DATE DUE

~~MY 11 '99~~ ~~OC 12'00~~			
~~ND 1'00~~ ~~NO 21'00~~			
~~DE 1'00~~			

DEMCO 38-296

Antibody Engineering

Breakthroughs in Molecular Biology

Antibody Engineering, second edition, is the fourth volume to appear in this exciting new series of high quality, affordable books in the fields of molecular biology and immunology. This series is dedicated to the rapid publication of the latest breakthroughs and cutting edge technologies as well as synthesis of major advances within molecular biology.

Other volumes in the series include:

*PCR Technology: Principles and Applications
for DNA Amplification*
edited by H. Erlich

DNA Fingerprinting: An Introduction
by L. T. Kirby

Adhesion: Its Role in Inflammatory Disease
edited by J. Harlan and D. Liu

Antibody Engineering

Second Edition

Edited by

Carl A. K. Borrebaeck

New York   Oxford
Oxford University Press
1995

Oxford University Press

Oxford   New York   Toronto
Delhi   Bombay   Calcutta   Madras   Karachi
Kuala Lumpur   Singapore   Hong Kong   Tokyo
Nairobi   Dar es Salaam   Cape Town
Melbourne   Auckland   Madrid

and associated companies in
Berlin   Ibadan

Published by Oxford University Press, Inc.,
200 Madison Avenue, New York, New York 10016

Oxford is a registered trademark of Oxford University Press

Library of Congress Cataloging-in-Publication Data

Antibody engineering / edited by Carl A.K. Borrebaeck. — 2nd ed.
p. cm. (Breakthroughs in molecular biology)
Includes bibliographical references and index.
ISBN 0-19-509150-7
1. Immunoglobulins   Biotechnology.
2. Protein engineering.
I. Borrebaeck, Carl A. K., 1948–
II. Series.
TP248.65.I49A57   1995
616.07'93   dc20   94-25809

1 3 5 7 9 8 6 9 2

Printed in the United States of America
on acid-free paper

Preface

The field of antibody engineering has continued to be an area of enormous growth and the scientific contributions since the first edition of *Antibody Engineering: A Practical Guide* (1991) have been very exciting and hold great promise for future development.

Today the polymerase chain reaction is a routine tool for everybody involved in antibody engineering and forms the basis in most technologies presented in this second edition. This approach allows the rapid isolation of V-genes from which antibody libraries are then being created and displayed on filamentous phage. Today we can also see the evolution of completely DNA-based approaches to handle many of the problems that were associated with conventional hybridoma technology, now being replaced by antibody engineering in more and more laboratories around the world. Recent examples of this evolution are the development of synthetic antibodies and affinity maturation by in vitro mutagenesis. However, before we have a stream-lined approach for the generation of human antibody fragments by phage display, the selection of phages in a library remains a major problem. Recent developments in this area will undoubtedly allow us to select out single phages from libraries containing 10^{9-10} members. Furthermore, the technologies for expressing proteins in *E. coli* are continuously being refined together with strategies to design many kinds of antibody fragments, which also include the recent developments in intracellular immunization discussed in this volume.

The aim of this series of antibody engineering books is to provide the most recent technologies to laboratories in the area or to laboratories just entering into the field of antibody engineering. With this second volume I believe that we are presenting a very up-to-date overview, by authorities in the field, of some of the "hottest" areas, thus allowing us to bring designer antibodies closer to pharmaceutical applications.

Lund, Sweden   C. A. K. Borrebaeck

Contents

Contributors

Viviana Amati
CNR Institute of Neurobiology
Rome, Italy

Carlos F. Barbas, III
Department of Molecular Biology
Scripps Research Institute
La Jolla, California

Eugenio Benvenuto
Dipartimento Innovazione
ENEA
Rome, Italy

Silvia Biocca
CNR Institute of Neurobiology
Rome, Italy

Carl A. K. Borrebaeck
Department of Immunotechnology
Lund University
Lund, Sweden

Andrew Bradbury
International School for Advanced
Studies
Trieste, Italy
Società Italiana per la Ricerca Scientifica
Trieste, Italy

Antonino Cattaneo
International School for Advanced
Studies
Trieste, Italy

Todd Chilton
Ixsys, Inc.
San Diego, California

M. Josefina Coloma
Department of Microbiology and
Molecular Genetics and the
Molecular Biology Institute
University of California at Los Angeles
Los Angeles, California

Anna Di Luzio
CNR Institute of Neurobiology
Rome, Italy

Stephen C. Emery
Scotgen Biopharmaceuticals, Inc.
Aberdeen, Scotland

Daniel Espinoza
Department of Microbiology and
Molecular Genetics and the
Molecular Biology Institute
University of California at Los Angeles
Los Angeles, California

Rosella Franconi
Dipartimento Innovazione
ENEA
Rome, Italy

Christian Freund
Department of Biochemistry
Universität Zürich
Zürich, Switzerland

Liming Ge
Department of Biochemistry
Universität Zürich
Zürich, Switzerland

Andrew J. T. George
Department of Immunology
Hammersmith Hospital
London, United Kingdom

Scott Glaser
Ixsys, Inc.
San Diego, California

Stefani Gonfloni
International School for Advanced
Studies
Trieste, Italy

William J. Harris
Scotgen Biopharmaceuticals, Inc.
Aberdeen, Scotland
Department of Molecular and Cell
Biology
University of Aberdeen
Aberdeen, Scotland

Alice Hastings
Department of Microbiology and
Molecular Genetics and the
Molecular Biology Institute
University of California at Los Angeles
Los Angeles, California

Andrew H. Henry
School of Biology and Biochemistry
University of Bath
Bath, United Kingdom

Hennie Hoogenboom
Cambridge Antibody Technology
Cambridgeshire, United Kingdom

William Huse
Ixsys, Inc.
San Diego, California

James S. Huston
Creative BioMolecules, Inc.
Hopkinton, Massachusetts

Donald Jin
Creative BioMolecules, Inc.
Hopkinton, Massachusetts

Peter Keck
Creative BioMolecules, Inc.
Hopkinton, Massachusetts

Achim Knappik
Department of Biochemistry
Universität Zürich
Zürich, Switzerland

Karin Kristensson
Ixsys, Inc.
San Diego, California

James D. Marks
Department of Anesthesia and
Pharmaceutical Chemistry
University of California at San Francisco
San Francisco, California

John E. McCartney
Creative BioMolecules, Inc.
Hopkinton, Massachusetts

Sherie L. Morrison
Department of Microbiology and
Molecular Genetics and the
Molecular Biology Institute
University of California at Los Angeles
Los Angeles, California

Hermann Oppermann
Creative BioMolecules, Inc.
Hopkinton, Massachusetts

Peter Pack
Department of Biochemistry
Universität Zürich
Zürich, Switzerland

Jan T. Pedersen
School of Biology and Biochemistry
University of Bath
Bath, United Kingdom

Patrizia Piccioli
CNR Institute of Neurobiology
Rome, Italy

Andreas Plückthun
Department of Biochemistry
Universität Zürich
Zürich, Switzerland

Anthony R. Rees
School of Biology and Biochemistry
University of Bath
Bath, United Kingdom

Jonathan S. Rosenblum
Department of Molecular Biology
Scripps Research Institute
La Jolla, California

Francesca Ruberti
International School for Advanced
Studies
Trieste, Italy

Stephen J. Searle
School of Biology and Biochemistry
University of Bath
Bath, United Kingdom

David M. Segal
National Cancer Institute
National Institutes of Health
Bethesda, Maryland

Seung-Uon Shin
Department of Microbiology and
Molecular Genetics and the
Molecular Biology Institute
University of California at
Los Angeles
Los Angeles, California

Bob Shopes
Tera Biotechnology, Inc.
La Jolla, California

Mei-Sheng Tai
Creative BioMolecules, Inc.
Hopkinton, Massachusetts

Paraskevi Tavladoraki
Dipartimento Innovazione
ENEA
Rome, Italy

David M. Webster
Department of Molecular Biology
Scripps Research Institute
La Jolla, California

Thomas Werge
CNR Institute of Neurobiology
Rome, Italy

Letitia A. Wims
Department of Microbiology and
Molecular Genetics and the
Molecular Biology Institute
University of California at
Los Angeles
Los Angeles, California

Ann Wright
Department of Microbiology and
Molecular Genetics and the
Molecular Biology Institute
University of California at
Los Angeles
Los Angeles, California

Antibody Engineering

CHAPTER 1

Antibody Structure and Function

Stephen J. Searle, Jan T. Pedersen,
Andrew H. Henry, David M. Webster,
and Anthony R. Rees

The remarkably diverse specificity of antibodies can now be explained both in genetic and structural terms. In the early 1970s Wu and Kabat[1,2] compared the amino acid sequences of immunoglobulin variable domains and showed that certain segments of sequence were more or less conserved while others exhibited high variability between one sequence and another. They proposed that the highly variable segments, or complementarity determining regions (CDRs), formed a contiguous structural element at one end of the antibody that was responsible for antigen recognition.

In 1973, this prediction was confirmed by Poljak and coworkers[3] who reported the first X-ray crystallographic structure of an antibody Fab fragment that contained the antigen binding region.

At the time of writing, 20 years on, only 23 such structures are available in the Brookhaven database of protein structures, while there are close to 2000 variable region sequences known. The rate of sequence acquisition continues to outstrip the rate of structure determination and, while X-ray crystallography will always be the preferred method for producing structures at atomic resolution, the need for good three-dimensional models has become essential. Such models will have increasing utility where the structural effects of point mutations within the antibody combining site need to be assessed, where minimal perturbation strategies for CDR grafting (humanization) are required, and even where the residues to be targeted for random mutation in the gene library methods now available (e.g., phage libraries) are selected in a more rational manner on the basis of their likely accessibility to the antigen.

To effectively identify the interaction interface between antibody and antigen, a three-dimensional structure is required. The fact that of the 23 antibody structures, only 7 are of complexes, suggests an ever more urgent need for algorithms that predict the nature of the antigen from a knowledge only of the antibody structure. This is a special case of the docking problem, a solution to which will become an increasing demand for antigens (e.g., membrane receptors) whose nature precludes the use of normal methods of structural analysis.

This chapter will attempt to address some of these issues, and antibody structure variable region modeling and its applications and antibody-antigen interactions will be reviewed. Also included is a step-by-step methodological approach to CDR modeling embodied in the computer program ABM.[4]

ANTIBODY STRUCTURE

General Features of Antibodies

The four-chain structure of antibodies, depicted in Fig. 1–1(a) for an IgG, was established by Porter[5] and Edelman and Poulik[6] in the early 1960s. The three-dimensional arrangement of the heavy and light chain sequences into constant and variable domains was established for the Fab fragment in 1973 by Poljak et al.[3] and for Fc by Diesenhofer et al. in 1976.[7] The spatial relationship of the domains was indicated in structures of intact IgG molecules: two hinge-deleted myeloma antibodies were solved by Silverton et al. in 1977[8] and Rajan et al. in 1983.[9] The effect of the hinge deletion was to constrain the Fab arms so that, overall, the antibody was T-shaped. Recently, the structure of a "normal" IgG was determined by Harris et al.[10] The murine antibody, raised against a canine lymphoma, was an asymmetric structure with two apparently independent two-fold axes. One relates the heavy chains in the Fc region while the other relates the constant domains of the Fab arms. The variable regions are not related by this second two-fold axis because the two Fabs exhibit different elbow angles (159° and 143°), as shown in Fig. 1–1(b). This asymmetric structure, which also shows different hinge angles between the Fc region and each of the Fab's (65° and 115°), is likely to be a dynamic ensemble of conformations in free solution. This arises out of the intrinsic flexibility of the antibody, which can be thought of as two Fab regions loosely tethered to a mobile Fc region via a flexible hinge. This flexibility may be important when the two combining sites located or the Fabs are undergoing dynamic binding to separate molecules of the same antigen.

The crystallographic information currently available for antibodies and their Fab fragments, either alone or complexed with antigen, are summarized in Table 1–1.

The Antibody Fold

When Poljak and coworkers[3] determined the first structure of an antibody Fab fragment, the nature of the three-dimensional fold was established. The patterns seen in the Fab consisted of two anti-parallel sheets, one characteristic of the N-terminal variable domains (V_L for light chain; V_H for heavy chain), and a related but distinctly different pattern for the constant domains

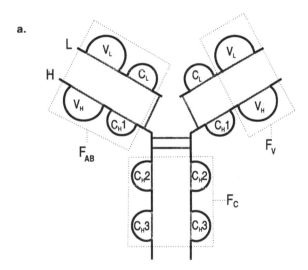

Figure 1–1. (a) An outline of a typical immunoglobulin illustrating the four-chain structure (2H + 2L). Each immunoglobulin light (L) or heavy (II) chain is structurally segmented into domains that are either of the variable type (V_L or V_H) or of the smaller constant type (C_H2 or C_H3). Larger combinations of these domains are also indicated. The Fab region comprises the variable F_V region (V_L + V_H) and part of the constant region ($C_{II}2 \times 2$), while the Fc region comprises the two pairs of constant domains. The example shown is typical of an immunoglobulin of the G class. Other classes have additional or different constant domains.

(continued)

b.

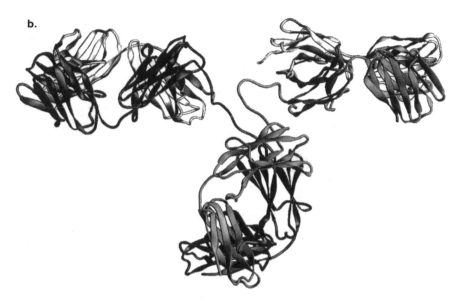

Figure 1–1 (*cont.*). (b) Ribbon representation of the X-ray structure of the murine antibody against canine lymphoma described by Harris et al.[10] The heavy chain is shown in grey and the light chains in white. The Fc region projects toward the viewer and assumes an asymmetric, oblique orientation with respect to the two Fabs (top left and right). This orientation illustrates the large difference in the hinge angles for the two Fabs (65° and 115°), each of which also has a different elbow angle (159° and 143°), suggesting a rather flexible hinge connection. The intact image was provided by the laboratory of Dr. Alexander McPherson. The immuniglobulin structure was determined by Harris et al.[10] Special thanks go to the Academic Computing Graphics and Visual Imaging Laboratory, University of California, Riverside, for help with the image, which was generated using the program RIBBONS (Carson and Bugg[101]).

(C_L for light chain; C_H1 for heavy chain). The characteristic β-sheet sandwich, now found with slight variations in a large superfamily of immunoglobulin-like molecules and known as the immunoglobulin fold[11], is depicted in Fig. 1–2(a). As indicated in the legend, the C_L/C_H1 fold lacks the two outer strands (C′ and C″). The hydrogen bonding pattern of these two types of fold are shown in Fig. 1–2(b) and a representation of the strand topology is shown in Fig. 1–2(c), to better illustrate why this type of fold is termed a "Greek key" motif. The two halves of the "sandwich" are held together by a disulphide bond, conserved in most but not all proteins containing the immunoglobulin fold.[11]

Inspection of Fig. 1–2 will also identify the positions of the three CDRs with respect to the β-sheet *framework*. Each CDR is a loop connecting two β-strands that has a fixed orientation on the framework depending on its

Table 1–1. List of antibodies and their fragments whose X-ray coordinates are currently available from the Brookhaven database of protein structures. Antibodies used for the β-barrel analysis (see section on Domain Packing) are marked with an asterisk

Brookhaven Entry	Name	Resolution (A̋)	Chain Types	Reference
2hfl*	HyHEL-5	2.54	κ/γII	71
3hfm*	HyHel-10	3.0	κ/γI	72
1bji/2bji	LOC	2.8	κ/κ	73
2fbj*	J539	1.95	κ/γIII	74
3fab/7fab*	NEW	2.0	κ/γII	75
4fab*	4-4-20	2.7	κ/γII	76
5fab/6fab*	36-71	1.9	κ/γI	77
1mcp/2mcp*	McPC603	3.0	κ/γIII	78
3mcg	MCG	2.0	$\lambda1/\lambda1$	79
1mcw	WEIR/MCG	3.5	$\lambda1/\lambda1$	80
2rhe	RHE	1.6	$\lambda1/\lambda1$	81
qrei	REI	2.0	κ/κ	82
2fb4/2ig2*	KOL	1.9	$\lambda1/\gamma$III	83
1f19*	R19.9	2.8	κ/γII	84
1fdl*	D1.3	2.5	κ/γII	85
1mam	YS*T9.1	2.5	NA/γII	86
8fab	HIL	1.8	λ/γI	87
1baf	ANO2	2.9	NA	88
1hil/1hin/1him	17/9	2.0	κ/γII	NA
(*)	Gloop2	2.8	κ/γI	89
ligf/2igf	B1312	2.8	κ/γI	90
1dfb*	3D6	2.7	κ/γI	91
1igm	POT	2.3	NA	92
1bbd	8F5	2.8	NA	93
1ncd	NC41	2.9	NA	NA
1igi	26-10	2.7	NA	94
1ggi	50.1	2.8	NA	95

NA = not available.

length and other characteristic sequence motifs (see later). Even at this stage it can be seen that, due to their close proximity, structural changes due to mutations within a CDR may be propagated to adjacent CDRs, or to the framework itself. A structure or model to guide such changes is therefore most important.

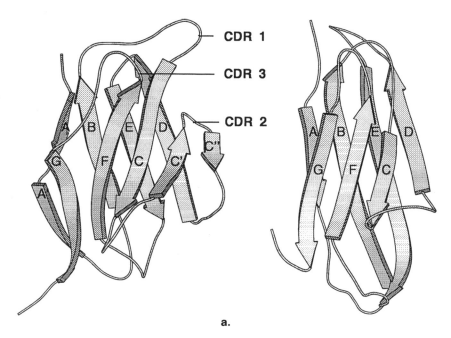

a.

Figure 1–2. (a) The domain structure of immunoglobulin showing the β-sheet sandwich structures characteristic of all members of the immunoglobulin super-family. (*left*) The variable (V) domain, consisting of nine β-strands packed to form a β-sheet sandwich. The loops connecting the strands form the complementarity determining regions (CDRs) shown generically as CDR1, CDR2 and CDR3. (*right*) The constant (C) domain, consisting of only seven β-strands packed in a similar manner to the V-domain, but lacking the two additional strands C′ and C″. (*continued*)

Domain Packing: Formation of the Framework

To generate the Fab structure, the V_L domain must pair with the V_H domain, and simultaneously the C_L domain must pair with the C_H1 domain. The manner in which these two sets of interactions occur is similar but not identical: The constant pair can be considered a rotational isomer of the variable pair. In this chapter, only the variable domain packing will be discussed in detail.

The V_L and V_H domains associate noncovalently to form a β-barrel structure (see Fig. 1–3). This places the six CDRs close to each other at the N-terminal end of the V_L–V_H dimer, or Fv region. This dimer association is spontaneous and has a binding constant, K_D, of about 10^{-10} M (Field, H., and Rees, A.R., unpublished). This high affinity is due almost entirely to the close complementarity of conserved residues at the dimer interface. Table 1–2 shows the distribution of amino acids found in human and mouse

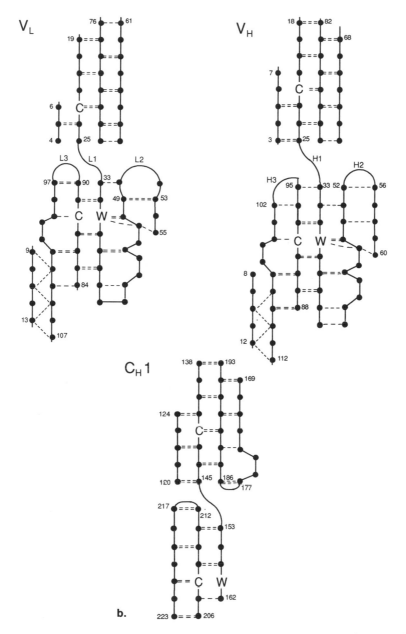

Figure 1–2 (*cont.*). (b) The hydrogen bonding patterns seen in V_L, V_H, and C_H1 domains. The hydrogen bonds are shown in double dotted lines and residue numbers indicated are after Kabat et al.[27] The positions of the conserved cysteines (C) and tryptophans (W) and the location of the CDRs (L1, L2, L3, H1, H2, and H3) are also shown. (*continued*)

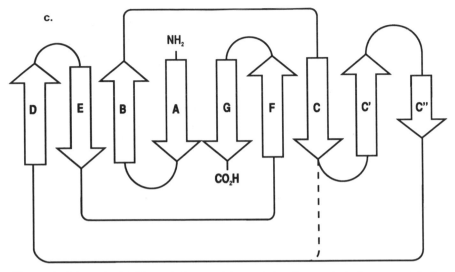

Figure 1–2 (*cont.*). (c) The relative positions of the β-strands in immunoglobulin constant and variable domains shown as a two-dimensional topology diagram. The type of connectivity seen in these types of domain is known as a "Greek key" motif (after early Greek cases).

V-region sequences for each of the interfacial residue positions. The manner in which these residues engage in close contact is shown in Fig. 1–3. The preponderance of aromatic and other hydrophobic residues is particularly striking in this view of the Fv region. Of course, heavy-chain and light-chain CDRs are also brought into contact by this dimer formation and may either add to or subtract from the binding energy. This latter possibility may explain the poor association of some V_L–V_H pairs (e.g., antibody B72.3)[12] although this antibody does not appear to have an unusual pattern of interface residues (Henry, A.H., and Rees, A.R., et al., unpublished).

From Table 1–2 it can be seen that many positions are conserved. There are variations, however, and when procedures that result in random pairings of V_L and V_H domain (e.g., gene libraries) are used, incompatibilities across the interface may be encountered, resulting in abortive pairing.

Although the three-dimensional structure of Fv regions is well conserved, not all are identical and care should be taken when modeling new sequences on existing X-ray structures. We have carried out an analysis of 12 antibody structures (starred in Table 1–1) to determine where these differences occur. The β-strands comprising the V-region β-barrel were defined according to their secondary structure and sequence conservation, rather than by excluded surface area as described by Chothia et al.[13] The barrels of 11 structures were then fitted onto the 12th (selected at random) and mean coordinates were calculated. The 12 structures were then fitted onto these mean

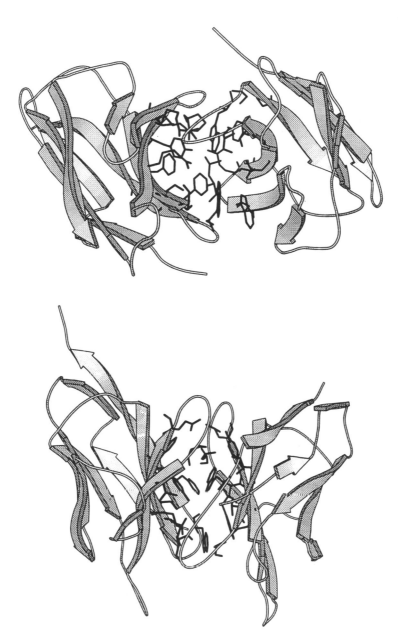

Figure 1–3. The pairing of V_L and V_H domains to form an F_v region. Plan and side views are shown to indicate the tight, noncovalent packing of aromatic and other hydrophobic residues at the V_L–V_H interface. Close inspection will reveal the manner in which the two conserved tryptophan residues, one from each domain, are packed.

Table 1–2. Distribution of residues found at the interface of (a) V_L and (b) V_H domains of mouse and human Fv regions*

Light-chain Position	Human	Mouse
	(a)	
41	W 99	W 98
42	Y 88, F 8	Y 74, F 12
43	Q 93, L 4	Q 74, L 22
44	Q 98	Q 88, E 5
45	K 58, H 13, L 10	K 81, R13
46	P 89, A 5	P 80, S 13
51	K 52, R 27	K 73, Q 12
52	L 72, V 10	L 70, R 9
53	L 75, I 8, V 11	L 81, W 15
54	L 92	I 91, V 4
55	V 86, F 6	V 83
89	E 42, F 37, V 9	A 21, I 13, L 26, F 9, V 13
90	A 90, G 8	A 67, E 28
91	D 38, T 22, V 28	D 7, I 11, T 36, V 30
92	Y 99	Y 99
93	Y 90, F 9	Y 67, F 30
94	C 99	C 99
106	F 93, Y 5	F 91
107	G 94	G 92
108	G 44, Q 34, T 10	G 56, A 21, S 12
109	G 95	G 95
110	T 95	T 92
	(b)	
155	W 96	W 98
156	V 70, I 23	V 86, I 10
157	R 90	R 53, K 44
158	Q 93	Q 90, K 5
164	G 83, A 6, S 6	G 58, R 20, K 8, S 7
165	L 65	L 98
166	E 88	E 96
167	W 98	W 92
168	V 46, I 22, M 17, L 13	I 61, V 18, L 11, M 9
169	G 58, A 22, S 15	G 68, A 29

Table 1–2. (*continued*)

Light-chain Position	Human	Mouse
215	Y 98	Y 96
216	V 90, F 9	Y 80, F 18
217	C 97	C 98
218	A 82, T 10	A 80
219	R 66, K 12, P 6	R 83
237	W 91	W 95
238	G 94	G 96

* The numbered positions are according to Pedersen.[24] The numbers following the one-letter code names represent percentage occurrences at each position in the aligned heavy- and light-chain sequences taken from the Kabat database.[27] Only frequencies greater than or equal to 5% are included.

coordinates and new mean coordinates calculated. This was repeated until the mean coordinate set converged (~ 10 cycles). The variance for the mean coordinates was then calculated for each N, Cα, C atom set in the barrel strands. This average barrel was fitted to the surface of a hyperboloid:

$$\frac{X^2}{A^2} + \frac{Y^2}{B^2} - \frac{Z^2}{C^2} = 1$$

where the parameters A, B, and C are taken from Novotny et al.[14,15] The strands of the mean barrel are shown in Fig. 1–4. The RMS derivation of the fit to the theoretical hyperboloid was 2.1 Å. Using this mean barrel, the deviations of 10 of the Fv region barrels were calculated. This analysis has pinpointed "hotspots" where residues deviate markedly in their backbone conformation from one antibody to another and from segments that are highly conserved. The most disordered residues ($> 3\sigma$ from mean coordinates) were found in strands F and G of the heavy chain, strand G is particularly variable with only two residues being retained in the fitting procedure. Interestingly, these two strands form the take-off points for CDR H3, the most conformationally variable of the six CDRs. The data for the conserved segments are plotted in Fig. 1–5(a) and (b) for each of the eight strands of the barrel, and show RMS deviation as a function of N, Cα, C position along the strand. Although some other strands show variability at their ends (e.g., the C-terminus of strand CL in Fig. 1–5(a)), the residues involved lie at the variable region–constant region interface and are unlikely to influence CDR

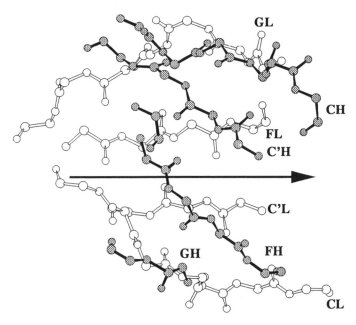

Figure 1–4. Plot of the average positions of the conserved segments of the β-barrel strands in a V-region, generated according to the multiple fitting procedure described in the text. The conjugate axis of the β-barrel is shown by the arrow, which points in the direction of the CDRs. The strands participating in the barrel are indicated by their normal letter names, as given in Fig. 1–2(c). The light-chain strands are in white and the heavy chain in black.

conformation. To summarize, of the eight strands in the Fv barrel, only parts of each strand exhibit high positional conservation. This mean barrel calculated on the basis of these conserved positions can be used to fit new sequences with high confidence (see later).

CDR Conformation

The combining site of an antibody is derived from the juxtaposition of six interstrand loops (or CDRs), three derived from the heavy chain (H1, H2, and H3) and three from the light chain (L1, L2, and L3). Four of the six are encoded in the V_L and V_H germline V-region genes (L1, L2, H1, H2) while the remaining two are formed at the junctions of the V-genes and their respective modifying gene segments (+J for light chain and +DJ for heavy chain). Despite considerable variability in these CDR sequences from one antibody to another,[1,2] there is some structural similarity for many of the CDRs. This was analysed in our laboratory in 1985 and 1986[16,17] when six antibody structures were compared. This analysis described the high

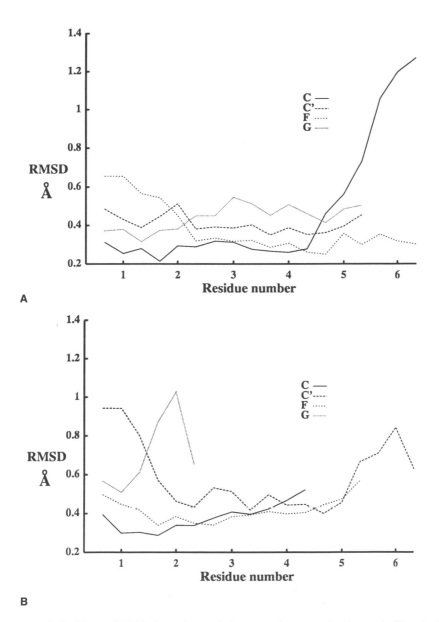

A

B

Figure 1–5. Plots of RMS deviations of the strand segments shown in Fig. 1–4 used to construct the mean barrel: (a) light-chain strands; (b) heavy-chain strands. The RMS deviation is calculated for N, Cα, C atoms at each residue position along the strands. Lettering of strands is as for Figs. 1–2(c) and 1–4. Residue numbering of the conserved strand segments is according to Kabat et al.[27] as follows: *Light chain*: C strand, 35–40; C′ strand, 45–49; F strand, 83–88. *Heavy chain*: C strand, 36–39; C′ strand, 44–49; F strand, 90–94; G strand, 103–104.

conservation of H1, H2, and L2 conformations and proposed a relationship between CDR length and conformation.

The present database of antibody structures (see Table 1–1) contains combining sites with various combinations of CDR length and sequence. The length ranges seen in the sequence database are:

CDR	No. of Residues
L1	10–17
L2	7 (highly conserved)
L3	7–11
H1	5–7
H2	9–12
H3	4–25

The structures of these CDRs often vary considerably with length. Four of the CDRs (L2, L3, H2, and H3) are hairpin loops. The preferred conformations of such loops have been extensively studied for other proteins[18,19] but the high variability in length of the antibody loops has complicated the development of a rigorous structural classification. For example, while H3 is by definition a hairpin loop since it interconnects two adjacent β-strands, its high length variation produces a large number of backbone conformations, the relationship between which has not been fully elucidated.

The most comprehensive structural classification of CDRs has been described by Chothia, Lesk, and coworkers[20–22] for both murine and human sequences. A full description of these "canonical" families (for CDRs L1, L2, L3, H1 and H2) is outside the scope of this review, but the reader should consult the references and the antibody modeling software, ABM[4] in which the known canonical families are tabulated. In summary, the canonical method assigns CDRs to a limited set of conformers based on the presence of residues that influence H-bonding, are involved in packing, or have preferred torsion angles. Should such a residue or residues be mutated by the antibody engineer, then of course it is likely that a conformational change will result, a further reason why molecular models to guide the engineering are desirable. An example of the canonical concept is shown in Fig. 1–6 for CDR L1.

While this method is useful for CDRs, where exact obedience to the "rules" is seen, further development is necessary in two areas. First, some CDRs are as yet structurally unclassified. That is to say, they lack the appropriate key residues in those positions that define a particular canonical class. This may reflect the fact that the conformational rules are somewhat less stringent in their requirements than proposed or that entirely new canonical classes are required. The second issue relates to the question of whether all permutations

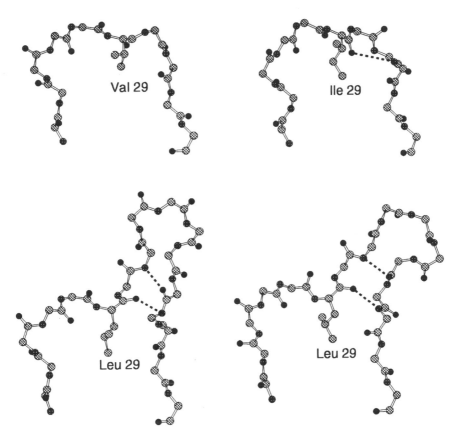

Figure 1–6. Four members of the light-chain CDR L1 canonical family, as defined by Chothia et al.[21] The conformation of this loop is defined by its length and the presence of a small hydrophobic residue at position 29 in the light chain sequence. This hydrophobic residue is indicated in the four examples and, if the entire V-region were shown, would be seen to play an important packing role with the framework. This packing leads to an arch-like conformation for short loops (upper two examples). For longer loops (lower two examples), the arch is retained and the extra residues form a bulge at the top of the loop.

of CDRs will be allowed. A canonical conformation in one V_L or V_H domain may be disallowed because of overriding energetic factors associated with domain association. CDR–CDR interactions across the V_L–V_H interface may prevent correct domain assembly, a possibility already discussed by Steipe et al.[23] Any preferences for particular CDR combinations have not yet been described, presumably because of the limited number of structures from which to generalize. Clearly, where mutations target canonical positions or convert one canonical structure to another, loss of antibody binding can be

a.

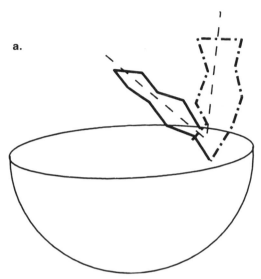

Figure 1–7. The effect of CDR take-off angle on the orientation of a CDR on the Fv framework. (a) A cartoon to illustrate the fact that the differences in CDR position on the framework result from the different angles at which the residues at the base of the CDR emerge from the β-strand of the framework; (b) the effect of this difference in take-off angle on the conformations of the seven different H3 CDR classes H3a–g, as defined in the text. The backbone positions of the framework β-strands are shown for each CDR to illustrate the identity of these positions, seen as overlapping white, light gray, dark gray, and black regions at the base of each of the loop clusters. Within classes similar take-off angles are seen whereas between classes the differences in take-off angles lead to markedly different CDR conformations. Each class contains antibody structures that differ in their take off angles by less than 35°. This is approximately half the maximum difference (77°) seen for all pairwise comparisons of antibody structures. The angular difference is calculated as the angle between the planes defined by the N terminal C-alpha, centre of geometry, and C terminal C-alpha of the H3 loops for each pair of structures after least squares fitting of the structures on conserved beta strands. The classes in all but one case have lower mean angular differences than the complete set of antibody structures. For class (a) the mean angular difference is 13.68° (5), for class (b) 25.11° (3), for class (c) 7.87° (2), for class (d) 5.98° (6), for class (e) 17.40° (5), for class (f) 12.13° (3), and for class (g) 13.54° (3). The mean for all structures is 20.39° (30). The figures in parentheses are the number of structures in each class. The Brookhaven codes of the structures used in the calculations are 1baf, 1bbd, 1dfb, 1fdl, 1fvc, 1ggi, 1hil, 1igf, 1iji, 1igm, 1mam, 1ncd, 2f19, 2fb4, 2fbj, 2hfl, 2mcp, 3hfm, 4fab, 6fab, 7fab, 8fab, 1bbj, 1dbm, 1ifh, 1igi, 1jhl, 1mfa, and 1tet. The undeposited structure gloop2 was also included. *(continued)*

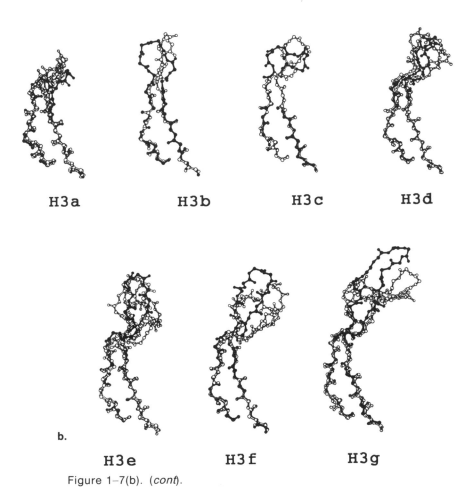

H3a   H3b   H3c   H3d

b.

H3e   H3f   H3g

Figure 1–7(b). (*cont*).

due to modification of antigen contact residues or to perturbation of the entire Fv domain structure.

The Special Problem of CDR H3

However comprehensive the canonical hypothesis turns out to be, it is presently incapable of defining the conformations of CDR H3 loops. This CDR occupies a central position in the combining site and has a critical effect on the combining site topography.

One of the factors determining CDR conformation is the loop take-off angle (Fig. 1–7(a)). This positional variability may be important for the binding of antigen when induced fit is required (see later). However, in the unbound state each CDR would be expected to exhibit a preference for one or a small number of energetically favorable conformers. We have analyzed the take-off angles of all six CDRs[24] (see Searle, S.J., Pedersen, J., and Rees, A.R., in preparation) which for CDR H3, can vary by up to 90° between different antibodies. This analysis suggested two structural classes of short H3s based on take-off angle alone. For longer CDRs the angles become more variable and specific loop features become important in defining the take-off angle. For example, a conserved Arg or Lys at position 94 interacts with an Asp, gly, or Ala at position 101 (Kabat numbering), fixing the conformation of this region of the CDR. Distinct classes have been defined for 8, 9 and 10, 11, and 12 residues, while for H3s of 13 residues or more a single class is defined with somewhat lower confidence in take-off angle. On the basis of the foregoing, we define seven classes of H3 loops, illustrated in Fig. 1–7(b).

H3a — loops shorter than 7 residues
H3b — loops of 7 residues
H3c–f — loops in which a conserved Arg or Lys at position 94 interacts with a Gly, Ala, or Asp at position 101 in 8, 9 and 10, 11 or 12 residues loops respectively
H3g — loops with the motif seen in c–f but of 13 or more residues

The legend to Fig. 1–7 gives the angular distributions for each of the seven classes.

Combining Site Topography

Wang et al.[25] have suggested three types of combining site: cavity, groove, and planar. This topographic classification is based on the analysis of 20 X-ray structures and is illustrated in Fig. 1–8. The CDR length combinations responsible for the surface shapes are shown in Table 1–3(a). Unfortunately, there is no totally satisfactory way of predicting which CDR combinations give rise to which surface topography, as indicated in Table 1–3(b), where three different antibodies are seen to give rise to the same types of topography as the antibodies in Table 1–3(a), but with different CDR length combinations. Clearly, more structures are required to place this type of classification on a firm footing.

Conservation of Variable Domain Surfaces

The classification "variable domain" when applied to antibodies suggests the existence of structural differences between antibody molecules. As has been shown, the β-sheet core is well conserved, and even where the greatest variability occurs, in the CDRs, patterns of conservation are also seen. Variable domains show structural conservation in one further respect: the positions and types of surface-exposed residues.

When the solvent accessibility is calculated for all framework residues (that is, without CDR residues) in the Fv region, using a set of 12 high-resolution crystal structures of Fab fragments, the alignment positions of those residues defined as "surface"[26] is conserved with 98% fidelity. Figure 1–9 shows the locations of these positions in the Fv region structure and Table 1–4 lists the residue positions and variability seen in human and mouse sequences.

a.

Figure 1–8. The three antibody-combining site topographies suggested by Wang et al.[25] and Webster et al.[102] (a) A *cavity* site, exemplified by antibody 4-4-20 (anti-fluorescein; Herron et al.[76]).   (*continued*)

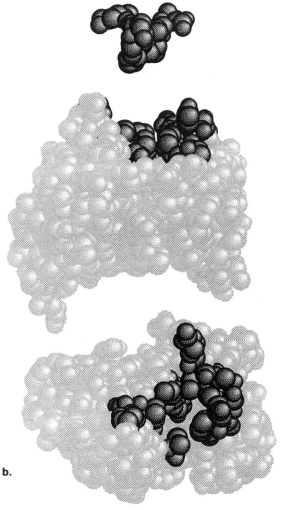

b.

Figure 1–8(b). (*cont.*) a *groove* site, exemplified by antibody B1312 (an anti-peptide antibody; Stanfield et al.[69]   (*continued*)

One of the most striking features of this analysis has been the fact that the identical V_L and V_H sequence families,[27] previously classified on the basis of contiguous sequences, can be generated using the surface residue profiles alone. This has suggested that the surface patterns of Fv regions may be more conserved than was previously thought. The main conclusions of the analysis are:

• V_L and V_H surface residue positions are conserved.

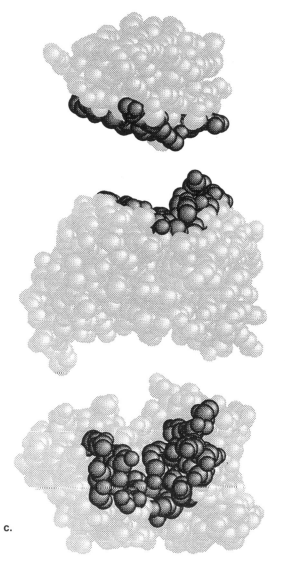

c.

Figure 1–8(c). (*cont.*) a *planar site*, exemplified by antibody HyHEL10; Padlan et al.[72]. The antigen (dark gray) is shown at the top of each figure; the center view is of the antibody-antigen complex, which has been pulled apart to reveal the complementary interaction surfaces (dark gray); the bottom view shows the antibody residues (in dark gray) that form the antigen contact region.

Table 1–3. CDR lengths found in antibodies exhibiting the three different topographic classes described in the text*

		L1	*L2*	*L3*	*H1*	*H2*	*H3*	*Topography*
(a)	HyHel 10	11	7	9	5	9	5	planar
	4-4-2	16	7	9	5	12	7	cavity
	B13i2	16	7	9	5	10	10	groove
(b)	D1.3	11	7	9	5	9	8	planar
	36-71	11	7	9	5	9	8	cavity
	Gloop2	11	7	9	5	10	5	groove

* The CDRs are defined according to Kabat et al.,[27] with the exception of H2, which is defined on a structural basis according to Pedersen et al.[42] References for the antibodies can be found in Table 1–1.

- Within each species there appears to be a preference for particular residues at certain positions.
- V_L and V_H sequences can be classified into families according to their surface residue profiles.
- Although recurring within a species, a particular surface residue profile is never seen in both mouse and human.

This study allowed the development of a novel method of humanization in which the surface pattern of a murine Fv region is replaced by its nearest human Fv surface.[26,28]

VARIABLE REGION MODELING

Methods and Test Example

The modeling of antibody combining sites has been the subject of intense development since the first, simple homology-based methods.[29,30] As the structural rules governing the conformations of short loops began to emerge,[18,19] improvements in prediction occurred. However, the loops that form the CDRs of antibodies were found to exhibit more structural variability, both because they were often much longer than loops typically found in other proteins and because of the effect of CDR–CDR contacts on conformation. Attempts to improve the prediction of CDRs can be grouped into knowledge-based categories[16,17,20–22,31–34] which include the canonical structures previously described, and *ab initio* employing either molecular

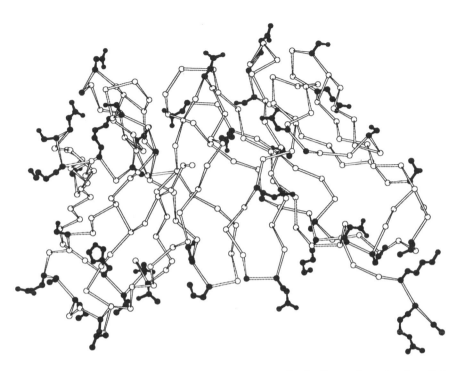

Figure 1–9. A backbone trace (white) for an antibody Fv region showing the locations of the surface accessible residues (black) listed in Table 1–4. The VL domain is on the right.

dynamics,[35] conformational search methods,[36–38] or simulated annealing.[39] None of the above methods has provided a method for modeling the *entire* combining site with reliability and reproducibility. The most successful homology-based method is that of Chothia and Lesk,[20–22] but even here, canonical structures are found for only 5 of the 6 CDRs at best, and occasionally for only 3 or 4 (see Table 1–5).

Considerable improvements are seen when homology methods and conformational search procedures, which incorporate energy screening, are combined. We have promoted this as the method of choice and have demonstrated its superiority over other methods for modeling all six CDRs with an accuracy approaching that of the medium-resolution X-ray structure.[4,40–42] The algorithm we have developed, CAMAL, is shown in outline in Fig. 1–10. The essential steps are outlined in the worked example below, which generates a model for the variable region of antibody D1.3 (see Table 1–1).

Modeling of Antibody D1.3 According to the A*b*M Protocol[4]

Sequence Entry
The V-domain sequences of D1.3 are entered and aligned against a database of antibodies with known crystal structures. The database is nonredundant with a total of 27 light chains and 22 heavy chains at the time of writing, most of which are derived from the Brookhaven Protein Data Bank. Alignment is based on sequence identity, with gaps corresponding to regions of highest structural variation within the CDRs.

Framework Construction
Due to their conserved nature, frameworks for the light- and heavy-chain domains are selected by similarity alone. For each database antibody, each residue is compared with the corresponding residue in the aligned D1.3 sequence. Exact identity adds 1 to the similarity score, while alignment

Table 1–4. Distribution of accessible residues in human and mouse (a) light-chain and (b) heavy-chain variable regions*

Light-chain Position	Human	Mouse
	(a)	
1	D 51, E 34, A 5, S 5	D 76, Q 9, E 6
3	V 38, Q 24, S 24, Y 6	V 63, Q 22, L 5
5	T 61, L 37	T 87
9	S 26, P 26, G 17, A 14, L 7	S 36, A 29, L 17, P 5
15	P 62, V 25, L 12	L 47, P 30, V 8, A 7
18	R 57, S 18, T 13, P 6	R 38, K 22, S 13, Q 12, T 9
46	P 94	P 82, S 9
47	G 89	G 71, D 18
51	K 43, R 31	K 70, Q 13, R 8, T 5
63	G 91	G 98
66	D 43, S 25, A 9	D 38, S 26, A 26
73	S 96	S 90, I 5
76	D 43, S 16, T 18, E 15	D 67, S 15, A 5, K 5
86	P 44, A 27, S 17, T 8	A 50, P 11, T 8, E 7, Q 6
87	E 71, D 11, G 7	E 91, D 6
111	K 74, R 12, N 6	K 93
115	K 54, L 40	K 87, L 5
116	R 60, G 33, S 5	R 89, G 9
117	Q 50, T 37, E 6, P 6	A 74, Q 14, P 5, R 5

Table 1–4. (*continued*)

Heavy-chain Position	Human	Mouse
	(b)	
118	E 47, Q 46	E 59, Q 29, D10
120	Q 83, T 7	Q 68, K 26
122	V 59, Q 13, L15	Q 57, V 27, L 5, K 5
126	G 54, A 23, P 18	G 36, P 30, A 29
127	G 53, E 22, A 14, D 7	E 45, G 43, S 6
128	L 61, V 31, F 7	L 96
130	K 46, Q 41, E 5	K 52, Q 27, R 17
131	P 95	P 91, A 5
132	G 74, S 16, T 7	G 82, S 17
136	R 53, K 23, S 17, T 7	K 66, S 17, R 13
143	G 96	G98
145	T 46, S 32, N 9, I 7	T 63, S 19, N 7, A 5, D 5
160	P 84, S 10	P 89, H 7
161	G 93	G 71, E 24
162	K 76, Q 10, R 8	K 50, Q 30, N 10, H 5
183	D 26, P 25, A 17, Q 10, T 7	E 31, P 22, D 17, Q 11, A 12
184	S 70, K 9, P 8	K 42, S 37, T 6
186	K 53, Q 22, R 7, N 7	K 83, Q 7
187	G 66, S 21, T 5	G 62, S 18, D 10
195	T 30, D 26, N 19, K 7	T 36, K 30, N 26, D 6
196	S 91	S 76, A 16
197	K 65, T 8, I 8, R 5	S 46, K 34, Q 11
208	R 46, T 18, K 17, D 6	T 55, R 26, K 8
209	A 50, P 21, S 13, T 8	S 67, A 14, T 11
210	E 46, A 18, D 13, S 9, V 5, Z 8	E 88, D 7
212	T 91	T 53, S 43

* CDR residues are excluded. Only the most commonly observed residues (greater than or equal to 5% frequency) are given for each position. The numbers after the one-letter code names indicate the percentage frequency of occurrence in the aligned light- and heavy-chain sequences, taken from the Kabat database[27] and from Tomlinson et al.[97] The definition of surface accessible is as given in Pedersen et al.[26] All residues are given in the single-letter code, Z = Glx.

gaps in the conserved framework away from chain termini introduce a a penalty of -200 for the database antibody being assessed. The antibodies REI and ANO2 (Table 1–2) supplied the most similar light and heavy frameworks for D1.3, with 67% and 59% similarity respectively.

The REI light-chain and ANO2 heavy-chain frameworks were paired by

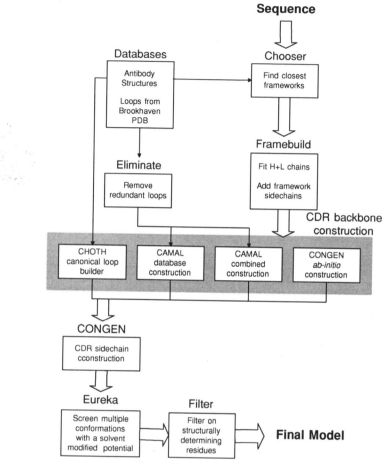

Figure 1–10. Flowchart of the antibody modeling algorithm, CAMAL, embodied in the program ABM (Rees et al.[4]) The shaded area contains those parts of the program that derive the CDR conformations.

fitting onto the average barrel described in the earlier section on domain packing. Sidechains were then replaced using a maximum overlap approach to match those of the D1.3 sequence.

CDR Modeling

The D1.3 sequence has 5 CDRs with canonical structures: L1, L2, L3, H1, and H2. The canonical classes are summarized in the table below. The most homologous candidate within each class, according to the Dayhoff mutation matrix, is identified for each CDR. For CDR H3, the most similar loop

Table 1–5. Results of modeling eight crystal structures of antibodies using the modeling program ABM[4]*

Structure	CDR	CDR Length	Global RMSD (N, Cα, C, O)	Canonical Structure Known
		(a)		
glb2	L1	11	1.161	+
	L2	7	0.647	+
	L3	9	1.031	+
	H1	5	1.785	+
	H2	10	1.609	+
	H3	4	1.273	−
	Total		1.251	
2hfl	L1	10	1.150	+
	L2	7	0.712	+
	L3	8	2.524	−
	H1	5	1.261	+
	H2	10	2.155	+
	H3	7	2.310	−
	Total		1.685	
2mcp	L1	17	0.784	+
	L2	7	0.538	+
	L3	9	0.739	+
	H1	5	1.004	+
	H2	12	2.014	+
	H3	11	2.306	−
	Total		1.231	
4fab	L1	16	2.470	−
	L2	7	0.792	+
	L3	9	1.255	+
	H1	5	0.721	+
	H2	12	2.028	+
	H3	7	2.132	−
	Total		1.566	
3hfm	L1	11	0.775	+
	L2	7	1.021	+
	L3	9	0.394	+
	H1	5	2.012	+
	H2	9	0.942	+
	H3	5	1.683	−
	Total		1.302	

(continued)

Table 1–5. (*cont.*)

Structure	CDR	CDR Length	Global RMSD (N, Cα, C, O)	Canonical Structure Known
1mam	L1	11	1.302	+
	L2	7	1.362	+
	L3	9	1.289	+
	H1	5	1.845	+
	H2	12	2.976	−
	H3	8	2.524	−
	Total		1.883	
bi3i	L1	16	2.667	
	L2	7	0.763	+
	L3	9	0.877	+
	H1	5	1.310	+
	H2	10	1.202	+
	H3	10	2.97	−
	Total		1.632	
		(b)		
d1.3(new)	L1	11	0.75	+
	L2	7	0.60	+
	L3	9	0.56	+
	H1	5	1.22	+
	H2	9	0.55	+
	H3	8	1.76	−
	Total		0.91	
d1.3(old)	L1	11	0.799	+
	L2	7	0.928	+
	L3	9	1.138	+
	H1	5	0.846	+
	H2	9	1.413	+
	H3	8	2.188	−
	Total		1.219	

* The RMSD figures are in ångstroms (Å) and are obtained after global least-squares fitting of the entire Fv structure of the crystal structure and the model. (a) Results obtained for seven antibodies using the algorithm shown in Fig. 1–10 in its current commercial version and without the inclusion of the improved take-off angle analysis described in the text. (b) Modeling of an additional antibody (d1.3) in the most recent research version of the algorithm, which incorporates new antibody structure data, allowing improved specification of CDR constraints, and either with (new) or without (old) implementation of the H3 take-off angle algorithm. Structure names are as found in Table 1–1.

within its class of take-off angle is identified:

CDR Loop	Canonical Class	CDR Template Identified
L1	2	REI, pdb code 1rei
L2	1	REI
L3	1	REI
H1	1	3D6, pdb code 1dfb
H2	1	ANO2, pdb code 1baf
H3		3D6

Loops within a canonical class have identical length so that only the sidechains require replacement by those of D1.3. Sidechains that are within a van der Waals interaction energy threshold, with respect to the surrounding framework and CDR residues, are added using the *Iterative* algorithm in CONGEN.[43]

One of the other methods available in CAMAL for constructing CDRs is to search an all protein database for loops of length and shape compatible with known CDR geometry. This is achieved by identifying loops that satisfy a set of α-carbon distances defined for each CDR, an established loop search technique developed by Jones and Thirup.[44] The α-carbon constraints used for each CDR were derived from crystal structure data of antibodies. The search, which is normally conducted on a nonredundant database of Protein Data Bank entries above 3 Å resolution, typically identified between a few (sometimes none) and several hundred thousand loops. A large number of hits is usually seen with short- to medium-length CDRs that have only a few α-carbon distance criteria. However, many of these will have similar conformations and, to avoid undue redundancy, the loops are clustered into groups so that the members of each group have backbone torsion angles for all residues that lie within 30° of one another.

By contrast, a low hit rate is seen when there are stringent α-carbon constraints. This is often the case for CDR L1, where few non-antibody proteins will contain loops that are both long enough and satisfy the constraints applied. Long L1s accommodate the length by incorporating a bulge in the N-terminal region of the arch, formed without disrupting the interaction between the conserved residues, 29 and 32 (Chothia et al.[20,21] and Fig. 1–6). When the database search method only identifies a small number of loops, conformational space will be inadequately saturated. Amplification is then achieved by regenerating a stretch of the loop with the conformational search procedure embodied in CONGEN. The region subjected to this search is normally the central part of the loop, since this is likely to be the most flexible. In addition, if the structural residues glycine and proline are present in a sequence, these may also be included.

CDR H3 of D1.3 is 8 residues long. Its construction requires a combination of database and conformational search. The database search identified 920 hits that clustered to 342 when the 30° torsional cut-off was applied. Subsequent conformational search gave 6,672 conformations, complete with sidechain atoms generated by the iterative method in CONGEN immediately following the conformational search. These 6,672 conformations were then subjected to energy screening.

Screening of Database Loops
Each candidate loop, generated either from the PDB database or CONGEN search, is grafted onto the D1.3 framework model and its energy evaluated using a solvent-modified *Eureka* potential.[45] High-energy conformers are discarded to leave a small number of low-energy loops, ranked using a structurally determining residue (SDR) algorithm that scores the resemblance of each loop residue conformation to known loop structures in the protein database. The final model is energy minimized to relax any poor contacts and torsion angles.

Results

The D1.3 model had a global N, Cα, C, O RMS deviation of 0.91 Å from the crystal coordinates (PDB entry 1fdl; Table 1–5(b)) when modeled by the above procedure, which includes the new take-off angle rules for H3 (RMSD 1.76 Å). When these H3 rules are not included, the RMSD is substantially worse (2.19 Å).

The results from modeling seven other high-resolution antibody structures are shown in Table 1–5(a) to illustrate the generality of the algorithm. In the example (D1.3) used to illustrate the protocol, the model was obtained before the X-ray coordinates were received. While "blind prediction" is always a good test of any modeling method, where an algorithm performs without modification for many different examples, as seen for ABM,[4] confidence in its predictive power can be high.

Accuracy, Limitations, and Future Developments

Methods for modeling antibody combining sites have made great advances since the earliest attempts. While for many antibodies the models will have an overall accuracy approaching that of medium-resolution X-ray structure, some parts of the variable region will be more accurately defined than others. For example, the framework regions generated by the docking of two previously unpaired V_L and V_H domains will normally have RMS derivations in the range 0.5–0.8 Å.[24,42] For CDRs conforming exactly to a canonical structure, the local backbone RMS deviations will be in the range 0.7–1.2 Å,[43] while global RMS deviations will be in the range 1–3 Å.[46] The larger global deviations are largely due to errors in take-off angles of the CDRs with respect to the framework.

Where CDRs do not conform to canonical classes they can be modeled by the CAMAL method. The four CDRs in the antibodies of Table 1–5 not conforming to canonical structures (e.g., L3 in HyHEL-5; L1 in 4-4-20; H2 in YS*T9.1; L1 in B13i2; see Tables 1–1 and 1–4), have global (N, Cα, C, O) deviations of 2.52 Å, 2.47 Å, 2.98 Å, and 2.67 Å respectively. These RMS figures account for both local backbone and take-off angle orientation and are thus comparable to those obtained for exact canonical structures.

For CDR H3, where no canonical structures exist other than the loosely defined classes we have described, RMS deviations will vary. The accuracy obtained will largely depend on the length of H3. The modeling protocol embodied in the program ABM[4] is satisfactory for H3 loops of 12 residues or less (see Table 1–5). For loops longer than 12 residues, the increased flexibility will make their prediction somewhat more difficult. This is no great disadvantage since, as Fig. 1–11 shows, the majority of H3 loops in the sequence database are less than 12 residues long.[47] However, it is a limitation

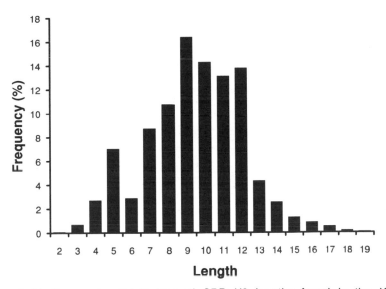

Length

Figure 1–11. Frequency distribution of CDR H3 lengths found in the Kabat database of immunoglobulin sequences (see also Kabat et al.[47]). The distribution shows two peaks at 5 residues and 9 residues, although the medium length peak is broad and asymmetric. Recent additions to the database, not represented in this figure, of antibodies showing specificity for viruses (e.g., 3D6; He et al.[91]) contain long (17–24 residues) H3 loops. This may be a particularly interesting characteristic of anti-viral antibodies.

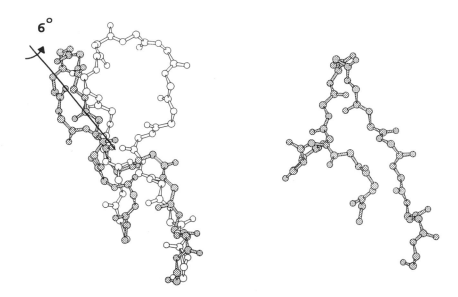

Figure 1–12. The conformations of CDR H3 loops from antibody 3D6 (left; He et al.[91]) and KOL (right; Marquart et al.[83]). For 3D6 the orientation of the CDR on the framework is shifted between the X-ray structure (gray) and the model (white) by a rotation in the upper part of the CDR of about 6°. The KOL structure is shown alongside to illustrate how two H3 loops of identical length (3D6 and KOL, both 17 residues) can differ in conformation. In the views shown, the orientation of the two CDRs is obtained after least-squares fitting the framework regions of the two antibodies (described in Martin et al.[41]).

that requires further development. To illustrate this difficulty, we modeled the 17 residue H3 of antibody 3D6.[92] The antibody KOL, for which an X-ray structure is known,[83] also has a 17 residue H3 but has only a single amino acid in common with 3D6. The predicted structure of the 3D6 H3 is shown in Fig. 1–12 with the X-ray structure for comparison. Alongside, the KOL conformation is shown. Although CAMAL correctly predicted an extended β-sheet for 3D6, the degree of twist in the predicted and X-ray structure differs by some 5° to 7° in the region of the extended β-sheet loop. Such a twist has also been observed between complexed and uncomplexed antibodies[48] and may reflect the flexibility required by some antibodies for effective antigen binding. Clearly, in this situation, it would be meaningless to assign a single conformation to such a CDR.

The correct prediction of sidechain conformation is essential where the identity of possible antigen contact residues is to be made. For canonical structures, sidechains can be taken directly from the CDR selected as the canonical template or, where a residue differs from this template, from another CDR that is in the same canonical class.[46] This clearly has limitations where no canonical CDR can be found in the structural database. Within the automated program A*B*M, the *Iterative* algorithm embedded in CONGEN[43] is used. This procedure searches the available conformational space for each sidechain and assigns the lowest energy conformation. In an attempt to improve the accuracy of sidechain placement, we have recently employed a Monte Carlo simulated annealing procedure,[42] which allows for effects due to sidechain–sidechain interaction to be incorporated and the addition of parameters that describe sidechain solvent accessibility. The latter is particularly important for the aromatic and other hydrophobic residues that may be surface exposed.

The issue of CDR flexibility is one that, in time, will need to be addressed. There is growing evidence that, for certain classes of antigen or hapten, conformational changes occur on binding (see later). Thus, it will become necessary to develop algorithms that specify "populations" of CDR conformations. This can only be achieved where some objective evaluation function (e.g., energy) is incorporated during, for example, molecular dynamics simulations

Applications

Site-Directed Mutations

When the first mutations of antibodies were carried out in our laboratory to improve antigen affinity, these were guided by crude models. Even so, the affinity of an anti-lysozyme antibody could be improved by a factor of 10 or so.[49] Since these early experiments, modeling has improved to the point where the effects of single-point mutations can be predicted with better reliability. Of course, if an X-ray structure is available, then the positions of CDR residues at or near the putative antigen binding region (see Fig. 1 8) can be more accurately specified. Examples of mutations to modify specificity or affinity have been described.[50–52].

The general rules to be observed when designing mutations are relatively elementary.

1. Where possible make a series of changes, from minimal to large effect; e.g.,

   lysine → arginine —maintains charge
   lysine → glutamine—loss of charge and some volume but retains the H-bonding potential
   lysine → alanine —loss of charge and H-bonding potential and large change in volume.

2. Avoid residues that may have a drastic structural effect (e.g., key residues defining a canonical class).
3. Target solvent-accessible residues since these are more likely to be involved in antigen contact, but see Near et al.[52]
4. Avoid CDR residues that are close to framework residues (see later). Some pairwise interactions of this sort may be important for maintaining CDR shape.
5. Make some changes in combination. Two residues individually may have a minimal effect, but their combined modification may allow a reorientation of antigen not possible where a single change is carried out (see Roberts et al.[49] and Sharon [50]).

Introduction of New Properties

In some circumstances it may be desirable to modify a combining site so that it is capable of binding a co-factor such as a metal ion. This can be approached in two ways. Existing residues in the combining site, spatially conforming to a template appropriate for a set of coordinating ligands, but sequentially discontinuous, may be changed to histidine, cysteine, or glutamate residues, or a combination of these, depending on the metal preferences. An example of this template-based approach can be found in Roberts et al.[53] and Wade et al.[54] An alternative method, which uses a computer program to search for sites within a protein that have a high potential for metal binding, can be used.[55] In one experimental example of this method, an anti-lysozyme antibody was modified within CDR L1 by introduction of 3 Zn^{2+} liganding residues. A fourth ligand was introduced at the N-terminus of the light chain. As predicted, the apo-antibody had the identical affinity for its antigen as the wild type, whereas when various metals were bound (Kas $\sim 10^8$–10^9 M^{-1}), antigen affinity could be modulated by up to 12-fold.[56] Although the introduction of metals has been hailed as a mechanism for improving rates by catalytic antibodies, there has so far been little more than meager enhancement of rates.[57] By contrast, it is likely that such metallo-antibodies may find applications as biosensors, affinity matrices, or pollution scavengers.

There has been one example of a mutation in an antibody that has directly led to catalytic activity. This, however, was in an antibody that had a close homologue that bound the same hapten but was catalytically inactive. It is to be hoped that, when the mechanisms by which catalytic antibodies act are further elaborated, mutational modification of antibodies that merely "bind" the transition state analog used to elicit them will lead to second-generation catalysts. The emergence of phage display methods may simplify their discovery.

Humanization of Antibodies

Humanization or "reshaping" of murine antibodies is an attempt to transfer the full antigen specificity and binding ability of the murine combining site to a human framework region by CDR grafting.[58] However, to preserve the binding affinity, the majority of CDR grafted antibodies require additional amino acid changes in the framework region, because such residues are either conformationally important or are in direct contact with the antigen.[58–60] Such necessary framework changes may introduce new antigenic epitopes and, if many changes are needed, the advantages of CDR grafting over chimeric antibody constructions[61] will be lost.

Minimization of the number of back mutations is therefore desirable. This can only be carried out in a rational manner using accurate methods to model the entire variable regions, requiring correct docking of V_L and V_H domains and accurate modeling of all six CDRs. This topic will be covered in detail elsewhere in this volume, but the following example illustrates the use of a modeling protocol in combination with a rapid experimental test.

Humanization by Resurfacing

In this method of humanization the surface residues only on the murine Fv are replaced. As discussed earlier, the surface residue positions are highly conserved. First, the V_L and V_H surface sets are searched against a human Fv sequence database and the nearest human V_L–V_H pair is selected.[26] Surprisingly, when this set is superimposed on the murine V_L–V_H set, very few changes are required to convert one to the other. In antibody N901[62] only three murine residues in V_L and seven residues in V_H required change to convert the murine surface to a human surface. The final "resurfaced" antibody was modeled by *ABM* and compared with a model of murine N901. Any residue in the framework region within 5 Å of a CDR residue atom was inspected. In one instance, shown in Fig. 1–13, leucine-3 of the murine framework has an interaction with residues arginine-24, serine-25, and serine-26 (numbering as in Roguska et al.[28]). In the human sequence selected for its surface identity to N901, position 3 is a valine. The two models indicated a major difference in the conformation of arginine-24 when valine was present. On conversion back to leucine, the murine conformation was reproduced. Synthesis and expression of this resurfaced N901, back-mutated at only this one position, led to a humanized antibody with the identical affinity to the murine parent. This example demonstrates two points:

1. Humanization can be carried out either by CDR grafting (the usual method) or by resurfacing.
2. Models of combining sites during either process are essential.

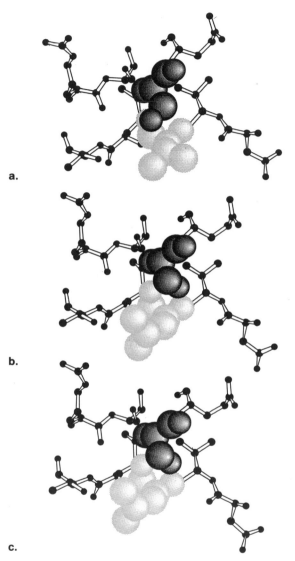

Figure 1–13. The interaction of the light-chain surface framework residue Leu-3 with residues Arg-24, Ser-25, and Ser-26 in antibody N901 (Griffin et al.[62]). (a) Conformation of the region around Leu-3 in the model of murine antibody N901. The leucine is shown in light gray and a neighboring serine (Ser-26) in dark gray. (b) Conformation of the same region in a resurfaced version of N901 where a human antibody has been selected to form the framework region. In this instance, the leucine has been replaced by a valine (light gray) while the other residues remain unchanged. Ser-26 (dark gray) can be seen to have changed its conformation. (c) Conformation of the same region after the valine has been replaced by leucine. All atoms have returned to their original positions (as in (a)).

ANTIBODY–ANTIGEN INTERACTIONS

Thermodynamic Perspectives

When two molecules associate, the "tightness" of the association, measured experimentally as the affinity (K_A), depends on a number of thermodynamic factors, both enthalpic and entropic.

The Negative Side

Each molecule when free in aqueous solution will have considerable configurational (translational and rotational) entropy that will be lost when association occurs. In addition, the surface residues of the two molecules will have H-bond, salt-bridge (with dissolved ions), and van der Waals contact with the water that will be lost when the antibody and antigen surfaces make contact.

The Positive Side

Bound water, when released from the two interacting surfaces, will lead to an increase in entropy of the solvent. Hydrophobic residues that become buried at the interface will similarly cause an increase in entropy of the solvent. H-bonds, salt bridges, and van der Waals interactions will form between the antibody and antigen.

The Net Effect

The general dogma for protein–protein interactions is that the free energy of association is related to the amount of surface area buried at the interface.[63] This in turn led to the notion that buried hydrophobic area would contribute the major driving force for association. While this notion seems reasonable for some large protein–protein complexes, it is inadequate to explain the high affinity some smaller molecules exhibit for antibodies.

In fact, the hydrophobicity model is somewhat overemphasized since, as Horton and Lewis[64] have shown, one measure of the hydrophobic potential, the solvation free energy, correlates poorly (12%) with the free energy of association (Fig. 1–14(a)). However, when the solvation term is broken down into its polar and apolar components, the correlation reappears (63% for polar and 77% for apolar; Fig. 1–14(b)). This suggests that polar interactions, measured by the enthalpy of association, will be at least as important as hydrophobic interactions in formation of a tight antibody–antigen complex. This is an important issue for antibody engineers since, to improve affinity, both hydrophobic and polar regions of the antibody combining site should be targets for mutation.

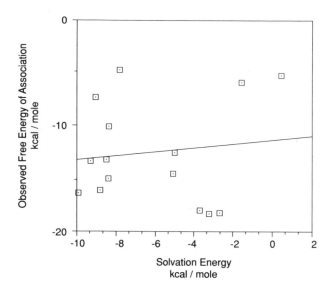

a.

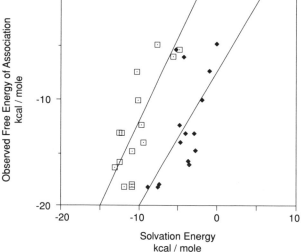

b.

Figure 1–14. (a) The observed free energy of association (kcal/mol) plotted against the calculated solvation energy. The open squares represent different proteins known to associate as rigid objects. (b) The observed free energy of association (kcal/mol) plotted against the calculated free energy of association where the polar and apolar components are calculated separately. Open squares represent the energy attributed to the apolar interactions and solid diamonds are the result of −1 times the polar contribution to the free energy. (Data in (a) and (b) reproduced from Horton and Lewis,[64] with permission).

Anti-Hapten Antibodies

X-ray crystallographic studies of hapten antibody complexes have demonstrated that, in general, the hapten tends to be buried in a cavity (see Fig. 1–8). Three examples are summarized in Table 1–6(a). The data from these hapten–antibody complexes suggest that the greater the surface area buried, the higher the affinity. However, it is likely that, for small molecules, their flexibility and associated entropy loss on binding will determine to a large extent the magnitude of the observed binding constant. In the three examples given, the order of increasing flexibility would be fluorescein < phenyloxazolone < phosphocholine. This entropy loss can be compensated for by numerous "good" contacts with the antibody such as van der Waals, H-bonds, or salt bridges.

Antibody engineering strategies should therefore attempt to make such changes by targeting residues likely to be in direct contact, or more distant residues whose modification may cause contact residues to form. To guide such strategies, either an X-ray structure or a model is required. In situations where conformational changes accompany the binding, structures of both the free and complexed antibody will be required to make complete sense of the interaction. Recently, Lascombe et al.[65] have suggested that an anti-arsenate antibody may undergo a large change in the conformation of CDR H3 on binding the hapten. Thus, a structure or model of the free Fab may not provide the perfect starting point for the antibody engineer. However, such examples, dramatically exemplified in anti-peptide antibodies (see later) provide for the interesting possibility of engineering a CDR so that its position is more rigidly fixed at or near its bound-state conformation, thus giving an entropic advantage to the system.

Anti-Peptide Antibodies

In 1984, Tainer et al.[66] predicted that antibodies directed against peptides would normally exhibit cross-reactivity with the parent protein if the corresponding regions in the protein were highly "mobile." Several alternative views have since been proposed,[67,68] but what has become clear is that the peptide in antibody–peptide complexes tends to be conformationally variable. The first antibody–peptide complex studied by X-ray crystallography[69] involved a 19-residue peptide derived from myohemerythrin. However, only nine peptide residues could be seen in the electron density, the remainder presumably being too mobile. This mobility in the bound state was studied directly by Rees and colleagues[70] using NMR and established that, while some residues of a lysozyme peptide become very tightly bound on complex formation, others are as mobile as in the free peptide. Recently, the structure of a further Fab–peptide complex has been reported that demonstrates an

Table 1-6. Some physicochemical characteristics of (a) antibody–hapten, (b) antibody–peptide, and (c) antibody–protein complexes*

Hapten	Buried Surface Area, $(Å)^2$	vdW Contacts	H-Bonds	Salt Bridges	K_a, M^{-1}	Antibody Name
(a)						
Phosphocholine	137	30	2	3	1.7×10^5	McPC603
Phenyloxazalone	170	>15**	>1	0	1.9×10^8	NQ10/12.5
Fluorescein	266	65	5	1	3.4×10^{10}	4-4-20
(b)						
Influenza haemagglutinin residues 100–108	400	74–81	13–15	1	5.0×10^7	anti-HA1
(c)						
Hen egg lysozyme	680	75	15	0	1×10^9	D1.3
Hen egg lysozyme	774	111	14	1	5×10^9	HyHEL10
Hen egg lysozyme	750	74	10	3	2×10^{10}	HyHEL5
Influenza neuraminidase	879	108	23	1	NA	NC41
D1.3	800	99***	9	1	NA	anti-D1.3

* Data for McPC603, 4-4-20, D1.3, HyHEL10, HyHEL5, and NC41 are taken from Davies et al.,[99] the D1.3–anti-D1.3 are taken from Bentley et al.,[99] and those for phenyloxalalone are from Alzari et al.[100]
** Not yet fully refined.
*** Author's estimate.

induced-fit mechanism for peptide binding.[48] While eight of the nine peptide residues can be located in the electron density, and are thus relatively fixed, CDR H3 is seen to undergo a major conformational change on binding of peptide. This change is defined as a twisting of the two strands about the long axis of the H3 loop, giving rise to residue movements of around 2 to 5 Å. The characteristics of the bound complex are shown in Table 1–6(b). These data show that, for a larger antigen, the number of contacts made with the antibody is proportional to the surface area buried. However, the affinity does not correlate in the same way. This may reflect the increased flexibility of a peptide compared with small molecules and the additional entropy loss accompanying immobilization of the antibody. Additionally, in the example shown, the loss in flexibility of CDR H3 on binding will also contribute to the entropy loss.

It would seem that antibodies to peptides and flexible ligands reach an upper affinity limit of about 10^8 M^{-1}. This is most likely to originate from the large loss of configurational entropy when they become bound. In contrast, it is possible to generate affinities of as much as 10^{12} M^{-1} for small, rigid haptens and 10^{10} M^{-1} for protein antigens. If high-affinity antibodies are required for therapeutic or diagnostic applications, it may be that peptides are not the ideal choice as immunogens.

Anti-Protein Antibodies

The great majority of antibodies have been raised against protein antigens. Of this large number, only seven high-resolution X-ray structures are known for Fab–antigen complexes. These are summarized in Table 1–6(c). Although the affinities of the NC41–neuraminidase and D1.3–anti-D1.3 complexes are not available, they are likely to be at least 10^9 M^{-1} (10^9 M^{-1} seems to be the minimum necessary for crystallization of protein complexes). What do these data tell us about protein antigen–antibody interactions? First, there is no obvious correlation between the number or type of interaction seen (van der Waals, H-bond, salt bridge) and affinity. Second, the buried surface area is not a good indicator of affinity. This might be the case if burial of hydrophobic residues and release of water molecules were the major contributors to the free energy of binding. In fact, recent work by Poljak et al.[96] suggest the reverse. During measurements of the enthalpic and entropic contributions to the association of HEL with D1.3 using micro-calorimetry, it was found that the binding was substantially enthalpically driven while the entropy actually decreased. While this may not always be the case, it is clear that the differing affinities observed in different complexes may have their origins in rather different profiles of interaction. A good fit of the two surfaces is clearly important for high affinity, but having achieved that, the thermodynamics may be driven by many different combinations of hydrophobic and electrostatic interactions. What is likely is that it will be more of a balance between the two than has been seen in other protein–protein complexes since, for much of its life, the antibody must retain stability in the *absence* of any bound antigen.

Thus, for the antibody engineer the problem may appear somewhat daunting, particularly as many different combinations of CDRs provide the interacting residues for the antigen:

MCPC603	L1		H1	H2		H3	
4-4-20		L1		L3	H1	H2	H3
NQ10/12.5	L1		L3	H1		H3	
D1.3	L1	L2	L3	H1	H2	H3	
HYHEL-5	L1	L2	L3	H1	H2	H3	
HYHEL-IU	L1	L2	L3	H1	H2	H3	
NC41		L2	L3	H1	H2	H3	
E225	L1	L2	L3		H2	H3	

It may be that, for the foreseeable future, the creation of libraries that scramble both CDR length and sequence will be the only effective way to engineer affinity and specificity in a timely fashion.

FUTURE PROSPECTS

The antibody is unique in the repertoire of protein molecules. The construction of the variable region allows for both rigid antibody–antigen docking and for induced fit mediated either by V_L–V_H domain rotational or sliding motions or by CDR conformational change. The V_L–V_H association is stable ($K_A \sim 10^{10}$ M^{-1}) and can be engineered by cross-linking to be even more kinetically stable, as discussed in another part of this volume. The enormous number of CDR length and sequence combinations provides a goldmine of binding molecules. Indeed, the antibody has become the "gold standard" in discussions of molecular recognition. All of these characteristics can be harnessed by tapping the natural repertoire. Why then should there be a role for antibody engineering? There are three reasons:

1. Antibodies are a recent evolutionary product. The genes encoding variable domains are governed by strict laws of natural selection and a somewhat slow evolutionary rate. The engineer is not restricted and, either by molecular library approaches or rational replacement of residues or whole CDRs, can create combining site shapes and chemical constitutions not yet discovered by nature (e.g., metallo antibodies).

2. Antibodies are large and contain constant regions without which the binding function would be pointless. As will be shown in this volume, these regions can be replaced with extraordinary ease to produce multifunctional proteins, or "magic bullets."

3. The rules that govern the interaction between antibodies and the enormous variety of antigenic shapes, sizes, and chemical character will only be deduced by a combination of X-ray crystallographic, molecular modeling, and protein-engineering studies. Only when such rules of recognition arrive will it be possible to generate antibodies with predetermined specificity and affinity by design.

Acknowledgments

We would like to thank the SERC (UK), the Danish Research Foundation, British Biotechnology Ltd., Wellcome Research Ltd., and Amersham International for financial support. We also thank Siew Quen Yeap of Oxford Molecular Ltd. for help with reproducing the protocol used by the program AbM for modeling the test antibody D1.3.

REFERENCES

1. Wu, T., and Kabat, E. 1970. Analysis of the sequences of the variable regions of Bence Jones proteins and myeloma light chains and their implications for antibody complementarity. *J. Exp. Med.* 132:211–50.

2. Wu, T., and Kabat, E. 1971. Attempts to locate complementarity-determining residues in the variable positions of light and heavy chains. *Ann. N.Y. Acad. Sci.* 190:382–93.

3. Poljak, R.J., Amzel, L.M., Avery, H., Chen, B.L., Phizackerley, R.P., and Saul, F. 1973. Three-dimensional structure of the Fab-fragment of a human immunoglobulin at 2.8 Å resolution. *Proc. Natl. Acad. Sci. USA* 70:3305–10.

4. Rees, A.R., Martin, A.C.R., Pedersen, J.T., and Searle, S.M.J. 1992. ABM™, a computer program for modeling variable regions of antibodies. Oxford, U.K.: Oxford Molecular Ltd.

5. Porter, R. 1962. In *Symposium on Basic Problems in Neoplastic Disease.* Gelhorn, A., and Hirschberg, E. eds. Columbia University Press. p. 177.

6. Edelman, G.M., and Poulik, M.D. 1961. Structural studies of the gamma globulins. *J. Exp. Med.* 113:867.

7. Deisenhofer, J., Colman, P.M., and Huber, R. 1976. Crystallographic structural studies of a human Fc fragment. I. An electron-density map at 4 Å resolution and a partial model. *Hoppe-Seyler's Z. Physiol. Chem.* 357:435–45.

8. Silverton, T.W., Naria, M.A., and Davies, D.R. 1977. Three-dimensional structure of an intact immunoglobulin. *Proc. Natl. Acad. Sci. USA* 74:5140–44.

9. Rajan, S.S., Ely, K.R., Abola, E.E., Wood, M.K., Colman, P.M., Athay, R.J., and Edmundson, A.B. 1983. Three-dimensional structure of the MCG IgC1 immunogloglobulin. *Mol. Immunol.* 20:787–799.

10. Harris, L.J., Larson, S.B., Hasel, K.W., Day, J., Greenwood, A., and McPherson, A. 1992. The three-dimensional structure of an intact monoclonal antibody for canine lymphoma. *Nature* 360:369–72.

11. Barclay, A.N., Birkeland, M.L., Brown, M.H., Beyers, A.D., Davis, S.J., Somoza, C., and Williams, A.F. 1993. *The Leucocyte Antigen Facts Book.* London, U.K.: Academic Press (ISBN 0-12-M8180-1).

12. Colcher, D., Horan Hand, P., Nuti, M., and Schlom, J. 1981. A spectrum of monoclonal antibodies reactive with human mammary tumor cells. *Proc. Natl. Acad. Sci. USA* 78:3199–203.

13. Chothia, C., Novotny, J., Bunccoleri, R.E., and Karplus, M. 1985. Domain association in immunoglobulin molecules—the packing of variable domains. *J. Mol. Biol.* 186:617–63.

14. Novotny, J., Bruccoleri, R., Newell, J., Murphy, D., Huber, E., and Karplus, M. 1983. Molecular anatomy of the antibody binding site. *J. Biol. Chem.* 258:14433–37.

15. Novotny, J., Bruccoleri, R., and Newell, J. 1984. Twisted hyperboloids (strophoid) as a model of β-barrels in proteins. *J. Mol. Biol.* 177:567–73.

16. Darsley, M.J., Phillips, B.C., Rees, A.R., Sutton, B.J., and de la Paz, P. 1985. An approach to the study of anti-protein antibody combining sites. In *Investigation and Exploitation of Antibody Combining Sites.* Plenum Press. pp. 63–68.

17. de la Paz, P., Sutton, B.J., Darsley, M.J., and Rees, A.R. 1986. Modelling of the combining sites of three anti-lysozyme monoclonal antibodies and the complex between one of the antibodies and its epitope. *EMBO J.* 5:415–25.

18. Wilmot, C.M., and Thornton, J.M. 1988. Analysis and prediction of the different types of β-turns in proteins. *J. Mol. Biol.* 203:221–32.

19. Rose, G.D., Gierasch, L.M., and Smith, J.A. 1985. Turns in peptides and proteins. *Advan. Protein Chem.* 37:1–109.

20. Chothia, C., and Lesk, A.M. 1987. Canonical structures for the hypervariable regions of immunoglobulins. *J. Mol. Biol.* 196:901–17.

21. Chothia, C., Lesk, A.M., Tvamontano, A., Levitt, M., Smith-Gill, S.J., Air, G., Sheriff, S., Padlan, E.A., Davies, D., and Tupil, M.R. 1989. The conformations of immunoglobulin hypervariable regions. *Nature* 342:877–83.

22. Chothia, C., Lesk, A.M., Gherardi, E., Tomlinson, I.M., Walter, G., Marks, J.D., Llewelyn, M.B., and Winter, G. 1992. Structural repertoire of the human VH segments. *J. Mol. Biol.* 227:799–817.

23. Steipe, B., Plückthun, A., and Huber, R. 1992. Refined crystal structure of a recombinant immunoglobulin domain and a complementarity-determining region 1-grafted mutant. *J. Mol. Biol.* 225:739–53.

24. Pedersen, J. 1993. Molecular modelling of antibody combining sites. Ph.D. Thesis, University of Bath, U.K.

25. Wang, D., Ligo, J., Mitra, D., Akolkar, P., Gruezo, F., and Kabat, B. 1991. The repertoire of antibodies to a single antigenic determinant. *Mol. Immunol.* 28:1387–97.

26. Pedersen, J., Henry, A.H., Searle, S.J., Guild, B.C., Roguska, M., and Rees, A.R. 1994. Comparison of surface accessible residues in human and murine

immunoglobulin Fv domains; implication for humanisation of murine anti-bodies. *J. Mol. Biol.* 235:959–73.

27. Kabat, E.A., Wu, T.T., Reid-Miller, M., Perry, H.M., and Gottesman, K.S. 1992. *Sequences of Proteins of Immunological Interest.* U.S. Department of Health and Human Services 5th Edition. USA.

28. Roguska, M.A., Pedersen, J.T., Keddy, C.A., Henry, A.H., Searle, S.M.J., Lambert, J.M., Goldmacher, V.S., Blatther, W.A., Rees, A.R., and Guild, B.C. 1994. Humanisation of murine monoclonal antibodies through variable domain resurfacing. *Proc. Natl. Acad. Sci.* 91:969–73.

29. Padlan, E.A., Daview, D.R., Pecht, I., Girol, D., and Wright, C. 1976. Model building studies of antigen binding sites: the hapten-binding site of MOPc315. *Cold Spring Harbor Symp. Quant. Biol.* 41:627–37.

30. Mainhart, C.R., Potter, M., and Feldman, R.M. 1984. A refined model for the variable domains of the J539 (1,6) D-galactan binding immunoglobulin. *Mol. Immunol.* 21:469–78.

31. Snow, M.E., and Anzel, L.M. 1986. Calculating three-dimensional changes in protein structure due to amino acid substitutions: the variable region of immunoglobulins. *Proteins Struct. Funct. Genet. J.* 267–79.

32. Smith-Gill, S.J., Mainhart, R., Lavote, T.B., Feldman, R.J., Drohan, W., and Brooks, B.R. 1987. A 3-dimensional model of an anti-lysozyme antibody. *J. Mol. Biol.* 194:713–24.

33. Padlan, E.A., and Kabat, E.A. 1988. Model building study of the combining sites of two antibodies to alpha (1 → 6) Dextran. *Proc. Natl. Acad. Sci. USA* 85:6885–89.

34. Feldmann, R.J., Potter, M., and Glaudemans, P.J. 1981. A hypothetical space-filling model of the V-regions of the galactan-binding myeloma immunoglobulin J539. *Mol. Immunol.* 18:683–98.

35. Fine, R.M., Wang, H., Sheritan, P.S., Yarmush, D.L., and Levinthal, C. 1986. Predicting antibody hypervariable loop conformations II: minimisation and molecular dynamics studies of MCPC505 from many randomly generated loop conformations. *Proteins Struct. Func. Genet.* 1:342–62.

36. Bruccleri, R.E., Heuber, E., and Novotny, J. 1988. Structure of antibody hypervariable loops reproduced by a conformational search algorithm. *Nature* 335:564–68.

37. Moult, J., and James, M.N.G. 1986. An algorithm which predicts the con-formation of short lengths of chains in proteins. *Proteins Struct. Funct. Genet.* 1:146–54.

38. Bassolino-Klimas, D., Bruccoleri, R.E., and Subraminian, S. 1992. Modeling the antigen combining site of an anti-dinthrophenyl antibody, AN01. *Protein Science* 1:1465–76.

39. Higo, J., Collura, V., and Garner, J. 1992. Development of an extended simulated annealing method—application to the modelling of complementarity determining regions of immunoglobulins. *Biopolymers* 32:33–43.

40. Martin, A.C.R., Cheetham, J.C., and Rees, A.R. 1989. Modeling antibody hyper-variable loops: a combined algorithm. *Proc. Natl. Acad. Sci. USA* 86:9268–72.

41. Martin, A.C.R., Cheetham, J.C., and Rees, A.R. 1991. Molecular modeling of antibody combining sites. *Methods Enzymol.* 203:121–52.

42. Pedersen, J.T., Searle, S.J., Henry, A.H., and Rees, A.R. 1992. Antibody modeling: beyond homology. *Immunomethods* 1:126–36.
43. Bruccoleri, R.E., and Karplus, M. 1987. Prediction of the folding of short polypeptide segments by uniform conformational sampling. *Biopolymers* 26: 137–68.
44. Jones, T.A., and Thirup, S. 1986. Using known structures in protein model building and crystallography. *EMBO J.* 5:819–22.
45. Dauber-Osguthorpe, P., Campbell, M.M., and Osguthorpe, D. 1991. Conformational analysis of peptide surrogates. *Int. J. Pept. Res.* 38:357–77.
46. Lesk, A.M., and Tramontano, A. 1991. Antibody structure and structural predictions useful in guiding antibody engineering. In *Antibody Engineering Manual*, New York: W.H. Freeman, 1st edition. pp. 1–38.
47. Tai, T., Wu, T., Johnson, G., and Kabat, E.A. 1993. Length distribution of CDR H3 in antibodies. *Proteins Struct. Funct. Genet.* 16:1–7.
48. Rini, J.M., Schulzegahmen, V., and Wilson, I.A. 1992. Structural evidence for induced fit as a mechanism for antibody-antigen recognition. *Science* 255:959–65.
49. Roberts. S., Cheetham, J.C., and Rees, A.R. 1987. Generation of an antibody with enhanced affinity and specificity for its antigen by protein engineering. *Nature* 328:731–34.
50. Sharon, D. 1990. Structural correlates of high antibody affinity: three engineered amino acid substitutions can increase the affinity of an anti-p-azophenylarsonate antibody 200-fold. *Proc. Natl. Acad. Sci. USA* 87:4814–17.
51. Denzin, L.K., Whitlow, M., and Voss, G.W. Jr. 1991. Single chain site-specific mutations of fluorescein amino acid contact residues in high affinity monoclonal antibody 4-4-20. *J. Biol. Chem.* 266:14095–103.
52. Near, R.I., Mudgeh-Hunter, M., Novotny, J., Bruccoleri, R. and Chung Ng, S. 1993. Characterisation of an anti-digoxin antibody binding site by site-directed *in vitro* mutagenesis. *Molec. Immunol.* 30:369–77.
53. Roberts, V.A., Iversen, B.L., Iversen, S.A., Benkovic, S.J., Lerner, R.A., Getzoff, E.D., and Tainer, J.A. 1990. Antibody remodeling: a general solution to the design of a metal-coordination site in an antibody building pocket. *Proc. Natl. Acad. Sci. USA* 87:6654–58.
54. Wade, W.A., Koh, J.S., Han, N., Hoekstra, D.M., and Lerner, R.A. 1993. Engineering metal coordination sites into the antibody light chain. *J. Am. Chem. Soc.* 115:4449–56.
55. Gregory, D.S., Martin, A.C.R., Cheetham, J.C., and Rees, A.R. 1993. The prediction and characterisation of metal binding sites in proteins. *Protein Eng.* 6:29–35.
56. Rees, A.R., Staunton, D., Webster, D.M., Searle, S.J., Henry, A.H., and Pedersen, J.T. 1994. Antibody design: beyond the limits. *Trends in Biotech.* 12:199–206.
57. Wade, W.S., Ashley, J.A., Jahangiri, G.K., McElhaney, G., Janda, K.D., and Lerner, R.A. 1993. A highly specific metal-activated catalytic antibody. *J. Am. Chem. Soc.* 115:4906–7.
58. Riechman, L., Clark, M., Waldmann, H., and Winter, G. 1988. Reshaping antibodies for therapy. *Nature* 332:323–27.
59. Kettleborough, C.A., Saldanha, J., Heath, V.J., Morrison, C.J., and Bendig, M.M.

1991. Humanization of a mouse monoclonal antibody by CDR-grafting: the importance of framework residues on loop conformation. *Protein Eng.* 4:773–83.

60. Foote, J., and Winter, C.T. 1992. Antibody framework residues affecting the conformations of the hypervariable loops. *J. Mol. Biol.* 224:487–99.

61. Morrison, S., Johnson, M.J., Herzenberg, S.A., and Oi, V.T. 1984. Chimeric human antibody molecules—mouse antigen binding domains with human constant region domains. *Proc. Natl. Acad. Sci. USA* 81:6851–55.

62. Griffin, J.D., Hercend, T., Beveridge, R., and Schlossman, S.F. 1983. Characterisation of an antigen expressed by human natural killer cells. *J. Immunol.* 130:2947–57.

63. Janin, J., and Chothia, C. 1990. The structure of protein–protein recognition sites. *J. Biol. Chem.* 265:16027–30.

64. Horton, N., and Lewis, M. 1992. Calculation of the free energy of association for protein complexes. *Protein Sci.* 1:169–81.

65. Lascombe, M-B., Alzari, P.M., Poljak, R.J., and Nisonoff, A. 1992. Three-dimensional structure of two crystal forms of Fab 12 19.9 from a monoclonal anti-arsonate antibody. *Proc. Natl. Acad. Sci. USA* 89:9429–33.

66. Tainer, J.A., Gretzoff, E.D., Alexander, H., Houtgen, R.A., Olson, A.J., and Lerner, R.A. 1984. The reactivity of anti-peptide antibodies is a function of the atomic mobility of sites in a protein. *Nature* 312:127–34.

67. Novotny, J., Handschumacher, M., Haber, E., Bruccoleri, R.E., Carlson, W.B., Fanning, D.W., Smith, J.A., and Rose, G.D. 1986. Antigenic determinants in proteins coincide with surface regions accessible to large probes (antibody domains). *Proc. Natl. Acad. Sci. USA* 83:226–30.

68. Thornton, J.M., Edwards, M.S., Taylor, W.R., and Barlow, D.J. 1986. Location of continuous antigenic determinants in the protruding regions of proteins. *EMBO J.* 5:409–13.

69. Stanfield, R.L., Fieser, T.M., Levner, R.A., and Wilson, I.A. 1990. Crystal structure of an antibody to a peptide and its complex with peptide antigen at 2.9 Å resolution. *Science* 248:712–19.

70. Cheetham, J.C., Raleigh, D.P., Griest, R.E., Redfield, C., Dobson, C.M., and Rees, A.R. 1990. Antigen mobility in the combining site of an anti-peptide antibody. *Proc. Natl. Acad. Sci. USA* 88:7968–72.

71. Sheriff, S., Silverton, E.W., Padlan, E.A., Chen, G.H., Smith-Gill, S.J., Finnel, B.C., and Davies, D.R. 1987. Three-dimensional structure of an antibody-antigen complex. *Proc. Natl. Acad. Sci. USA* 84:8075–79.

72. Padlan, E.A., Silverton, E., Sheriff, S., Cohen, G., Smith-Gill, S.J., and Davies, D.R. 1989. Structure of an antibody-antigen complex: crystal structure of the HyHEL10 Fab-lysozyme complex. *Proc. Natl. Acad. Sci. USA* 86:5938–42.

73. Schiffer, M., Ainsworth, C., Xu, B., Carperos, K., Olsen, A., Soloman, F., Stevens, C., and Cahng, H. 1989. Structure of a second crystal form of Bence-Jones protein LOC: strikingly different domain association in two crystal forms of a single protein. *Biochemistry* 28:4066–72.

74. Mainhart, C.R., Potter, M., and Feldman, R.M. 1984. A refined model for the variable domains of the J539 $\beta(1,6)$ D-galactan binding immunoglobulin. *Mol. Immunol.* 21:469–78.

75. Saul, F.A., Amzel, L.M., and Poljak, R.J. 1978. The preliminary refinement and structural analysis of the Fab fragment from human immunoglobulin NEW at 2 Å resolution. *J. Biol. Chem.* 253:585–97.

76. Herron, J., He, X., Mason, M., Voss, E., and Edmundson, A. 1989. Three-dimensional structure of a fluorescein-Fab complex crystallised in 2-methyl-2,4-pentanediol. *Proteins. Struct. Func. Genet.* 5:271–80.

77. Rose, D.R., Strong, K.R., Margolis, M.N., Gefter, M.L., and Petsgo, G.A. 1990. Crystal structure of the antigen binding fragment of the murine anti-arsenate antibody 36-71 at 2.9 Å resolution. *Proc. Natl. Acad. Sci. USA* 87:338–42.

78. Segal, D., Padlan, E.A., Cohen, G., Rudikoff, S., Potter, M., and Davies, D.R. 1974. The three-dimensional structure of a phosphorylcholine binding mouse immunoglobulin Fab and the nature of the binding site. *Proc. Natl. Acad. Sci. USA* 71:4298–302.

79. Ely, K., Herron, J., Harker, A., and Edmundson, A. 1989. Three-dimensional structure of a light chain dimer crystallised in water. *J. Mol. Biol.* 210:601–15.

80. Ely, K., Wood, M., Rajan, S., Hodson, J., Abola, E., Deutsch, H., and Edmundson, A. 1985. Unexpected similarities in the crystal structures of the MCG light chain dimer and its hybrid with the WEIR protein. *Mol. Immunol.* 22:93–100.

81. Furey-jnr, W., Wang, B., Yoo, C., and Sax, M. 1983. Structure of a novel Bence-Jones protein (RHE) fragment at 1.6 Å. *J. Mol. Biol.* 167:661–92.

82. Epp, O., Lattmann, E.E., Schiffer, M., Huber, R., and Palm, W. 1975. The molecular structure of a dimer composed of the variable portions of the Bence-Jones protein REI refined at 2.0 Å resolution. *Biochemistry* 14:4943–52.

83. Marquart, M., Diesenhofer, J., and Huber, R. 1980. Crystallographic refinement and atomic models of the intact immunoglobulin molecule KOL and its antigen binding fragment at 3.0 Å and 1.9 Å resolution. *J. Mol. Biol.* 141:369–91.

84. Lascombe, M-B., Alzari, P., Boulot, G., Salujian, P., Tougard, P., Berek, C., Haba, S., Rosen, E., Nissonof, A., and Poljak, R. 1989. Three-dimensional structure of FabR19.9, a monoclonal antibody specific for the p-azobenzene-arsonate group. *Proc. Natl. Acad. Sci. USA* 86:607–11.

85. Amit, A.G., Mariuzza, R.A., Phillips, S.E.V., and Poljak, R.J. 1986. The three-dimensional structure of an antibody-antigen complex at 2.8 Å resolution. *Science* 233:747–53.

86. Rose, D.R., Przybyska, M., To, R.J., Kayden, C.S., Oomen, R.P., Vorberg, E., Young, M.N., and Bundle, D.R. 1993. Crystal structure at 2.45 Å resolution of a monoclonal Fab specific for the Brucella-A cell wall polysaccharide moiety. *Protein Sci.* 2:1106–13.

87. Saul, F.A., and Poljak, R.J. 1992. Crystal structure of the Fab fragment from the myeloma immunoglobulin IgG HIL at 1.8 Å resolution. In preparation. Preliminary structure entry deposited in the Brookhaven protein data bank.

88. Brunger, A., Leahy, D., Hynes, T., and Fox, R. 1991. 2.9 Å resolution structure of an anti-dinitrophenyl spin labelled monoclonal antibody Fab fragment with bound hapten. *J. Mol. Biol.* 221:239–56.

89. Jeffrey, P.D. 1989. The structure and specificity of immunoglobulins. Ph.D. Thesis, University of Oxford, U.K.

90. Stanfield, R.L., Fieser, T.M., Levner, R.A., and Wilson, I.A. 1990. Crystal

structure of an antibody to a peptide and its complex with peptide antigen at 2.8 Å resolution. *Science* 248:712–19.

91. He, X., Ruker, F., Casale, E., and Carter, D. 1992. Structure of a human monoclonal antibody Fab fragment against gp41 of HIV1. *Proc. Natl. Acad. Sci. USA* 89:7154–58.

92. Fan, Z.C., Shan, L., Guddat, L.W., He, X.M., Gray, W.R., Raison, R.L., and Edmundson, A.B. 1992. 3-dimensional structure of an Fv from human IgM. *J. Mol. Biol.* 228:188–207.

93. Tormo, J., Stadler, E., Skern, T., Auer, H., Kanzler, O., Betzel, C., Blaas, D., and Fita, I. 1992. 3-dimensional structure of the Fab fragment of a neutralising antibody to human rhinovirus serotype-2. *Protein Sci.* 1:1154–61.

94. Jeffrey, P.D., Strong, R.K., Sieker, L.C., Chang, C.Y., Campbell, R.L., Petsko, G.A., Haber, E., Margolies, M.N., and Sheriff, S. 1993. 26-10 Fab-digoxin complex—affinity and specificity due to surface complementarity. *Proc. Natl. Acad. Sci. USA* 90:10310–14.

95. Rini, J.M., Stanfield, R.L., Stura, R.L., Salinas, E.A., Profy, P.A., and Wilson, I.A. 1993. Crystal structure of an HIV type 1 neutralising antibody, 50.1, in complex with its V3 loop peptide antigen. *Proc. Natl. Acad. Sci. USA* 90:6325–29.

96. Tello, D., Goldbaum, F.A., Mariuzza, R.A., Ysern, X., Schwarz, F.P., and Poljak, R.J. 1993. 3-D structure and thermodynamics of antigen biding by anti-lysozyme antibodies. *Biochem. Soc. Transac.* 21:943–46.

97. Tomlinson, I., Walter, G., Marks, J., Llewelyn, M., and Winter, G. 1992. The repertoire of human germline VH sequences reveals about 50 groups of VH segments with different hypervariable loops. *J. Mol. Biol.* 227:776–98.

98. Davies, D.R., Padlan, E.A., and Sheriff, S. 1990. Antibody-antigen complexes. *Ann. Rev. Biochem.* 59:439–74.

99. Bentley, G.A., Boulot, G., Riottot, M.M., and Poljak, R.J. 1990. 3-dimensional structure of an idiotope-anti-idiotope complex. *Nature* 348:254–57.

100. Alzari, P.M., Spinelli, S., Mariuzza, R.A., Boulot, G., Poljak, R.J., Jarvis, J.M., and Milstein, C. 1990. 3-dimensional structure of an anti-phenyloxazolone antibody—the role of somatic mutation and heavy–light chain pairing in the maturation of the immune response. *EMBO J.* 9:3807–14.

101. Carson, M., and Bugg, C.E. 1988. BSRIBBON—a program for producing 3-D ribbon models of macromolecules suitable for interactive graphics display. *J. Appl. Cryst.* 21:578.

102. Webster, D.M., Henry, A.H., and Rees, A.R. 1994. Antibody-antigen interactions. *Curr. Opin. Struct. Biol.* 4:123–29.

CHAPTER 2

Human Monoclonal Antibodies from V-Gene Repertoires Expressed on Bacteriophage

James D. Marks

Monoclonal antibodies have proven to be valuable laboratory and diagnostic reagents and have great potential as human therapeutic agents. Typically, mice are immunized to produce murine monoclonal antibodies. The process is inefficient, however, and only a relatively small number of antibodies are produced. The inability to characterize a large number of antibodies may result in failure to isolate the precise specificity desired. In addition, the murine immune system may completely fail to recognize important human antigens or epitopes, particularly when purified proteins are not available for immunization. Finally, murine monoclonal antibodies are immunogenic when administered to humans, resulting in decreased efficacy over time and the risk of allergic reactions. As a result, human antibodies are preferred for therapy in humans. It has proven extremely difficult, however, to make human monoclonal antibodies using conventional hybridoma technology.[1] In most instances, immunization is not possible due to the toxicity of the immunogen, and even with immunization, few antibodies are produced due to the inefficiency of the fusion process.

In vivo, the humoral immune system produced high-affinity antibodies in an antigen-driven selection process. The process has three key features: (1) the generation of a vast array of diverse antibody molecule genes; (2) the expression of this antibody gene repertoire on the surface of B-lymphocytes, and (3) antigen-driven selection of rare antigen-binding B-lymphocytes for proliferation and differentiation.[2] Higher-affinity antibodies are generated by mutation of the antibody genes of binding B-cells and further antigen selection.

Recently, it has proven possible to mimic the key features of the humoral immune system *in vitro* by expressing antibody fragment gene repertoires on the surface of bacteriophage (phage display). As a result, high-affinity human antibodies can be produced without prior immunization or the use of conventional monoclonal antibody technology.[3] This approach is based on three technical advances in molecular biology: the ability to express antigen binding antibody fragments in *E. coli*,[4–5] the application of the polymerase chain reaction (PCR) to create very large antibody gene repertoires,[6–9] and the ability to express antibody fragments on the surface of filamentous bacteriophage.[10–11] These technical innovations will be explained in detail in the following paragraphs.

REVIEW OF ANTIBODY STRUCTURE

The following sections are more clearly understood if one briefly reviews the structure of antibodies presented in Chapter 1. An antibody is composed of two heavy chains and two light chains. Each light chain is composed of an N-terminal variable (V) domain (V_L) and a constant (C) domain (C_L). Each heavy chain is composed of an N-terminal V domain, three or four C domains, and a hinge region. The V_H and V_L domains consist of four regions of relatively conserved sequence called framework regions (FR1, FR2, FR3, and FR4), which form a scaffold for three regions of hypervariable sequence (complementarity determining regions, CDRs). The CDRs contain most of the antigen-binding residues. The smallest antigen-binding fragment is the Fv, which consists of the V_H and V_L domains (Fig. 2–1). The Fab fragment consists of the V_H–C_H1 and V_L–C_L domains covalently linked by a disulfide bond (Fig. 2–1).

EXPRESSION OF ANTIBODY FRAGMENTS IN BACTERIA

Since the mid 1980s, it has been possible to express recombinant antibodies in eukaryotic cells. The process is inefficient and requires large quantities of DNA due to low transformation efficiencies. This inefficiency prohibits creating and analyzing a large number of clones. The process is also relatively

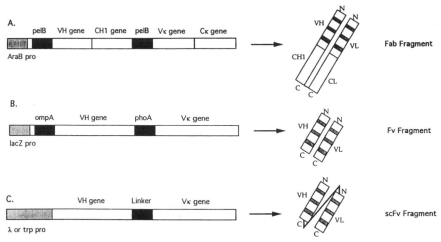

Figure 2–1. Expression of antibody fragments in bacteria. The relevant part of the plasmid gene construct is shown. Expression is driven by a promoter and the expressed protein directed to the periplasmic space by the appropriate signal sequence. A. Construct for expression of Fab fragments.[5] araB pro = arabonse promoter; pelB = pectate lyase leader sequence. B. Construct for expression of Fv fragments.[5] lacZ pro = lacZ promoter; ompA = ompA signal sequence; phoA = alkaline phosphatase signal sequence. C. Construct for expression of scFv fragments.[14–15] λ or trp pro = lambda or tryptophan promoter; linker = synthetic DNA encoding a polypeptide sequence linking the V_H domain to the V_L domain.

slow, taking days to weeks before enough antibody is produced for analysis. The ability to express antibodies in bacteria is ideal for creating, expressing, and analyzing large numbers of different antibodies due to the ease of genetic manipulation, efficient transformation efficiencies, rapid growth, simple fermentation, and favorable economics. For example, a microgram of vector DNA can be used to create more than 100,000,000 recombinant bacteria. Initial attempts to express antibodies in *E. coli* produced very low yields of functional antibody when the heavy and light chains were expressed intracellularly.[12–13] The proteins were contained in "inclusion bodies," and generation of functional native antibody required *in vitro* denaturation and refolding, and the yields were very low. It was presumed that the highly reducing intracellular environment of *E. coli* was not conducive to disulfide bond formation.

A significant breakthrough was achieved in 1988 when two groups simultaneously described the expression of native, correctly folded antibody fragments in *E. coli* in high yields. One group[4] attached the *E. coli* signal sequences ompA (outer membrane protein A) and phoA (alkaline phosphatase) to the V_H and V_L domains of an antibody (Fig. 2–1). The signal sequences directed the expressed domains into the periplasmic space where they folded correctly into functional heterodimeric Fv fragments the V_H and V_κ domains non-covalently linked). Another group[5] used two copies of a different signal sequence, pelB (pectate lyase), to express a functional Fab antibody fragment. More recently, the scFv antibody fragment (the V_H and V_L domains linked by a flexible peptide chain, Fig. 2–1) was developed to overcome the tendency of the non-covalently linked V_H and V_L domains of Fvs to dissociate.[14–15] The polypeptide can link either the C-terminus of the V_H to the N-terminus of the V_L or the C-terminus of the V_L to the N-terminus of the V_H. As long as the linker is of adequate length, a wide range of different linker sequences can be tolerated.[16] scFvs typically have affinities similar to the antibody they are derived from.[17]

In these initial reports of bacterial expression, antibody fragments were harvested from the bacterial periplasm. Subsequently, it has been determined that scFv and Fab antibody fragments are present in large quantities in the bacterial supernatant, and hence unpurified bacterial supernatants can be screened directly for binding to antigen.

GENERATION OF ANTIBODY GENE REPERTOIRES USING THE POLYMERASE CHAIN REACTION (PCR)

Until the advent of PCR, the cloning of antibody genes was a laborious and time-consuming process involving the creation and screening of genomic or cDNA libraries. PCR has greatly simplified the task of obtaining and cloning immunoglobulin DNA and has made it possible to reproduce antibody gene repertoires present *in vivo*. PCR produces amplification of a specific DNA sequence through the repetition of a process involving three steps.[18–19] First, the template DNA is denatured and then synthetic complementary oligo-nucleotide primers are annealed to the 5′ ends of the template. In the final step, DNA polymerase extends the primers, thus replicating the template sequence. The use of a thermal stable DNA polymerase makes it possible to repeat this cycle without the enzyme being denatured during the denatura-tion or annealing steps.[20] By using 30 to 40 cycles, amplification factors of greater than 10^{10} can be achieved.[20–21] After amplification, the DNA can be easily cloned, particularly if restriction sites have been incorporated into the primers.

Optimal PCR results are obtained when amplification is performed by annealing two oligonucleotide primers, one at each end of the target DNA sequence. Design of primers for the 3′ end of immunoglobulin variable region genes was straightforward since primers could be based on the constant regions, all of which have been sequenced. Design of primers for the 5′ end of the V gene was thought to be less straightforward due to the sequence variability of different V-genes, although it was known that primers did not have to be a perfect match for the template DNA.[22] In the earliest attempt to use PCR to amplify V-genes, N terminal protein sequencing was done on purified antibody from a hybridoma and the sequence used to assign the V_H and V_L gene families.[23] The V_H and V_L gene assignments were used to design degenerate primers based on FR1.

A generally applicable approach was taken by Orlandi et al.[6] The nucleotide sequences of murine V_H and V_L genes were extracted from the Kabat database,[24] aligned, and the frequency of the most common nucleotide plotted for each position. Conserved regions were identified at the 5' and 3' regions of the V_H and V_L genes, and the sequences of these regions were used to design oligonucleotide primers. Restriction sites were incorporated into these primers to permit cloning directly into vectors for sequencing and expression in eukaryotic cells. V_H and V_L genes amplified from a hybridoma were expressed in eukaryotic cells, and functional recombinant antibody was produced, verifying this method.[6] PCR with immunoglobulin = specific primers could also be used to amplify genomic V_H and V_L gene repertoires from DNA or RNA prepared from mouse spleens.[7,25–26] Sequence analysis of the amplified V_H and V_L genes indicated that a diverse gene repertoire could be produced.[25]

The general approach described above can be used to reproduce the V_H and V_L gene repertoires present in humans. To maximize gene repertoire diversity and the efficiency of amplification, PCR primers were designed based on the consensus sequence of each V_H and V_L gene family (Table 2–1). For design of PCR primers at the 5' end of the V_H gene (V_HBack primers), 66 human V_H sequences were extracted from the Kabat database,[24] the EMBL database, and all of the V_H sequences published in the literature as of 1988. For design of PCR primers at the 5' end of the V_κ gene (V_κ Back primers), 42 germline and rearranged human V_κ sequences were extracted from the Kabat database,[24] the EMBL database, and the literature. For design of PCR primers at the 5' end of the V_λ gene (V_λBack primers), 36 germline and rearranged human V_λ sequences were extracted from the Kabat database,[24] the EMBL database, and the literature. The V-gene sequences were classified according to family and the frequency of the most common nucleotide at each position used to derive the sequences of a set of family-based primers.[8–9] Amplification primers at the 3' end of the gene (forward primers) were designed based on each of the sequences of the $6J_H$, $5J_\kappa$, or $4J_\lambda$ genes that comprise the 3' terminal portion of the V-gene. Restriction sites were incorporated into the forward and back primers to facilitate cloning. When used together, the forward and back primers resulted in the amplification of the entire V_H, V_κ, or V_λ gene (Fig. 2–1).

To verify the applicability of the primers, RNA was prepared from human peripheral blood lymphocytes (PBLs) and used as a template for first strand

Table 2–1. Oligonucleotide primers used for PCR of human immunoglobulin genes

1. Primers for 1st strand cDNA synthesis

Human Heavy Chain Constant Region Primers
HuIgG1-4CH1FOR 5'-GTC CAC CTT GGT GTT GCT GGG CTT-3'
HuIgMFOR   5'-TGG AAG AGG CAC GTT CTT TTC TTT-3'

Human κ Constant Region Primers
HuC$_\kappa$FOR   5'-AGA CTC TCC CCT GTT GAA GCT CTT-3'

Human λ Constant Region Primers
HcC$_\lambda$FOR   5'-TGA AGA TTC TGT AGG GGC CAC TGT CTT-3'

2. Primers for primary amplifications of V_H, V_κ, and V_λ genes

Human VH Back Primers

HuVH1aBACK   5'-CAG GTG CAG CTG GTG CAG TCT GG-3'
HuVH2aBACK   5'-CAG GTC AAC TTA AGG GAG TCT GG-3'
HuVH3aBACK   5'-GAG GTG CAG CTG GTG GAG TCT GG-3'
HuVH4aBACK   5'-CAG GTG CAG CTG CAG GAG TCG GG-3'
HuVH5aBACK   5'-GAG GTG CAG CTG TTG CAG TCT GC-3'
HuVH6aBACK   5'-CAG GTA CAG CTG CAG CAG TCA GG-3'

Human V$_\kappa$ Back Primers

HuV$_\kappa$1a BACK   5'-GAC ATC CAG ATG ACC CAG TCT CC-3'
HuV$_\kappa$2aBACK   5'-GAT GTT GTG ATH ACT CAG TCT CC-3'
HuV$_\kappa$3aBACK   5'-GAA ATT GTG TTG ACG CAG TCT CC-3'
HuV$_\kappa$4aBACK   5'-GAC ATC GTG ATG ACC CAG TCT CC-3'
HuV$_\kappa$5aBACK   5'-GAA ACT ACA CTC ACG CAG TCT CC-3'
HuV$_\kappa$6aBACK   5'-GAA ATT GTG CTG ACT CAG TCT CC-3'

Human V$_\lambda$ Back primers

Huλ1BACK   5'-CAG TCT GTG TTG ACG CAG CCG CC-3'
Huλ2BACK   5'-CAG TCT GCC CTG ACT CAG CCT GC-3'
Huλ3aBACK   5'-TCC TAT GTG CTG ACT CAG CCA CC-3'
Huλ3bBACK   5'-TCT TCT GAG CTG ACT CAG GAC CC-3'
Huλ4BACK   5'-CAC GTT ATA CTG ACT CAA CCG CC-3'
Huλ5BACK   5'-CAG GCT GTG CTC ACT CAG CCG TC-3'
Huλ6aBACK   5'-AAT TTT ATG CTG ACT CAG CCC CA-3'

Human J$_H$ Forward Primers
HuJH1–2FOR	5'-TGA GGA GAC GGT GAC CAG GGT GCC-3'
HuJH3FOR	5'-TGA AGA GAC GGT GAC CAT TGT CCC-3'
HuJH4–5FOR	5'-TGA GGA GAC GGT GAC CAG GGT TCC-3'
HuJH6FOR	5'-TGA GGA GAC GGT GAC CGT GGT CCC-3'

Human J$_\kappa$ Forward Primers
HuJ$_\kappa$1FOR	5'-ACG TTT GAT TTC CAC CTT GGT CCC-3'
HuJ$_\kappa$2FOR	5'-ACG TTT GAT CTC CAG CTT GGT CCC-3'
HuJ$_\kappa$3FOR	5'-ACG TTT GAT ATC CAC TTT GGT CCC-3'
HuJ$_\kappa$4FOR	5'-ACG TTT GAT CTC CAC CTT GGT CCC-3'
HuJ$_\kappa$5FOR	5'-ACG TTT AAT CTC CAG TCG TGT CCC-3'

Human J$_\lambda$ Forward Primers
HuJ$_\lambda$1FOR	5'-ACC TAG GAC GGT GAC CTT GGT CCC-3'
HuJ$_\lambda$2–3FOR	5'-ACC TAG GAC GGT CAG CTT GGT CCC-3'
HuJ$_\lambda$4–5FOR	5'-ACC TAA AAC GGT GAG CTG GGT CCC-3'

3. Primers to create scFv linker DNA

Reverse J$_H$ primers
RHuJH1–2	5'GCA CCC TGG TCA CCG TCT CCT CAG GTG G-3'
RHuJH3	5'-GGA CAA TGG TCA CCG TCT CTT CAG GTG G-3'
RHuJH4–5	5'-GAA CCC TGG TCA CCG TCT CCT CAG GTG G-3'
RHuJH6	5'-GGA CCA CGG TCA CCG TCT CCT CAG GTG C-3'

Reverse V$_\kappa$ for scFv linker
RHuV$_\kappa$1aBACKFv	5'-GGA GAC TGG GTC ATC TGG ATG TCC GAT CCG CC-3'
RHuV$_\kappa$2aBACKFv	5'-GGA GAC TGA GTC ATC ACA ACC GAT CCG CC-3'
RHuV$_\kappa$3aBACKFv	5'-GGA GAC TGC GTC AAC ACA ATT TCC GAT CCG CC-3'
RHuV$_\kappa$4aBACKFv	5'-GGA GAC TGG GTC ATC ACG ATG TCC GAT CCG CC-3'
RHuV$_\kappa$5aBACKFv	5'-GGA GAC TGC GTG AGT GTC GTT TCC GAT CCG CC-3'
RHuV$_\kappa$6aBACKFv	5'-GGA GAC TGA GTC AGC ACA ATT TCC GAT CCG CC-3'

(*Continued*)

Table 2–1. (*Continued*)

Reverse V$_\lambda$ for svFv linker

RHuV$_\lambda$BACK1Fv	5'-GGC GGC TGC GTC AAC ACA GAC TGC GAT CCG CCA GAG-3'
RHuV$_\lambda$BACK2Fv	5'-GCA GGC TGA GTC AGA GCA GAC TGC GAT CCG CCA GAG-3'
RHuV$_\lambda$BACK3aFv	5'-GGT GGC TGA GTC AGC ACA TAG GAC GAT CCG CCA GAG-3'
RHuV$_\lambda$BACK3bFv	5'-GGG TCC TGA GTC AGC TCA GAA GAC GAT CCG CCA GAG-3'
RHuV$_\lambda$BACK4Fv	5'-GGC GGT TGA GTC AGT ATA ACG TGC GAT CCG CCA GAG-3'
RHuV$_\lambda$BACK5Fv	5'-GAC GGC TGA GTC AGC ACA GAC TGC GAT CCG CCA GAG-3'
RHuV$_\lambda$BACK6Fv	5'-TGG GGC TGA GTC AGC ATA AAA TTC GAT CCG CCA CCG GAG-3'

4. Primers with appended restriction sites for reamplification of scFv gene repertoires

HuVH1aBACKSfi	5'-GTC CTC GCA ACT GCG GCC CAG CCG GCC ATG GCC CAG GTG CAG CTG GTG CAG TCT GG-3'
HuVH2aBACKSfi	5'-GTC CTC GCA ACT GCG GCC CAG CCG GCC ATG GCC CAG GTC AAC TTA AGG GAG TCT GG-3'
HuVH3aBACKSfi	5'-GTC CTC GCA ACT GCG GCC CAG CCG GCC ATG GCC GAG GTG CAG CTG GTG GAG TCT GG-3'
HuVH4aBACKSfi	5'-GTC CTC GCA ACT GCG GCC CAG CCG GCC ATG GCC CAG GTG CAG CTG CAG GAG TCG GG-3'
HuVH5aBACKSfi	5'-GTC CTC GCA ACT GCG GCC CAG CCG GCC ATG GCC GAG GTG CAG CTG TTG CAG TCT GC-3'
HuVH6aBACKSfi	5'-GTC CTC GCA ACT GCG GCC CAG CCG GCC ATG GCC CAG GTA CAG CTG CAG CAG TCA GG-3'
HuJ$_\kappa$1BACKNot	5'-GAG TCA TTC TCG ACT TGC GGC CGC ACG TTT GAT TTC CAC CTT GGT CCC-3'
HuJ$_\kappa$2BACKNot	5'-GAG TCA TTC TCG ACT TGC GGC CGC ACG TTT GAT CTC CAG CTT GGT CCC-3'
HuJ$_\kappa$3BACKNot	5'-GAG TCA TTC TCG ACT TGC GGC CGC ACG TTT GAT ATC CAC TTT GGT CCC-3'
HuJ$_\kappa$4BACKNot	5'-GAG TCA TTC TCG ACT TGC GGC CGC ACG TTT GAT CTC CAC CTT GGT CCC-3'
HuJ$_\kappa$5BACKNot	5'-GAG TCA TTC TCG ACT TGC GGC CGC ACG TTT AAT CTC CAG TCG TGT CCC-3'
HuJ$_\lambda$1FORNot	5'-GAG TCA TTC TCG ACT TGC GGC CGC ACC TAG GAC GGT GAC CTT GGT CCC-3'
HuJ$_\lambda$2–3FORNot	5'-GAG TCA TTC TCG ACT TGC GGC CGC ACC TAG GAC GGT CAG CTT GGT CCC-3'
HuJ$_\lambda$4–5FORNot	5'-GAG TCA TTC TCG ACT TGC GGC CGC ACY TAA AAC GGT GAG CTG GGT CCC-3'

5. Primers for DNA sequencing and PCR fingerprinting

LMB3	5'-CAG GAA ACA GCT ATG AC-3'
fdseq	5'-GAA TTT TCT GTA TGA GG-3'
Linkseq	5'-CGA TCC GCC ACC GCC AGA-3'
pHENseq	5'-CTA TGC GGC CCC ATT CA-3'

cDNA synthesis. First strand cDNA was then amplified by PCR using the appropriate forward primer and one of the 5 family-based V_H Back primers, one of the 6 family-based V_κ Back primers, or one of the 7 V_λ Back primers. The resulting PCR products were cloned and the sequences of more than 100 clones determined. Each clone had a unique sequence. For the V_H genes, 5 of the 6 V_H families were represented; for the V_κ genes, all 6 of the V_κ families were represented; and for the V_λ genes, 4 of the 7 V_λ families were represented.[8–9] Thus, the 5′ and 3′ end of rearranged human V_H and V_L genes proved to be sufficiently conserved to design "universal" primers for PCR amplification of the genes from RNA. The resulting V-gene repertoires are diverse, as determined by DNA sequencing, indicating that it is is possible to reproduce the human antibody V-gene repertoires present *in vivo*.

CREATION OF scFv ANTIBODY FRAGMENT GENE REPERTOIRES

The amplified V_H and V_L gene repertoires can be spliced together using PCR to create scFv antibody fragment gene repertoires that can be cloned in one step for expression in bacteria.[9] To create the spliced scFv repertoires, the amplified V_H, V_κ, and V_λ genes are purified and combined in a PCR reaction containing "linker" DNA, the reactants temperature cycled to join the fragments, and the spliced V-genes amplified by addition of the flanking primers (Fig. 2–2). The linker DNA codes for the peptide $((G_4S)_3)$[15] is complementary to 24 nucleotides at the 3′ end of the V_H gene and 24 nucleotides at the 5′ end of the V_L gene. After purification, the spliced scFv gene repertoire is reamplified with flanking primers containing appended restriction sites NcoI and NotI (Fig. 2 2). These restriction sites are chosen because they do not occur internally in human V_H or V_L germline genes.

The resulting scFv gene repertoire can be digested with NcoI and NotI, ligated into a bacterial expression vector and libraries of greater than 100,000,000 potentially different recombinant clones obtained after transformation of *E. coli*.[9] Available data indicate that the scFv libraries are diverse. Digestion of the scFv insert with a frequently cutting restriction enzyme indicates DNA sequence diversity, and sequencing of greater than 30 clones demonstrates diversity with respect to V_H–V_L pairing and DNA sequence. Moreover, at least 30% of the recombinant clones express high levels of scFv protein as determined by blotting.

Thus the V_H and V_L gene repertoires present *in vivo* can be reproduced by using PCR and spliced together to create scFv antibody fragment gene repertoires. The gene repertoires can be used to create large (> 100,000,000 members) antibody libraries in *E. coli* that express scFv protein. These developments created the possibility of producing human monoclonal

A. 1st strand cDNA synthesis

mRNA

HuIgG1-4FOR
HuIgMFOR

1st Strand VH-CH1 cDNA

mRNA

HuCLFOR

1st Strand VL-CL cDNA

B. Primary PCRs

HuVHBACK

1st Strand VH cDNA

HuJHFOR

VH cDNA

HuVLBACK

1st Strand VL-CL cDNA

HuJLFOR

VL cDNA

C. PCR Assembly

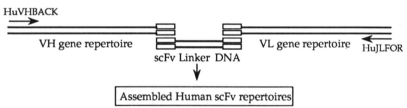

HuVHBACK

VH gene repertoire

scFv Linker DNA

VL gene repertoire

HuJLFOR

Assembled Human scFv repertoires

D. Reamplification with primers containing restriction sites

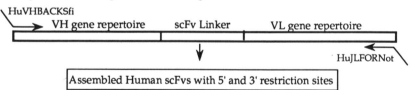

HuVHBACKSfi

VH gene repertoire   scFv Linker   VL gene repertoire

HuJLFORNot

Assembled Human scFvs with 5' and 3' restriction sites

Figure 2–2. Creation of human scFv gene repertoires using PCR splicing. A. mRNA is primed with immunoglobulin heavy- and light-chain constant region specific primers and 1st-strand cDNA synthesized. B. V_H and V_V gene repertoires are amplified from 1st-strand cDNA using family-specific V-gene primers and J-gene primers. C. Purified V_H and V_L gene repertoires are spliced together in a PCR reaction mixture containing DNA, which encodes the scFv linker that is complementary to the 3' end of the V_H gene and the 5' end of the V_L gene. D. The resulting scFv gene repertoire is reamplified with primers containing appended restriction sites. Adapted from Markets et al.[9] with permission.

antibody fragments in *E. coli*, once a method was found for isolating clones expressing binding scFvs from nonbinders.

EXPRESSION OF ANTIBODY FRAGMENTS ON THE SURFACE OF BACTERIOPHAGE

The ability to express antibody fragments on the surface of viruses (bacteriophage, phage) which infect bacteria (phage display) makes it possible to isolate a single binding antibody fragment from a library of greater than 10^8 nonbinding clones. In the first example of phage display, filamentous bacteriophage fd was used to express small peptides fused to the N-terminus of the phage minor coat protein (pIII) by inserting synthetic DNA encoding the peptide into the 5′ end of gene III.[27–28] There are three to five copies of pIII located at the tip of the phage,[29] three of which can be visualized by electron microscopy.[30] During infection, the phage is attached to the bacterial F-pilus by pIII, which appears to form pores through which the phage DNA passes.[31] Despite this critical role of pIII, phages tolerate the peptide insertions without loss of infectivity.[28] More recently, peptides have been fused to the major coat protein, pVIII, coded for by gene VIII.[32–33] There are approximately 2,500 copies of pVIII that make up the body of the phage.

Despite the successful display of peptides on phage, it was not clear whether large proteins such as antibody fragments could be functionally expressed on the surface of phage. McCafferty et al.,[10] however, demonstrated that when the gene encoding an anti-lysozyme scFv antibody fragment was inserted into gene III, the scFv–pIII fusion was incorporated into the phage (Fig. 2–3), allowing the phage to bind lysozyme.[10] In addition, the scFv fusion did not affect the infectivity of the phage. Two groups also demonstrated that heterodimeric Fab fragments could be displayed on the surface as pIII[11] or pIII[34] fusions by linking either the heavy or light chain to a coat protein and secreting the other chain into the bacterial periplasm where the two chains would associate.

Since the antibody fragments on the surface of the phage are functional, phage bearing antigen-binding antibody fragments can be separated from nonbinding phage by antigen affinity chromatography.[10,34–37] Mixtures of phage expressing different antigen specificities are allowed to bind to an affinity matrix, nonbinding phage are removed by washing, and bound phage are eluted by treatment with acid or alkali. Several formats have been used for affinity chromatography of phage, including antigen coupled to columns, dishes, tubes, or on the surface of cells. Enrichment factors of 20 fold to 1,000,000 fold are obtained for a single round of affinity selection.[9,36] By infecting bacteria with the eluted phage, however, more phage can be grown

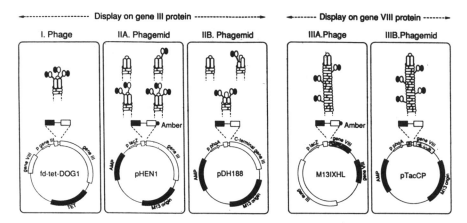

Figure 2–3. Examples of different types of vectors for phage display of scFv or Fab antibody fragments. For pIII fusions, phage vectors (I) result in display of three copies of the antibody pIII fusion, while phagemid vectors (IIA and IIB) result in 0 to 3 copies of fusion protein. Many more copies of antibody fragment are displayed with pVIII fusions. Other details are explained in the text. Antibody fragments are depicted as dark (heavy chain) or light (light chain) shaded ovals; antibody genes are depicted as a dark shaded bar (heavy chain) and a light shaded bar (light chain). For displayed antibody fragments, only infectious phage is illustrated. p gene III = gene 3 promoter, placZ = lacZ promoter, p phoA = alkaline phosphatase promoter, AMP = ampicillin resistance, TET = tetracycline resistance. Adapted from Marks et al.[3] with permission.

and subjected to another round of selection. In this way, an enrichment of 1,000 fold in one round can become 1,000,000 fold in two rounds of selection.[10] Thus even when enrichments are low,[9] multiple rounds of affinity selection can lead to the isolation of rare phage and the genetic material contained within which encodes the sequence of the binding antibody.

The avidity of phage binding to a solid phase coated with antigen depends on the affinity of each molecule of displayed antibody and on the number of antibody fragments per phage that can engage in binding. The number of antibody fragments per phage is largely determined by the choice of either pIII or pVIII coat proteins for fusion and the use of phage or phagemid vectors. Other factors may also contribute, including density of the coating antigen, any association of antibody fragments on the surface of phage as dimers,[38] and proteolysis of the fusion protein.

VECTORS FOR PHAGE DISPLAY OF ANTIBODY FRAGMENTS

A number of different phage and phagemid vectors have been described for display of antibody fragments (Fig. 2–4). The vectors differ mainly in the

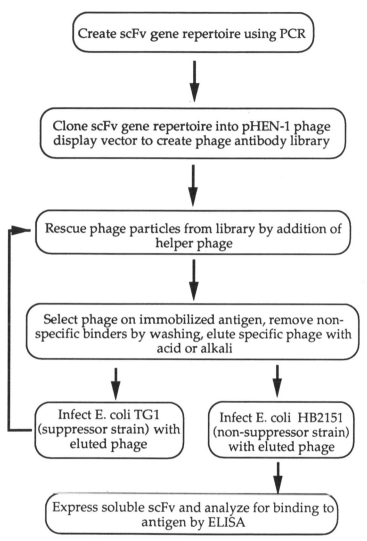

Figure 2–4. Overview of methodology for producing human antibody fragments using phage display. (1) An scFv gene repertoire with appended restriction sites is created using PCR as shown in figure 1. (2) The repertoire is digested with appropriate restriction enzymes and ligated into the phage display vector pHEN-1. The ligation mix is used to transform *E. coli* TG1 and a phase antibody library is produced. (3) Phage particles are rescued from the library by the addition of helper phage. (4) Phage antibodies are incubated with immobilized antigen, nonspecific phage removed by washing, and antigen-specific phage eluted by the addition of acid or alkali. (5) Eluted phage are used to infect *E. coli* TG1 to produce more phage for selection on antigen. (6) After 3 to 4 rounds of selection on antigen, eluted phage are used to infect *E. coli* HB2151 to produce soluble scFv for analysis for binding to antigen using ELISA.

number of antibody fragments per phage. For pIII fusions, phage vectors[10,39] result in three to five copies of antibody fragment per phage. For pVIII fusions, phage[40] or phagemid[34,41] vectors result in many more copies of the fusion protein, for example up to 24 antibody molecules per phage.[34] Multivalent display results in a greater avidity of binding and theoretically might help retain lower affinity phage on antigen during washing, especially for phage with rapid dissociation rate constants. Discrimination between phage with different affinities, however, may be more difficult with multivalent display.[42] Indeed, antibody fragments isolated from V-gene repertoires expressed as pVIII fusions have very low affinity constants and poor antigen specificity.[43]

For pIII fusions, phagemid vectors[11,36,44] result in phage with less than three copies of fusion per phage. This is because phagemid vectors do not carry the genetic material to make phage directly and thus require superinfection with helper phage for successful packaging of phage particles; pIII from the helper phage competes with pIII-antibody fusion from the phagemid vector for incorporation into the phage. The pIII protein has two domains, of which the N-terminal domain is required for infectivity. In phage vectors, fusions must be made to the N-terminal domain or the phage will not be infective. With phagemid vectors, fusions can be made to the N-terminal domain, or the N-terminal domain can be removed and fusions made to the second domain. For fusions to the second domain, however, at least one copy of wild type pIII must be present for the phage to be infectious. Regardless of which domain is used for fusion, with phagemid vectors there is less than one fusion protein per phage on average.[36] Such "monovalent" display appears to enhance the discrimination between phage with different affinities.[37]

Analysis for binding is simplified by including an amber codon between the antibody fragment gene and gene III (see pHEN-1, Figs. 2–3 and 2–5). This makes it possible to easily switch between displayed and soluble antibody fragment simply by changing the host bacterial strain. When phage is grown in a supE suppressor strain of *E. coli*, the amber stop codon between the antibody gene and gene III is read as glutamine and the antibody fragment is displayed on the surface of the phage. When eluted phage is used to infect a nonsuppressor strain, the amber codon is read as a stop codon and soluble antibody is secreted from the bacteria.[11]

PHAGE DISPLAY CAN BE USED INSTEAD OF CONVENTIONAL HYBRIDOMA TECHNOLOGY TO PRODUCE ANTIBODIES

After immunization of an animal or human with antigen, antibodies can be isolated using phage display instead of conventional hybridoma technology.

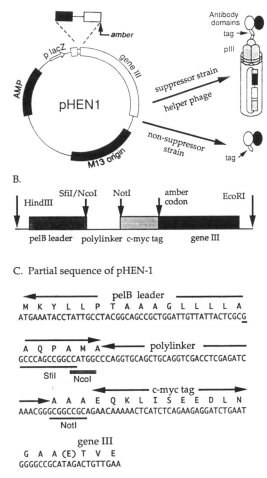

B.

SfiI/NcoI　　　NotI　　　amber　　　　EcoRI
　HindIII　　　　　　　　　　codon

pelB leader　polylinker　c-myc tag　　gene III

C. Partial sequence of pHEN-1

◄──────── pelB leader ────────
M K Y L L P T A A A G L L L L A
ATGAAATACCTATTGCCTACGGCAGCCGCTGGATTGTTATTACTCGCG

───►　　　◄──── polylinker ────
A Q P A M A
GCCCAGCCGGCCATGGCCCAGGTGCAGCTGCAGGTCGACCTCGAGATC
　　　─────　─────
　　　SfiI　　NcoI

　　　　　　　　　　◄──── c-myc tag ────►
───►A A A E Q K L I S E E D L N
AAACGGGCGGCCGCAGAACAAAAACTCATCTCAGAAGAGGATCTGAAT
　　　─────
　　　NotI

　　　　gene III
G A A (E) T V E
GGGGCCGCATAGACTGTTGAA

Figure 2–5. The phage display vector pHEN-1. This phage display vector permits both the display of antibody fragments on the surface of phage (when the host is a suppressor strain of *E. coli*) or soluble expression of antibody fragment (when the host is a suppressor strain of *E. coli*). The vector backbone is pUC119, and the sequence between the HindIII and EcoRI sites of the pUC polylinker is graphically illustrated in section B. The pelB leader sequence directs expressed protein to the periplasmic space. The restriction sites SfiI or NcoI, and NotI, are used to clone the antibody fragment gene repertoire. The c-myc peptide tag permits easy detection of binding in ELISA-based assays via the use of a monoclonal antibody, which recognizes the peptide sequence. The tag can also be used for purification of antibody fragment using affinity chromatography. The presence of an amber codon between the tag and gene III makes it possible to easily switch between antibody fragment displayed on the phage surface and soluble antibody fragment by use of either a suppressor or nonsuppressor *E. coli* host. A partial sequence of pHEN-1 is shown in section C. Further details can be found in Hoogenboom et al.[11]

Repertoires of antibody genes are amplified from immunized mice or humans using PCR and cloned for expression as scFv or Fab antibody fragments on the surface of bacteriophage. The antibody gene repertoires are amplified from lymphocyte or spleen RNA using PCR and oligonucleotide primers optimized for the amplification of murine[6] or human immunoglobulin variable region genes.[8–9] Libraries of at least 10^6 different phage antibodies are produced. Rare phage expressing binding antibody fragments are isolated by affinity selection as described above. Using this approach, high affinity murine antibodies have been made against the hapten phenyloxazolone[39] and human antibodies have been made against tetanus,[44] hepatitis B surface antigen,[45] HIV-1, and respiratory syncytial virus.[46] Many different antibody fragments are produced with affinities (10 nM to 1 nM) that compare favorably to the affinities of monoclonal antibodies produced using conventional hybridoma technology.[39]

PHAGE DISPLAY CAN BE USED TO MAKE ANTIBODIES WITHOUT IMMUNIZATION

Phage display can be used to make antibodies without prior immunization by displaying very large and diverse V-gene repertoires on phage.[9] The natural V-gene repertoires present in human peripheral blood lymphocytes were isolated from unimmunized donors by PCR amplification of the V_H genes from IgM mRNA and the V_κ and V_λ genes from κ or λ mRNA. The V_H and V_L genes were spliced together at random using PCR to create a scFv gene repertoire which was cloned into a phage vector to create a library of 30 million phage, each expressing a different antibody on its surface.[9]

From this single "naive" phage antibody library, binding antibody fragments have been isolated against 17 different antigens, including a hapten, 3 different polysaccharides, and 13 proteins, by panning the phage on immobilized antigen.[9, 38] Overall, at least one binder can be isolated against a protein antigen 70% of the time. Antibodies have been produced against multiple epitopes on the same protein (bovine thyroglobulin, human immunoglobulin, and human tumor necrosis factor) and against self proteins (human thyroglobulin, immunoglobulin, and tumor necrosis factor). The antibody fragments are highly specific for the antigen used for selection and have affinities in the 1 μM to 100 nM range.[9, 38] The affinities are typical for hybridoma antibodies produced in the primary immune response (after a single immunization with antigen).

Recently, antibody fragments against erythrocyte cell surface antigens have been produced without immunization by selecting naive phage antibody libraries directly on cells.[47] Antibodies have been produced against the blood group B antigen (a polysaccharide antigen with 500,000 sites/cell), Rh(D) and Rh(E) antigens (20,000 to 30,000 sites/cell), and the Kpb antigen (5,000 sites/cell). The antibody fragments are highly specific to the antigen used for selection, are functional in agglutination and immunofluorescence assays, and have affinities of 1000 nM to 1 μM.

The above results indicate that highly specific human antibodies with affinities typical of the primary immune response can be produced *in vitro* without prior immunization by creating and expressing diverse antibody fragment repertoires on the surface of bacteriophage. Phage expressing binding antibody fragments are separated from nonbinders by affinity selection on antigen.

PHAGE DISPLAY CAN BE USED TO INCREASE ANTIBODY AFFINITY

Phage display can also be used to increase antibody affinity. The most successful approach has been to combine the original V_H gene with a repertoire of V_L chains to create new partners (chain shuffling).[39] Using light-chain shuffling and phage display, we improved the affinity of an antibody that bound the hapten phenyloxazolone (phOx) from 300 nM to 15 nM (20 fold).[48] Shuffling of the V_H gene, while leaving the V_H CDR3 and new light chain intact, further improved the affinity from 15 nM to 1 nM. In both instances, the shuffled repertoires were derived without immunization.[48] These affinities are typical of hybridoma antibodies produced against the same hapten after three immunizations. DNA sequence analysis indicated that the V_H and V_L genes of the higher-affinity binders were derived from the same germline genes as the original antibody fragment, but had 10 to 23 point mutations. Chain shuffling has proven more effective for increasing antibody affinity than random mutagenesis,[49] probably because the V-genes encoding the shuffled chains are derived from the mRNA of B-lymphocytes and are more likely to be functional, particularly when multiple mutations are introduced.

In conclusion, phage display technology offers a new approach for producing high-affinity human antibodies. When human immunization is possible, limitations of conventional hybridoma technology can be overcome by preparing antibody fragment gene repertoires from mRNA derived from IgG producing B lymphocytes or plasma cells. The repertoires are cloned for expression on phage and the resulting phage library subjected to affinity chromatography on immobilized antigen. In the many instances where immunization is not possible, very large and diverse antibody fragment gene repertoires derived from healthy humans are expressed on phage. Antigen-specific antibody fragments can then be isolated using antigen affinity chromatography. The affinities of the antibody fragments can be increased by creating mutant gene repertoires, expressing them on phage, and subjecting the phage library to rounds of affinity chromatography.

METHODS

The following methods section will detail the techniques and protocols used to produce monoclonal human scFv antibody fragments using phage display. An overview of the steps required is shown in Fig. 2–4. The first methods section describes the generation of immune or "naive" scFv gene repertoires. The second section describes the cloning of the repertoires into the phagemid expression vector pHEN-1 to create a phage antibody library. The third and fourth sections describe how to isolate and characterize antigen-specific binders from the library. The final section briefly describes approaches to increase the affinity of scFv antibody fragments.

Method 1: Generation of Human scFv Gene Repertoires

Overview

This section will describe the protocols used to generate human scFv gene repertoires. The overall scheme is detailed in Fig. 2–2. RNA is prepared from the peripheral blood lymphocytes of either a normal healthy human or an immunized human, depending on whether a naive or immune phage antibody library will be made. cDNA is prepared from the RNA and used as a template for PCR amplification of the V_H and V_L genes. The V_H and V_L genes are spliced together using PCR to create scFv gene repertoires. Finally, restriction sites are appended to the scFv gene repertoires to permit forced cloning into a phage display vector.

Specifics of Method 1

Total RNA is prepared from approximately 10^7 B-lymphocytes using a modified method of Cathala[50] as shown in Protocol 1. The immunoglobulin mRNA present in the total RNA is used to produce first strand cDNA as shown in Protocol 2. For creation of "naive" scFv gene repertoires, RNA is primed with an IgM constant region primer. The IgM V_H gene repertoire will be more diverse than the IgG since it is more likely to represent B-cells that have not been selected and amplified by antigen. For creation of "immune" scFv gene repertoires, RNA is primed with an IgG constant region primer.

Human antibody V_H and V_L gene repertoires are generated from first strand cDNA using PCR as described in Protocol 3.1. The diversity of the gene repertoires is maximized by using both V_κ and V_λ light-chain genes. Diversity is further increased by using PCR primers based on each of the human V_H and V_L gene families and each of the J gene segments (Table 2–1). The purified V_H and V_L genes (V_κ or V_λ) are combined in a second PCR reaction mixture containing DNA which encodes the scFv linker and which is complementary to the 3' end of the V_H gene and the 5' end of the V_L gene (Protocol 3.2). The V_H and V_L genes are first temperature cycled without primers to join the fragments, then flanking primers are added to amplify the scFv gene repertoires. This two-step process increases diversity by avoiding amplification of only a small number of spliced fragments. Finally, restriction sites are added to the scFv gene repertoires by reamplification using primers containing appended restriction sites (Protocol 3.3). Multiple reamplification reactions should be performed to ensure adequate material for digestion and ligation. The products are gel purified to remove primers and nonspecific products that decrease the efficiency of digestion.

To make the linker DNA, 52 separate 50 μl PCR reactions are performed using each of the 4 reverse J_H primers in combination with each of the 6 V_κ or 7 V_λ reverse primers (Table 2–1). The template is approximately 1 ng of pSW2scFvD1.3[10] containing DNA encoding the short peptide $(Gly_4Ser)_3$.[15] The PCR reaction mixtures are as described in Protocol 3.3, except for the use of different primers. The cycle is 94°C for 1 minute, 45°C for 1 minute, and 72°C for 1 minute. The linker DNA is purified on a 2% agarose gel and extracted from the gel slice by spinning through a Spin-X column (Costar).

Points to Consider

1. Production of high-quality mRNA requires strict attention to detail. Designate "RNA only" reagents and wear gloves at all times to avoid contamination with RNAses on skin. Follow other precautions as detailed in Protocol 1.

2. Avoidance of contamination is critical when performing PCR, particularly when the products are to be cloned for phage display. Contamination with a single copy of an scFv gene encoding an antigen binding fragment can result in the reisolation of this gene after selection of the phage antibody library on antigen. The probability of contamination can be reduced by aliquoting all PCR reagents (water, dNTPs, buffer) and using aliquots only once. A dedicated PCR pipetman should be used in combination with aerosol resistant pipet tips. Reactions are best set up in a laminar flow hood or in a designated room in which no cloned DNA is permitted. Be sure to include negative controls in each amplification step. While these measures may seem extreme, it is very frustrating, and a large waste of time, to make

and select a phage display library, only to find that the binders represent an already cloned scFv.

3. In all PCR amplification steps, the enzyme is added after the reaction mixes have been heated to 94°C. This improves amplification results by preventing extension of primers that have misannealed.

4. Vent DNA polymerase is used for primary amplifications and PCR splicing because its proof reading exonuclease activity results in a lower error rate. The exonuclease also removes random bases that can be added during the PCR process at the 3′ end of the gene. If not removed, the additional base causes mismatching and decreases efficiency of the splicing process. Taq DNA polymerase is used for the reamplification of the scFv gene repertoires with primers containing appended restriction sites and for creation of the scFv linker DNA due to poor results obtained with Vent DNA polymerase.

5. Amplification of the V_H and V_L gene repertoires is generally straightforward. Failure to obtain a product at this stage is usually the result of poor quality or absent mRNA.

6. For PCR splicing of the V_H and V_L genes to create the scFv repertoires, a high annealing temperature is used to prevent spurious annealing. Some optimization of the annealing temperature may be required for different brands of thermal cyclers.

Method 2: Cloning of Human scFv Gene Repertoires for Phage Display

Overview

The next step is to use the scFv gene repertoires to create a large phage antibody library of at least 10^7 clones. The gel purified scFv gene repertoires are digested with restriction enzymes and ligated into the phage display vector pHEN-1 whcih has been digested with the same restriction enzymes. The ligation mixes are used to transform electrocompetent *E. coli*[51] resulting in the production of a phage antibody library.

pHEN-1[11] is a phagemid vector based on pUC119 that contains cloning sites for scFv antibody genes, the c-myc peptide tag, and gene III (Fig. 2–5). By including an amber codon between the scFv gene and gene III, antibody fragments can be expressed either for display on the surface of phage as a pIII fusion (by growth in a suppressor strain of *E. coli*) or as a c-myc peptide tagged soluble fragment (by growth in a nonsuppressor strain of *E. coli*). The ability to express soluble antibody fragment merely by switching bacterial strains greatly simplifies analysis for binding to antigen using ELISA. Inclusion of the c-myc peptide tag makes it possible to detect binding by using a monoclonal antibody (9E10) that recognizes the peptide.[52]

Expression is under control of the lac promoter, and the pelB leader sequence directs expressed protein to the periplasm.

Specifics of Method 2

The gel purified scFv gene repertoires are digested with the restriction enzymes NcoI and NotI as described in Protocol 4.1. Overdigestion is essential due to the poor efficiency with which PCR fragments are digested. The vector pHEN-1 is prepared using CsCl and also digested with the restriction enzymes NcoI and NotI as described in Protocol 4.2.

Points to Consider

1. Although the description of digestion and ligation conditions might seem unnecessary to many readers, it must be emphasized that creating large libraries from DNA fragments generated using PCR is not straightforward. Acceptable results will only be achieved by strict attention to details.
2. Overdigestion of the PCR generated scFv repertoires is necessary due to the poor efficiency with which PCR fragments are digested. Gel purification of the repertoires prior to digestion removes remaining primers that may inhibit digestion. The use of the optimal buffer for each enzyme also improves digestion efficiency. Large quantities of fragment and vector are required, and this requires scaling up digestions as indicated.
3. Vector should be prepared using CsCl to maximize digestion efficiency and minimize background in the library.
4. Cleaning up the ligation mixture by phenol/chloroform extraction and ethanol precipitation increases transformation efficiencies 10 to 100 fold. The ligation mixture must be well washed with 70% ethanol after precipitation to maximize the efficiency of electroporation.

Method 3: Selection of Binders from the Phage Antibody Library

Overview

The next stage in the process is to isolate phage expressing binding antibodies from the background of phage expressing nonbinders. First, phage particles are rescued by superinfecting the phage library with helper phage. The phage are incubated with immobilized antigen, nonbinding phage are removed by washing, and bound phage eluted by the addition of alkali. The eluted phage are used to infect *E. coli* TG1 to produce more phage for the next round of selection. A portion of the eluted phage is used to infect *E. coli* HB2151[53] to produce soluble scFv antibody fragment for binding analysis by ELISA.

Specifics of Method 3

Phage particles are rescued from the phage antibody library and concentrated by precipitation with polyethylene glycol as described in Protocol 5 to give a titer of approximately 10^{12} to 10^{13} phage/ml. Immunotubes (Nunc) are coated overnight with antigen as described in Protocol 6. The next day, tubes are blocked for two hours with 2% skimmed milk powder in PBS and then 10^{12} to 10^{13} phage are added. After a two-hour incubation, nonbinding phage are removed from the tubes by washing with buffer and bound phage eluted by the addition of alkali (triethylamine). The alkali is neutralized by the addition of buffered Tris-HCl. Eluted phage are used to infect *E. coli* TG1, which are then plated on culture media containing 100 µg/ml ampicillin and 1% glucose. The next day the colonies are scraped from the plate and used to prepare phage for the next round of selection. For production of soluble scFv protein for analysis for binding by ELISA, eluted phage are used to infect *E. coli* HB2151, a nonsuppressor strain. Dilutions of the infected HB2151 are plated to obtain single colonies. The rescue–selection–elution–infection process is repeated a total of four times (rounds). Each round can be completed in 2 days, so the entire selection process can be completed in slightly longer than a week. After four rounds of selection, 10% to 90% of the clones should bind antigen.

Points to Consider

1. For successful superinfection of *E. coli* by helper phage, production of pIII from pHEN-1 must be prevented by the inclusion of glucose in the culture media. Failure to include glucose will result in superinfection of only a few clones and failure to adequately rescue the library. For this reason, it is essential to titer the number of superinfection events on kanamycin plates to ensure that the entire library has been rescued.

2. After superinfection, the glucose must be removed by washing so that antibody–pIII fusion will be expressed. Failure to remove the glucose will result in the production of phage without antibody fragments on the surface.

3. Superinfections and reinfections must be performed at 37°C.

4. Since enrichment ratios for low-affinity phage antibodies may be as low as 20 fold, the number of phage eluted after the first round of selection should be at least $\frac{1}{20}$ of the original library size (10^6 eluted phage for a 10^7 library). If smaller numbers are eluted, the washing

conditions during the first round of selection should be made less stringent.

5. The inclusion of 2% skimmed milk powder in the phage solution during selection significantly decreases the isolation of scFv antibody fragments that bind nonspecifically to many different antigens.

Method 4: Detection and Characterization of Binders

Overview

After four rounds of selection, clones should be analyzed for binding to antigen by ELISA. The specificity of binding clones is then confirmed by ELISA using the relevant antigen and a panel of irrelevant antigens. The number of different specific binders can then be determined by PCR fingerprinting and confirmed by DNA sequencing of the scFv gene. Unique binding scFvs can then be purified and affinities measured using an appropriate method.

Specifics of Method 4

After the third or fourth round of selection, the ability of soluble scFv to bind antigen should be determined. Bacteria are grown and induced in microtiter plates exactly as described in Protocol 7. The use of microtiter plates makes it simple to analyze several hundred clones. Soluble scFv leaks out of the bacterial cells during expression and thus is present in the bacterial supernatant. After induction, the bacteria are removed by centrifugation and the supernatant used for ELISA as described in Protocol 8. Typically, 10% to 90% of the clones screened bind antigen. The specificity of each binder should be determined by ELISA on wells coated with the relevant antigen and a panel of irrelevant antigens.

The next step is to determine the number of unique scFv binders. This process is simplified by the use of PCR fingerprinting.[9, 39] The scFv gene is amplified directly from the bacterial glycerol stocks of binders using primers that flank the gene (LMB3 and fdseq (Table 2–1)). PCR fingerprinting is performed in microtiter plates in 20 µl reaction volumes as described in Protocol 9. The scFv genes are then digested with the restriction enzyme BstN1 which cuts frequently in human V genes. The products are analyzed on a 4% Nusieve agarose gel. We have found that from a "naive library" each unique restriction pattern typically represents only one unique sequence.[9] From an immune library, a single restriction pattern may represent many different scFv genes.[39] Therefore, if the binders are from a "naive" library, the DNA from 2 clones of each restriction pattern should be sequenced. If the binders are from an immune library, multiple clones of each pattern should be sequenced. The primer pHENseq, which anneals to the myc tag sequence, can be used to sequence the V_L gene and the primer linkseq, which anneals to the scFv linker sequence, used to sequence the V_H gene.

Finally, the affinity of each unique scFv should be determined using an appropriate technique. When necessary, scFv protein can be purified to homogeneity by taking advantage of the c-myc peptide tag. The 9E10 antibody that binds the c-myc tag is coupled to Sepharose and the affinity column is used to purify scFv protein directly from bacterial supernatant.[9] A 5 ml column will typically purify 1 to 2 mg of scFv. The scFv can be expressed from bacteria grown in shaker flasks. Yields of scFv antibody fragments produced using phage display are typically 5 to 10 mg/L of bacterial supernatant.

CONCLUSIONS

This chapter has described how human monoclonal antibody fragments can be produced using phage display. The process requires neither conventional hybridoma technology nor immunization. The antibody fragments produced are very specific for the antigen used for selection and have proven useful for agglutination-based assays and immunochemistry.[47] In many instances, however, the affinities will be inadequate for the desired application. Fortunately, the affinities can be easily increased using the same approach and the protocols described above. The V_H gene of a binding scFv is combined with a repertoire of V_L genes using PCR to create a mutant scFv phage antibody library (chain shuffling).[39,48] The mutant phage are then selected on antigen to isolate higher-affinity binders. Affinity can be increased further by shuffling the V_H gene, while leaving the V_H CDR3 and new light chain intact.[48]

PROTOCOL 1. RNA PURIFICATION FROM PERIPHERAL BLOOD LYMPHOCYTES (PBLs)

1. Collect fresh human blood and separate white blood cells over Ficoll immediately.
2. Wash the PBLs three times with ice cold phosphate buffered saline (PBS).
3. Collect the PBLs in a 50 ml Falcon tube and add 7 ml of lysis buffer (will lyse up to 5×10^8 PBLs) and vortex vigorously to lyse the cells.
4. Add 7 volumes (49 ml) of 4M lithium chloride and incubate at 4°C for 15 to 20 hours (overnight).

5. Transfer the suspension to 30 ml Corex tubes (acid-washed and silated) and spin at 6,500 rpm for 2 hours at 4°C in a swinging bucket rotor.
6. Pour off the supernatant and wipe lips of tubes with kimwipes. Pool pellets by resuspending in 3M lithium chloride (approximately 15 ml). Centrifuge 1 hour at 6,500 rpm.
7. Pour off the supernatant and dissolve pellets in 2 ml of RNA solubilization buffer. Freeze suspension thoroughly at −20°C.
8. Thaw by vortexing for 20 seconds every 10 minutes for 45 minutes.
9. Extract once with an equal volume of phenol and once with an equal volume of chloroform.
10. Precipitate the RNA by adding $\frac{1}{10}$ volume 3M sodium acetate, pH 4.8, and 2 volumes of −20°C ethanol. Mix thoroughly and leave overnight at −20°C.
11. Spin RNA at 12,000 rpm for 30 minutes in a swinging bucket rotor. Suspend pellet in 0.2 ml of DEPC (diethylpyrocarbonate, see notes below) treated water. Transfer to a 1.5 ml microcentrifuge tube and reprecipitate. Store as ethanol precipitate until ready to use.

Note 1. Use disposable plasticware when possible. All glassware, including Corex tubes, should be baked overnight at 180°C. Use separate reagents (phenol, chloroform, ethanol) for RNA work. Specific reagents should be prepared as below.

Note 2. Reagents:
1. Lysis buffer: 5M guanidine monothiocyanate, 10 mM EDTA, 50 mM Tris-HCl, pH 7.5, 1 mM DTT, filter with Millipore 0.45 micron filter.
2. 4M and 3M lithium chloride, autoclaved.
3. RNA solubilization buffer: 0.1% SDS, 1 mM EDTA, 10 mM Tris-HCl, pH 7.5, autoclaved.
4. 3M sodium acetate, pH 4.8, DEPC treated (add 0.2 ml DEPC/100 ml solution) and autoclaved.
5. DEPC treated water (0.2 ml DEPC/100 ml water) autoclaved.

Note 3. Reference: Cathala et al., *DNA* 2: 329–35, 1983.

PROTOCOL 2. FIRST STRAND cDNA SYNTHESIS

1. RNA is prepared from PBLs as described in Protocol 1.
2. For first strand cDNA synthesis, prepare the following reaction mixture in a silated 1.5 ml microcentrifuge tube:

10 × RT buffer	5 µl
20 × dNTPs (each 5 mM)	5 µl
0.1 M DTT	5 µl

HuIgMFOR primer	2 µl
HuC$_\kappa$FOR primer	2 µl
HuC$_\lambda$FOR primer	2 µl
RNAsin	80U (2 µl)

All primers are 10 pmol/µl.

3. Take an aliquot (1 to 4 µg) of RNA in ethanol, place in a sterile 1.5 ml microcentrifuge tube and spin 5 minutes in microcentrifuge. Wash once with 70% ethanol, dry, and resuspend in 25.5 µl of DEPC (see Protocol 1) treated water.

4. Heat to 65°C for 3 minutes to denature the RNA, quench on ice 2 minutes, and add to first strand reaction mixture. Add 2.5 µl AMV reverse transcriptase (Super RT, Anglian Biotech) and incubate at 42°C for 1 hour.

Note: 10 × RT buffer is 1.4 M KCl, 500 mM Tris-HCl, pH 8.1 at 42°C, 80 mM NgCl$_2$.

PROTOCOL 3. GENERATION OF HUMAN scFv ANTIBODY GENE REPERTOIRES

3.1. Amplification of V$_H$ and V$_L$ Genes from First Strand cDNA

1. Prepare first strand cDNA as described in Protocol 2.
2. Make up 50 µl PCR reaction mixes in 0.5 ml microcentrifuge tubes containing:

Water	31.5 µl
10 × Vent buffer	5.0 µl
20 × dNTPs	2.0 µl
Acetylated BSA (10/mg/ml)	0.5 µl
Forward primer	2.0 µl
Back primer	2.0 µl
cDNA reaction mix	5.0 µl

3. Overlay with paraffin oil.
4. Heat to 94°C for 5 minutes in a thermal cycling block.
5. Add 2.0 µl (2 units) Vent DNA polymerase under the oil.
6. Cycle 30 times to amplify the V-genes at 94°C for 1 minute, 60°C for 1 minute, and 72°C for 1 minute.
7. Purify the PCR fragments by electrophoresis on a 1.5% agarose gel, extract from the gel using Geneclean (glassmilk).

Note 1. $10 \times$ Vent PCR buffer is 100 mM KCl, 10 mM $(NH_4)_2SO_4$, 200 mM Tris-HCl, pH 8.8, 20 mM $MgSO_4$, 1% Triton X-100.

Note 2. $20 \times$ dNTPs is 5 mM of each deoxynucleotide.

Note 3. V_H, V_κ, and V_λ genes are amplified in separate PCR reactions using the appropriate Back and Forward primers. Back primers are an equimolar mixture of either the 6 V_H Back, 6 V_κ Back, or 7 V_λ Back primers (final total concentration of primer mixture equals 10 pmol/µl). Forward primers are an equimolar mixture of either the 4 J_H, 5 J_κ, or 4 J_λ primers (final total concentration of primer mixture equals 10 pmol/µl).

3.2. PCR Splicing of V_H and V_L Genes to Create scFv Gene Repertoire

1. Make up 25 µl PCR reaction mixes in 0.5 ml microcentrifuge tubes containing:

Water	7.5 µl
$10 \times$ Vent buffer	2.6 µl
$20 \times$ dNTPs	1.0 µl
scFv linker (~ 100 ng)	2.0 µl
V_H primary DNA (~ 500 ng)	5.0 µl
V_L primary DNA (~ 500 ng)	5.0 µl

2. Overlay with paraffin oil.
3. Heat to 94°C for 5 minutes in a thermal cycling block.
4. Add 2.0 µl (2 units) Vent DNA polymerase under the oil.
5. Cycle 7 times without amplification at 94°C for 1 minute, 72°C for 2.5 minutes to randomly join the fragments.
6. After 7 cycles, hold at 94°C while adding 1 µl of each flanking primer (an equimolar mixture of the 6 V_H Back primers and an equimolar mixture of the 5 V_κ FOR or 4 V_λ FOR primers (final total concentration of each primer mix equals 10 pmol/µl).
7. Cycle 25 times to amplify the fragments at 4°C for 1 minute, 72°C for 2.5 minutes.
8. Purify the PCR spliced scFv gene repertoires as described above and resuspend in 25 µl of water. The scFv gene should be approximately 0.8 to 0.9 kb.

3.3. Reamplification of scFv Gene Repertoires with Primers Containing Appended Restriction Sites

1. Make up 50 µl PCR reaction mixes in 0.5 ml microcentrifuge tubes containing:

Water	37 µl
10 × Taq buffer	5.0 µl
20 × dNTPs	2.0 µl
Forward primer	2.0 µl
Back primer	2.0 µl
scFv gene repertoire ($\sim$10 ng)	1.0 µl

2. Overlay with paraffin oil.
3. Heat to 94°C for 5 minutes in a thermal cycling block.
4. Add 1.0 µl (5 units) Taq DNA polymerase under the oil.
5. Cycle 25 times to amplify the V-genes at 94°C for 1 minute, 55°C for 1 minute, and 72°C for 1 minute.
6. Purify the PCR fragments by electrophoresis on a 1.5% agarose gel, electroelute, and precipitate with ethanol.

Note 1. Any brand of Taq polymerase in its appropriate buffer can be used.
Note 2. Back primers are an equimolar mixture of the 6 V_H BackSfi primers; Forward primers are an equimolar mixture of either the 5 J_κ Not, or 4 J_λ Not primers (final total concentration of primer mixture equals 10 pmol/µl).

PROTOCOL 4. RESTRICTION AND LIGATION OF scFv GENE REPERTOIRES

4.1. Restriction of scFv Gene Repertoires (examples are given using New England Biolabs (NEB) buffer)

1. Make up 200 µl reaction mix to digest scFv repertoires with NcoI:

scFv DNA (1 to 4 µg in 100 µl water)	100 µl
Water	74 µl
10 × NEB 4 buffer	20 µl
NcoI (10 U/µl)	6.0 µl

2. Incubate at 37°C overnight.
3. Phenol/chloroform extract with a half volume of each, ethanol precipitate, wash with 70% ethanol, dry, and resuspend in 100 µl water.
4. Make up 200 µl reaction mix to digest scFv repertoires with NotI:

scFv DNA	100 µl
Water	72 µl
Acetylated BSA (10 mg/ml)	2 µl
10 × NEB 3 buffer	20 µl
NotI (10 U/µl)	6.0 µl

5. Incubate overnight at 37°C.
6. The digested products are extracted with phenol/chloroform, ethanol precipitated, and are then ready for ligation.

4.2. Ligation of scFv Gene Repertoire into pHEN-1

1. 40 µg of cesium chloride purified pHEN-1 are digested with the restriction enzymes NcoI and NotI exactly as described above. The digested vector DNA is purified on a 0.8% agarose gel, extracted from the gel by electroelution, and ethanol precipitated.
2. The following 100 µl ligation mixture is set up:

10 × ligation buffer	10 µl
Water	52 µl
pHEN-1 (100 mg/µl)	10 µl
scFv gene repertoire (50 ng/µl)	20 µl
T4 DNA ligase (400 U/µl, NEB)	8.0 µl

3. Ligate overnight at 16 to 20°C.
4. Bring volume to 200 µl, extract once with phenol/chloroform, twice with ether, ethanol precipitate and wash with 1.0 ml of 70% ethanol (failure to adequately wash will decrease transformation efficiency). Resuspend in 10 µl of water and use 2 µl to transform electrocompetent *E. coli* TG1.

PROTOCOL 5. RESCUE OF PHAGEMID PARTICLES FOR SELECTION ON ANTIGEN

1. Inoculate 50 ml of 2 × TY containing 100 µg/µl of ampicillin and 1% glucose (2 × TY-AMP-GLU) with 5×10^8 bacterial cells from the library glycerol stock and grow with shaking at 37°C until the OD_{600} is 0.9.
2. Add 5 ml of bacteria to 50 ml of 2 × TY-AMP-GLU prewarmed to 37°C. Add 2×10^{10} plaque forming units of VCSM13 (Stratagene) and incubate at 37°C for 1 hour without shaking.
3. Determine the number of bacteria infected with VCSM13 by plating sensible dilutions of the culture on TYE plates containing 25 µg/µl kanamycin. The number of infected cells should exceed the library size by an order of magnitude to ensure rescue of all clones in the library.
4. Centrifuge cells at 4000 rpm for 10 minutes to remove glucose, resuspend in 250 ml of 2 × TY containing 100 µg/µl of ampicillin and 25 µg/µl kanamycin and grow overnight at 37°C with shaking.
5. Remove bacteria by centrifugation, and precipitate phage by adding $\frac{1}{5}$ volume of PEG/NaCl (20% polyethylene glycol 8000, 2.5 M NaCl) to the supernatant. Mix well and incubate at 4°C for 1 hour.

6. Centrifuge 20 minutes at 4000 rpm, pour off supernatant and resuspend phage in 10 ml of phosphate buffered saline (PBS).
7. Centrifuge 20 minutes at 4000 rpm to remove any remaining cellular debris, remove supernatant to a new tube, add $\frac{1}{5}$ volume PEG/NaCl to precipitate phage and immediately centrifuge 20 minutes at 4000 rpm.
8. Resuspend phage in 2 ml of PBS and filter through a 0.45 micron filter. Phage are now ready for use in selections.

Note: 2 × TY media and TYE plates are as described in Miller, J. H., *Experiments in Molecular Genetics*, Cold Spring Harbor Laboratory Press, Cold Spring Harbor, New York, 1972.

PROTOCOL 6. SELECTION OF PHAGE ANTIBODIES ON IMMOBILIZED ANTIGEN

1. Coat a 75 × 12 mm Nunc immunotube (Maxisorb, catalog number 4-44202) with 4 ml of antigen (10 to 1000 µg/ml) in PBS by incubating overnight at room temperature.
2. Wash the tube 3 times with PBS and block the tube with 2% skim milk power in PBS (MPBS) at 37°C for 2 hours.
3. Wash the tube 3 times with PBS.
4. Add 10^{12} to 10^{13} phage in 4 ml of 2% MPBS and incubate 30 minutes at room temperature while tumbling and a further 1.5 hours without tumbling.
5. Remove nonspecific phage by washing the tubes twenty times with PBS containing 0.1% Tween 20 followed by 20 times with PBS. Each wash is performed by pouring buffer in and out of the tube.
6. Elute bound phage by adding 1.0 ml of 100 mM triethylamine and tumbling the tube on a turntable for 10 minutes.
7. Remove the eluant to a new tube and neutralize immediately by adding 0.5 ml of 1.0 M Tris-HCl, pH 7.4, and mixing.
8. To prepare phage for the next round of selection, use $\frac{1}{2}$ (0.75 ml) of the eluted phage to infect 10 ml of exponentially growing *E. coli* TG1 by incubating the phage with the cells for 30 minutes at 37°C. Centrifuge at 4000 rpm for 15 minutes, resuspend in 1.5 ml 2 × TY and plate on 2 150 mm TYE plates containing 100 µg/ml ampicillin and 1% glucose (TYE-AMP-GLU). Grow overnight at 37°C and then scrape the colonies off the plates into 4 ml of 2 × TY-AMP-GLU containing 10% glycerol and store at −70°C. Phagemid particles can then be rescued from this bacterial stock as described in protocol 5 for the next round of selection. Also plate sensible dilutions of the culture prior to centrifugation on TYE-AMP-GLU to determine the number of phage eluted.

9. To produce soluble scFv fragments for ELISA, use 0.25 ml of the eluted phage to infect 10 ml of exponentially growing *E. coli* strain HB2151 (a suppressor strain). Plate sensible dilutions of the culture on TYE-AMP-GLU to obtain single colonies.

PROTOCOL 7. EXPRESSION OF SOLUBLE scFv FROM pHEN-1 IN *E. coli* HB2151

1. Single colonies of *E. coli* HB2151 infected with phagemid pHEN-1 are obtained as described in Protocol 6, step 9.
2. Thd single colonies are used to inoculate 150 µl of 2 × TY-AMP-GLU in 96-well microtitre plates. The bacteria are grown overnight at 37°C with shaking.
3. A 96-well replicator is used to transfer a small amount of inocula from each well to a new microtiter plate containing 150 µl/well of 2 × TY-AMP-0.1% glucose. The bacteria are grown at 37°C, shaking until an O.D. 600 nm is reached (approximately 3 to 4 hours). Meanwhile add 50 µl/well of 40% glycerol to the wells of the original plate and store at −70°C.
4. When the cells have reached the desired O.D., add 50 µl/well of 2 × TY-AMP containing 4 mM IPTG. Grow overnight at +30°C.
5. Spin the microtitre plate at 4000 rpm for 15 minutes and use the supernatant for ELISA as described in Protocol 8.

Note: This method is based on that of DeBellis and Schwarz[54] and relies on the low levels of the glucose repressor present in the starting media being metabolized by the time the inducer (IPTG) is added.

PROTOCOL 8. ELISA FOR DETECTION OF BINDING OF SOLUBLE scFv TO ANTIGEN

1. Coat a microtiter plate (Falcon 3912) overnight at room temperature with 50 µl/well of antigen at 10 to 100 µg/ml in PBS.
2. Wash wells 3 times with PBS and block with 200 µl/well of 2% skim milk powder in PBS (MPBS) for 2 hours at 37°C.
3. Wash wells 3 times with PBS and add 50 µl/well of bacterial supernatant containing expressed scFv protein. Incubate for 1 hour at room temperature.
4. Wash wells 3 times with PBS containing 0.1% Tween 20 (TPBS) and 3 times with PBS. Add 50 µl/well of 9E10 monoclonal antibody at 1 µg/ml in 2% MPBS. Incubate at room temperature for 1 hour.

5. Wash wells 3 times with TPBS and 3 times with PBS. Add 50 µl/well of horseradish peroxidase conjugated anti-mouse IgG Fc monoclonal antibody (Sigma A-2554) at 1 µg/ml in 2% MPBS. Incubate at room temperature for 1 hour.

6. Wash wells 3 times with TPBS and 3 times with PBS. Prepare developing solution by adding 1 10 mg ABTS (2,2'-azino bis(3-ethylbenzthiazoline-6-sulphonic acid)) tablet (Sigma) to 20 ml of 50 mM citrate buffer, pH 4.5. Add 20 µl of 30% hydrogen peroxide to the ABTS solution and mix. Add 100 µl/well of developing solution. Leave at room temperature for 20 to 30 minutes.

7. Stop reaction by adding 50 µl/well of 3.2 mg/ml sodium fluoride. Read plate at 405 nm.

PROTOCOL 9. PCR FINGERPRINTING OF scFv GENES IN pHEN-1

1. Make up the following PCR reaction mixture where n = the number of different clones to be screened:

Water	$14.8 \times (n + 1)$ µl
$10 \times$ Taq buffer	$2 \times (n + 1)$ µl
$20 \times$ dNTPs	$1 \times (n + 1)$ µl
LMB3 primer	$1 \times (n + 1)$ µl
fdseq primer	$1 \times (n + 1)$ µl
Taq polymerase	$0.2 \times (n + 1)$ µl

2. Aliquot 20 µl of PCR reaction mix into wells of a 96-well PCR compatible microtiter plate.

3. Touch a toothpick to a bacterial glycerol stock or bacterial colony and then twist the toothpick in the PCR reaction mix to transfer a small amount of bacteria.

4. Overlay the PCR reaction mixes with mineral oil, insert the microtiter plate into a thermal cycler preheated to 94°C, and hold at this temperature for 10 minutes to lyse the bacteria.

5. Cycle 25 times at 94°C for a 1minute; 50°C for 1 minute; 72°C for 1 minute.

6. Make a restriction enzyme mixture where n = the number of PCR reactions performed:

Water	$17.8 \times (n + 1)$ µl
$10 \times$ NEB buffer 2	$2 \times (n + 1)$ µl
Bs5N1 (10 units/ml, NEB)	$0.2 \times (n + 1)$ µl

7. Add 20 µl of the above mix to each well containing a PCR mix under the mineral oil. Incubate at 60°C for 2 hours.
8. Analyze the results on a 4% Nusieve agarose gel.

Note: NEB = New England Biolabs.

REFERENCES

1. James, K., and Bell, G.T. 1987. Human monoclonal antibody production: current status and future prospects. *J. Immunol. Methods* 100:5.
2. Winter, G., and Milstein, C. 1991. Man-made antibodies. *Nature* 349:293.
3. Marks, J.D., Hoogenboom, H.R., Griffiths, A.D., and Winter, G. 1992. Molecular evolution of proteins on filamentous phage: mimicking the strategy of the immune system. *J. Biol. Chem.* 267:16007.
4. Skerra, A., and Pluckthun, A. 1988. Assembly of a functional immunoglobulin Fv fragment in *Escherichia coli. Science* 240:1038.
5. Better, M., Chang, C.P., Robinson, R.R., and Horwitz, A.H. 1988. *Escherichia coli* secretion of an active chimeric antibody fragment. *Science* 240:1041.
6. Orlandi, R., Gussow, D.H., Jones, P.T., and Winter, G. 1989. Cloning immunoglobulin variable domains for expression by the polymerase chain reaction. *Proc. Natl. Acad. Sci. US.A* 86:3833.
7. Huse, W.D., Sastry, L., Iverson, S.A., Kang, A.S., Alting, M.M., Burton, D.R., Benkovic, S.J., and Lerner, R.A. 1989. Generation of a large combinatorial library of the immunoglobulin repertoire in phage lambda. *Science* 246:1275.
8. Marks, J.D., Tristrem, M., Karpas, A., and Winter, G. 1991. Oligonucleotide primers for polymerase chain reaction amplification of human immunoglobulin variable genes and design of family-specific oligonucleotide probes. *Eur. J. Immunol.* 21:985.
9. Marks, J.D., Hoogenboom, H.R., Bonnert, T.P., McCafferty, J., Griffiths, A.D., and Winter, G. 1991. By-passing immunization: Human antibodies from V-gene libraries displayed on phage. *J. Mol. Biol.* 222:581.
10. McCafferty, J., Griffiths, A.D., Winter, G., and Chiswell, D.J. 1990. Phage antibodies: filamentous phage displaying antibody variable domains. *Nature* 348:552.
11. Hoogenboom, H.R., Griffiths, A.D., Johnson, K.S., Chiswell, D.J., Hudson, P., and Winter, G., 1991. Multi-subunit proteins on the surface of filamentous phage: methodologies for displaying antibody (Fab) heavy and light chains. *Nucleic Acids Res.* 19:4133.
12. Cabilly, S., Riggs, A.D., Pande, H., Shively, J.E., Holmes, W.E., Rey, M., Perry, L.J., Wetzel, R., and Heyneker, H.L. 1984. Generation of antibody activity from immunoglobulin polypeptide chains produced in *Escherichia coli. Proc. Natl. Acad. Sci. USA* 81:3273.
13. Boss, M.A., Kenten, J.H., Wood, C.R., and Emtage, J.S. 1984. Assembly of functional antibodies from immunoglobulin heavy and light chains synthesised in *E. coli. Nucleic Acids Res.* 12:3791.

14. Bird, R.E., Hardman, K.D., Jacobson, J.W., Johnson, S., Kaufman, B.M., Lee, S.M., Lee, T., Pope, S.H., Riordan, G.S., and Whitlow, M. 1988. Single-chain antigen-binding proteins. *Science* 242:423.

15. Huston, J.S., Levinson, D., Mudgett, H.M., Tai, M.S., Novotny, J., Margolies, M.N., Ridge, R.J., Bruccoleri, R.E., Haber, E., Crea, R., and Oppermann, H. 1988. Protein engineering of antibody binding sites: recovery of specific activity in an anti-digoxin single-chain Fv analogue produced in *Escherichia coli*. *Proc. Natl. Acad. Sci. USA* 85:5879.

16. Huston, J. S., Mudgett-Hunter, M., Mei-Sheng, T., McCartney, J., Warren, F., Haber, E., and Opperman, H. 1991. Protein engineering of single-chain Fv analogs and fusion proteins. *Meth. Enzymol.* 203:46.

17. Bird, R.E., and Walker, B.W. 1991. Single chain antibody variable regions. *Trends Biotech.* 9:132.

18. Saiki, R.K., Scharf, S., Faloona, F., Mullis, K.B., Horn, G.T., Erlich, H.A., and Arnheim, N. 1985. Enzymatic amplification of beta-globin genomic sequences and restriction site analysis for diagnosis of sickle cell anemia. *Science* 230:1350.

19. Mullis, K., Faloona, F., Scharf, S., Saiki, R., Horn, G., and Erlich, H. 1986. Specific enzymatic amplification of DNA *in vitro*: the polymerase chain reaction. *Cold Spring Harb. Symp. Quant. Biol.* 1:263.

20. Saiki, R.K., Gelfand, D.H., Stoffel, S., Scharf, S.J., Higuchi, R., Horn, G.T., Mullis, K.B., and Erlich, H.A. 1988. Primer-directed enzymatic amplification of DNA with a thermostable DNA polymerase. *Science* 239:487.

21. Li, H.H., Gyllensten, U.B., Cui, X.F., Saiki, R.K., Erlich, H.A., and Arnheim, N. 1988. Amplification and analysis of DNA sequences in single human sperm and diploid cells. *Nature* 335:414.

22. Scharf, S.J., Horn, G.T., and Erlich, H.A. 1986. Direct cloning and sequence analysis of enzymatically amplified genomic sequences. *Science* 233:1076.

23. Larrick, J.W., Chiang, Y.L., Sheng-Dong, R., Senck, G., and Casali, P. 1988. Generation of specific human monoclonal antibodies by *in vitro* expansion of human B cells: a novel recombinant DNA approach. In *In vitro immunisation in hybridoma technology*. Borrebaeck, ed. Amsterdam: Elsevier Science Publishers.

24. Kabat, E.A., Wu, T.T., Reid-Miller, M., Perry, H.M., and Gottesman, K.S. 1987. *Sequences of proteins of immunological interest*. U.S. Department of Health and Human Services, U.S. Government Printing Office.

25. Gussow, D., Ward, E.S., Griffiths, A.D., Jones, P.T., and Winter, G. 1989. Generating binding activities from *Escherichia coli* by expression of a repertoire of immunoglobulin variable domains. *Cold Spring Harb. Symp. Quant. Biol.* 1:265.

26. Ward, E.S., Gussow, D., Griffiths, A.D., Jones, P.T., and Winter, G. 1989. Binding activities of a repertoire of single immunoglobulin varible domains secreted from *Escherichia coli*. *Nature* 341:544.

27. Smith, G.P. 1985. Filamentous fusion phage: novel expression vectors that display cloned antigens on the virion surface. *Science* 228:1315.

28. Parmley, S.F., and Smith, G.P. 1988. Antibody-selectable filamentous fd phage vectors: affinity purification of target genes. *Gene* 73:305.

29. Goldsmith, M.E., and Konigsberg, W.H. 1977. Adsorption protein of the bacteriophage fd: isolation, molecular properties, and location in the virus. *Biochemistry* 16:2686.

30. Gray, C.W., Brown, R.S., and Marvin, D.A. 1981. Adsorption complex of the filamentous fd virus. *J. Mol. Biol.* 146:621.

31. Glaser-Wuttke, G., Keppner, J., and Rasched, I. 1989. Pore-forming properties of the adsorption protein of filamentous phage fd. *Biochim. Biophys. Acta.* 985:239.

32. Il'ichev, A.A., Minenkova, O.O., Tat'kov, S.I., Karpyshev, N.N., Eroshkin, A.M., Ofitserov, V.I., Akimenko, Z.A., Petrenko, V.A., and Sandakhchiev, L.S. 1990. [The use of filamentous phage M13 in protein engineering]. *Mol. Biol. Mosk.* 24:530.

33. Greenwood, J., Willis, A.E., and Perham, R.N. 1991. Multiple display of foreign peptides on a filamentous bacteriophage: Peptides from Plasmodium falciparum circumsporozoite protein as antigens. *J. Mol. Biol.* 220:821.

34. Kang, A.S., Barbas, C.F., Janda, K.D., Benkovic, S.J. and Lerner, R.A. 1991. Linkage of recognition and replication functions by assembling combinatorial antibody Fab libraries along phage surfaces. *Proc. Natl. Acad. Sci. USA* 88:4363.

35. Breitling, S.D., Seehaus, T., Klewinghaus, I., and Little, M. 1991. A surface expression vector for antibody screening. *Gene* 104:147.

36. Garrard, L.J., Yang, M., O'Connell, M.P., Kelley, R.F., and Henner, D.J. 1991. Fab assembly and enrichment in a monovalent phage display system. *Bio/technology* 9:1373.

37. Bass, S., Greene, R., and Wells, J.A. 1990. Hormone phage: an enrichment method for variant proteins with altered binding properties. *Proteins* 8:309.

38. Griffiths, A.D., Malmqvist, M., Marks, J.D., Bye, J.M., Embleton, M.J., McCafferty, J., Baier, M., Hollinger, K.P., Gorick, B.D., Hughes-Jones, N.C., Hoogenboom, H.R., and Winter, G. 1993. Human anti-self antibodies with high specificity from phage display libraries. *EMBO J.* 12:725.

39. Clackson, T., Hoogenboom, H.R., Griffiths, A.D., and Winter, G. 1991. Making antibody fragments using phage display libraries. *Nature* 352:624.

40. Huse, W.D. 1991. Combinatorial antibody expression libraries in filamentous phage. In *Antibody Engineering. A Practical Approach.* Borrebaeck, ed. (New York: W. H. Freeman), 103–120.

41. Chang, C.N., Landolfi, N.F., and Queen, C. 1991. Expression of antibody Fab domains on bacteriophage surfaces. *J. Immunol.* 147:3610.

42. Cwirla, S.E., Peters, E.A., Barrett, R.W., and Dower, W.J. 1900. Peptides on phage: a vast library of peptides for identifying ligands. *Proc. Natl. Acad. Sci. USA* **87**:6378.

43. Gram, H., Marconi, L., Barbas, C.F., Collet, T.A., Lerner, R.A., and Kang, A.S. 1992. *In vitro* selection and affinity maturation of antibodies from a naive combinatorial immunoglobulin library. *Proc. Natl. Acad. Sci. USA* 89:3576.

44. Barbas, C.F., Kang, A.S., Lerner, R.A., and Benkovic, S.J. 1991. Assembly of combinatorial antibody libraries on phage surfaces: The gene III site. *Proc. Natl. Acad. Sci. USA* 88:7978.

45. Zebedee, S.L., Barbas, C.F., Hom, Y., Caothien, R.H., Graff, R., Degraw, J., Pyati, J., LaPolla, R., Burton, D.R., Lerner, R.A., and Thronton, G.B. 1992.

Human combinatorial antibody libraries to hepatitis B surface antigen. *Proc. Natl. Acad. Sci. USA* 89:3175.

46. Burton, D.R., Barbas, C.F., Persson, M.A.A., Koenig, S., Chanock, R.M., and Lerner, R.A. 1991. A large array of human monoclonal antibodies to type 1 human immunodeficiency virus from combinatorial libraries of asymptomatic individuals. *Proc. Natl. Acad. Sci. USA* 88:10134.

47. Marks, J.D., Ouwehand, W.H., Bye, J.M., Finnern, R., Gorick, B.D., Voak, D., Thorpe, S., Hughes-Jones, N.C., and Winter, G. 1993. Human antibody fragments specific for blood group antigens from a phage display library. *Bio/Technology*. In press.

48. Marks, J.D., Griffiths, A.D., Malmqvist, M., Clackson, T., Bye, J.M., and Winter, G., 1992. Bypassing immunisation: high affinity human antibodies by chain shuffling. *Bio/Technology* 10:779.

49. Hawkins, R.E., Russell, S.J., and Winter, G. 1992. Selection of phage antibodies by binding affinity: mimicking affinity maturation. *J. Mol. Biol.* 226:889.

50. Cathala, G., Savouret, J., Mendez, B., Wesr, B.L., Karin, M., Martial, J.A., and Baxter, J.D. 1983. *DNA* 2:329.

51. Gibson, T.J. 1984. Studies on the Epstein-Barr virus genome. Ph.D. thesis, Cambridge University.

52. Munro, S., and Pelham, H.R.B. 1986. An Hsp-like protein in the ER: Identity with the 78kd glucose regulated protein and immunoglobulin heavy chain binding protein. *Cell* 46:291.

53. Carter, P., Bedouelle, H., and Winter, G. 1985. Improved oligonucleotide site-directed mutagenesis using M13 vectors. *Nucleic Acids Res.* 13:4431.

54. DeBellis, D., and Pelham, H.R.B. 1990. Regulated expressions of foreign genes fused to Iac control by glucose levels in growth medium. *Nucleic Acids Res.* 18:1311.

CHAPTER 3

Synthetic Antibodies

Jonathan S. Rosenblum and Carlos F. Barbas, III

Antibodies are the immune system's solution to the problem of molecular recognition. They are generated and evolved by the immune system to bind with high affinity to molecules the host has never before encountered. In an organism, an antigenic challenge is met with an order of 100 million different antibody molecules. From this primary response, the highest-affinity binders are selected for maturation through somatic mutation. Antibodies with infinitely many specificities can be produced in this manner. Indeed, the survival of the host may depend on it. A primary response typically has members with micromolar dissociation constants, while somatic mutation expeditiously increases binding by three to five orders of magnitude.

The aspects of antibody structure that permit rapid evolution of high-affinity receptors *in vivo* also make antibodies excellent targets for protein engineering. Antigen binding sites are composed of six hypervariable loops or complementary determining regions (CDRs), three each from the heavy and light chains (Figs. 3–1 and 3–2). Variations in the lengths and

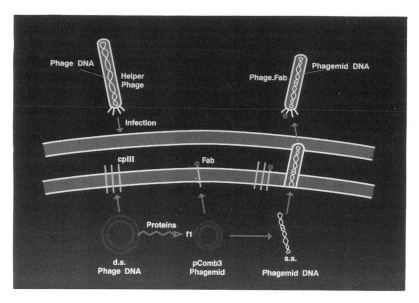

Figure 3–1. The three-dimensional structure of a Fab. This view highlights the disposition of the six CDRs (three each from the heavy and light chains labeled H1 through H3 and L1 through L3, respectively) around the binding site. The framework and constant domains are in gray. Molecular models were constructed by Mike Pique using AVS software.

compositions of the CDRs lend greatly to the plasticity of the binding site. Antibodies to small haptens typically have binding sites that resemble pockets or grooves, whereas antibodies to proteins often have extensive, undulating surfaces for binding. Surface areas of antibodies buried on binding antigens range from 161 Å^2 on binding phosphorylcholine (169 g/mole) to 916 Å^2 on binding neuraminidase (50,000 g/mole).[1] In contrast to the diversity displayed in the CDRs, the β-sheet structure that holds the loops is invariant structurally. This combination of loop diversity and framework invariance allows rapid evolution *in vivo* and facile engineering *in vitro*.

THE DOMINANT ROLE OF CDRs

Prior to the determination of a high-resolution structure of an antibody–antigen complex, Wu and Kabat[2] predicted the antibody domains responsible for recognition based on amino acid sequence analysis. They proposed that the domains that displayed the highest diversity would be the complementarity determining regions and would contribute the most to the broad specificities of the antibody repertoire. The importance of CDRs has been

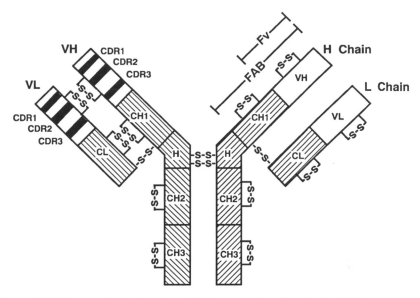

Figure 3–2. The covalent structure of a human IgG antibody. An IgG is composed of two chains, the H (for heavy) and the L (for light) chains. Much of the structure of the antibody molecule comes from specific disulfide bonds both within and between chains. Such whole antibodies are bivalent—two heavy and two light chains fold to make two distinct antigen-binding sites. Smaller, monovalent antibody fragments, Fab and Fv, can be made. Though naturally heterodimers, these fragments can be made as single-chain molecules by installing a linker between the H and L chain of the Fv fragment.

borne out through structural studies, humanization of rodent Abs through CDR grafting, and biological activity of CDR-based peptides.

X-ray crystallography has shed much light on the mechanisms of antibody–antigen interaction.[3,4] Indeed, the six CDRs provide most of the contacts necessary to bind antigen. However, all six CDRs do not contribute equally, just as the heavy and light chains do not contribute equally. In structures thus far determined, four or more CDRs are used to bind antigen, with the heavy chain generally contributing more buried surface area than the light chain.[1] High-resolution structures are available for two antibodies that share very similar light chains, the anti-DNA antibody BVO4[5] and the anti-fluorescein antibody 4-4-20.[6] These reports and others (see below) suggest that CDRs, particularly those of the heavy chain, can determine the specificity of the antibody. A specific CDR, H3, often makes essential contacts and has even been shown to undergo a major rearrangement on binding in an anti-peptide antibody.[7] With respect to sequence and length, and therefore structure, H3 is the greatest source of molecular diversity within the antibody binding site. H3 can be as short as 2 residues or greater than 26.[8] It has been estimated that humans have the potential to generate more than 10^{14} peptides in this region.[9]

Although murine monoclonal antibodies are generally accessible through the application of hybridoma technology,[10] the elicitation of a human anti-murine antibody (HAMA) response on multiple administration precludes their use as general human therapeutics.[11–13] Humanization of rodent antibodies[14] is the process whereby the CDRs of a high-affinity rodent antibody are transplanted into the framework of a human antibody, minimizing the rodent component, and possibly abrogating the HAMA response. Several important issues have been raised through the humanization of antibodies to small molecules[14] and proteins (human epidermal growth factor receptor,[2,20] lysozyme,[15,16] IL-2 receptor, receptor[17,18]). First, not all of the CDRs are essential to instill the rodent antibody specificity on the human antibody framework. Thus, not only are CDRs important for specificity, but individual CDRs can dominate. More importantly, several humanization studies have noted that it was essential to replace one[19] or more[20,21] of the human framework residues with corresponding murine ones. For example, in the humanization of an anti-p185HER2 antibody,[20] a CDR-transplanted antibody had a K_d for the antigen of 25 nM, whereas a variant containing five murine framework residues and two adjustments in the transplanted CDRs had a K_d of 0.1 nM (i.e., it binds 250-fold tighter than the CDR-transplant-only antibody). Further, this variant with framework substitutions bound three times more tightly than the parent murine antibody, suggesting that the increase in binding may come from factors other than more closely approximating the murine structure. Such essential framework residues could either directly contact antigen or affect CDR conformation. The individualized nature of essential framework residues highlights the need for high-resolution structures (or accurate models) as starting points for humanization. Presently, there are no general rules regarding which framework residues might be necessary; their identification seems to depend on the particular antibody. However, a number of framework residues have been proposed to play critical roles in modulating the conformation of particular CDRs.[20,22] That antibodies can be humanized at all is a testament to the unique recognition capacity inherent in the CDRs.

Antibodies to protein antigens generally have broad, undulating surfaces that can even protrude into the binding site of a protein antigen (for example, HyHEL-10, whose surface bulges into the lysozyme binding site[23]). It is therefore not surprising that antibodies to receptors have been produced that can compete with ligand for receptor binding. These include antibodies to reovirus receptor,[24] fibrinogen receptor,[25,26] and thyrotropin receptor.[27] In such cases it is plausible that a single CDR loop bulges into the receptor, structurally mimicking the natural ligand. In support of this hypothesis, Taub et al.[26] noted sequence similarity between the unusually long H3 of the anti-fibrinogen receptor antibody PAC1 and fibrinogen. The R–G–D sequence of fibrinogen is apparently mimicked by an R–Y–D motif in PAC1. Additionally, three of five residues on the amino terminal side of the R–G/Y–D motif are identical between PAC1 and fibrinogen. A 21-residue peptide encompassing H3 was synthesized that could compete with either PAC1 ($K_i = 10 \, \mu M$) or fibrinogen ($K_i = 5 \, mM$) for binding to activated platelets. More recently it was shown that a cyclic peptide whose sequence is derived from H3 of an antibody to the immunodominant V3 loop of the human immunodeficiency virus-1 exhibits antiviral activity.[28] These reports, indicating that in some cases CDRs do not need an antibody framework for biological activity, demonstrate the unusual versatility of the molecular recognition capacity inherent in the CDRs.

CAPTURING MOLECULAR DIVERSITY IN A TEST-TUBE: ESSENTIALS OF BACTERIAL EXPRESSION AND FILAMENTOUS PHAGE BIOLOGY

Creating synthetic antibodies requires knowledge of how nature's molecules work and the tools to manipulate them with the same dexterity exhibited by the immune system. *In vivo*, selection and maturation of antibodies is achieved by linking recognition and replication. This is accomplished by the display of antibody on the cell surface, which allows for proliferative stimulation on binding antigen. Reproducing this linkage outside the B-cell relies on an understanding of *E. coli* antibody expression and of the life cycle of phage, a virus that infects this prokaryote.

Facile methods of antibody engineering build on the demonstration that antibodies could be produced in *E. coli*[29,30] that retain the functional and structural[31] characteristics of hybridoma-produced antibodies. In order to produce the antibodies' native disulfide bonds (Fig. 3–1), and thus correctly folded antibodies, the individual protein chains must be transported to the oxidizing environment of the periplasm (the region between the inner and outer bacterial membranes). Fusion of the antibody sequence with a prokaryotic leader sequence results in secretion into the periplasmic space where functional antibodies assemble. This is directly analogous to the production of antibodies in eukaryotic cells where the two chains are transported from the cytoplasm to the lumen of the endoplasmic reticulum. Importantly, this is the same compartment in which filamentous phage are assembled. It is this feature common to antibody folding and phage morphogenesis that allows us to link antibody binding and replication without a B-cell.

As a result of a wealth of knowledge regarding the fundamental biology of filamentous (Ff) bacteriophage[32–35] they have recently become useful

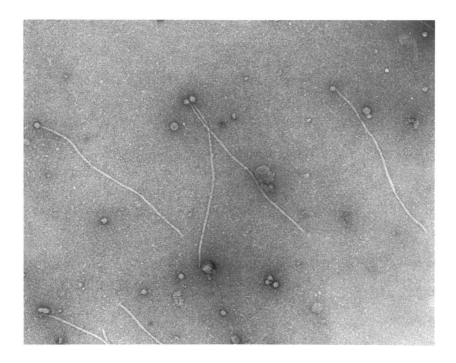

Figure 3–3. Filamentous phage morphology. Fab phage are approximately 900 nm long and 6 to 10 nm in diameter. This electronmicrograph shows that Fab phage selected for binding to hepatitis B surface antigen (HBsAg) bind HBsAg only at the very tip of the phage.[48] As such, Fab phage express functional Fab in the correct location.

tools for protein engineering. Ff phage, most notably f1, fd, and M13 (the three are almost identical) infect gram-negative bacteria by virtue of a specific interaction between the phage and a receptor on the bacterial pilus. Ff phage are long and thin (ca. 900 × 6–10 nm, Fig. 3–3); essentially they are protein sheaths coating a single-stranded closed circular DNA genome.

The protein component of Ff phage is made up of five proteins. Approximately 2,700 copies of the gene 8 protein (g8p) form the bulk of the sheath. Minor coat proteins (g3p, g6p, g7p, g9p) are present in about five copies each. The proteins are arranged so that g3p and g6p are expressed at one end of the phage, g7p and g9p at the other. The phage protein g3p makes specific contact with the bacterial receptor. It has two domains, an amino terminal infectivity domain and a carboxy terminal capping domain.

The genome of Ff phage is about 6.4 kb long, coding for ten proteins. In addition to the coat proteins is an endonuclease/topoisomerase pg2p), a single-stranded DNA binding protein (g5p), and several poorly understood

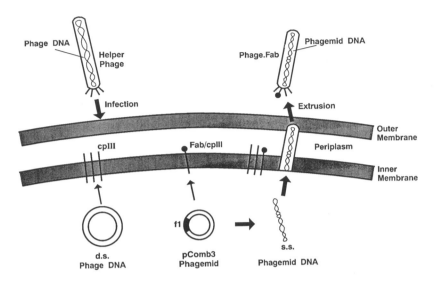

Figure 3–4. Helper phage rescue to create Fab-expressing phage. Fab fusion phage are produced in several steps. An *E. coli* cell harboring a phagemid is first infected by a helper phage. Helper phage DNA directs the production of normal phage proteins while pComb3 directs the synthesis of a variation of the gene 3 protein (cpIII) to which a Fab is appended. Phagemid DNA is packaged and extruded with the aid of helper phage-encoded proteins. Fab-cpIII fusion and wild type cpIII compete for inclusion in the assembled phage.

proteins: g1p, X, g4p. An important regulatory element, the intergenic region (IR), is also present. The IR contains a DNA origin of replication and a DNA packaging signal. When included on a bacterial plasmid (a plasmid with an Ff IR is known as a phagemid), the IR is sufficient to allow helper phage-mediated replication and packaging of resulting single-stranded DNA. Such packaged phagemids can infect bacteria and are known as transducing particles (Fig. 3–4). Note that the phagemid need not contain any phage related sequence other than the IR to be packaged.

Unlike most bacterial viruses, the Ff phage are not assembled in the cytoplasm. Likewise, they are not released by cell lysis. The only phage-related structure present in the cytoplasm is the flexible rod formed from the sequestration of single-stranded viral DNA by g5p. The Ff phage are secreted as they are being assembled in the bacterial membrane. This occurs after the rod of g5p coated genomic (or phagemid) DNA migrates to the inner membrane. Coat proteins then replace g5p in a vectorial fashion, g7p and g9p are first, followed by polymerization of g8p onto the DNA rod, and finally g6p and g3p cap the particle. At least 300 phage can be produced by a single bacterium in one generation. Ff phage can easily be prepared in concentrations exceeding 10^9 virions per microliter.

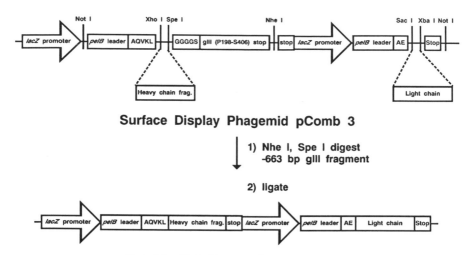

Surface Display Phagemid pComb 3

1) **Nhe I, Spe I digest**
   -663 bp gIII fragment

2) **ligate**

Soluble Fab Expressing Phagemid

Figure 3–5. pComb3: A prototypical phage display vector.[47] The surface display version of pComb3 contains a heavy-chain Fab fragment fused to a truncated gene 3 protein. A light-chain gene is expressed separately. Both the H chain/gIII and the L chain are targeted to the periplasm by a pelB leader sequence. gIII is an integral membrane protein and thus acts as a membrane anchor for the Fab. Digestion of pComb3 with the restriction enzymes Nhe1 and Spe1 and self-ligation of the vector removes the gIII gene. Removal of gene III causes the Fab to be targeted to the periplasm without attachment to gIII. As such the Fab is no longer associated with the bacterial membrane—it is produced in soluble form.

   Since g3p and g8p are secreted into the periplasmic space of *E. coli*, and they are assembled to form the surface of the phage, they are ideal targets for fusion with antibody domains. Fusion phage systems[36–39] have been developed to capitalize on a number of the above features of normal phage morphogenesis. Genomic g3p fusion systems allow the polyvalent display of foreign peptides, whereas phagemid g3p (Fig. 3–5) and g8p fusion permits monovalent display of foreign proteins. Such fusion phage systems readily allow sorting of enormous libraries (of the order of 10^8 unique members present in thousands of copies each). This rapid sorting is a result of the linkage of recognition and replication. Only those phage that bear a functional peptide or protein expressed on their surface will be permitted to propagate through a procedure known as panning (Fig. 3–6). As such, useful members of the library are *selected* from the vast, useless background. Previous methods to access large libraries (e.g., the phage lambda Fab system,[40]) involved *screening* methods. Screening methods entail the individual assessment of each member of the library, through plaque lifts or other assays. A diverse (ca. 5×10^8) fusion phage library can be selected

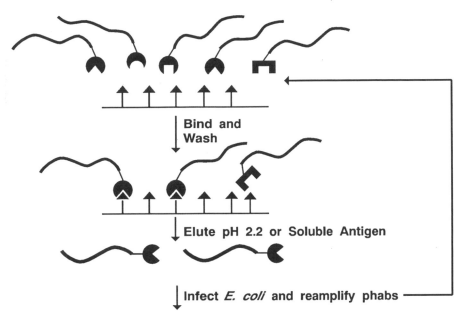

Figure 3–6. The panning process. A diverse library of Fab-expressing phage (or phabs) is applied to an immobilized target (panning antigen, or "panigen"). Several hours are allowed for a binding equilibrium to be reached, then nonbinding phage are washed from the solid support. Specifically bound phage can be eluted at low pH or by the addition of soluble panigen. Eluted phage are amplified for further rounds of selection by infection of *E. coli* followed by helper phage rescue. One "round" of panning (from isolation of a pool of phage to amplification of selected clones) takes one day. Several rounds are generally performed to minimize the presence of nonspecific binding clones.

from a 50 microliter well, whereas screening the same library would involve analysis of at least 10,000 petri plates (at 50,000 colonies per plate). The advantage of selection over screening are obvious.

The final aspect necessary for the production of synthetic antibodies was the procurement of starting materials—the cloning of the immune response. With the advent of PCR[41] and the wealth of sequences of antibodies,[42] it was soon possible to clone single antibodies from hybridomas,[43] a V_H library from spleen,[44] and finally an entire immunoglobulin repertoire.[40]

Antibody fusion phage, or Phabs, can be thought of as *in vitro* immune systems on several levels. First, like the immune system, Phabs are combinatorial. The random pairing of light and heavy chains mimics nature's random pairing. This random pairing initially concerned some scientists. They believed that it was tremendously unlikely that high-affinity antibodies could be isolated without knowledge of the chain pairing *in vivo*. Of course, such information is lost in the stepwise cloning of a repertoire of heavy and light chains. This has recently been shown not to be a weakness of the combinatorial approach; rather it is a strength. Not only does it mimic some of the aspects of the natural mechanisms for the diversification of the immune response, but it allows for additional generations of phage-based antibody development. Next, and most importantly, recognition and replication are linked. In a eukaryote, B-cells expressing antibodies that bind to an antigen are stimulated to proliferate. In the phage systems, this stimulation, or amplification, is carried out by the Ff host *E. coli*, working in concert with a helper phage. Somatic mutation optimizes primary binders in the immune system, and, as will be discussed later, there are a variety of methods that allow similar optimization of primary binders via phage display.

The mechanisms of immunological diversity generation are complex and not entirely understood. As such, it is unlikely, and perhaps unnecessary, that these same mechanisms will be transferred to bacteria for the purpose of antibody engineering. A current challenge for phage-based antibody engineering systems is to mimic the variety present in the immune system in a manner consistent with the biology of a prokaryotic host.

One method is to clone the repertoire of an immunized or infected host.[45–53] Following display on phage, the highest-affinity antibodies can be selected by panning (Fig. 3–6). In this manner, human monoclonal Abs to tetanus toxoid, hepatitis B surface antigen, gp120 of the human immunodeficiency virus, respiratory syncitial virus, as well as many other immunologic targets, have been isolated. Notably, many of these antibodies have been isolated from asymptomatic individuals exhibiting low titer. An important outcome of these studies was the proof of the validity of the combinatorial approach. The chain pairings isolated as a result of Fab affinity for antigen *in vitro* were not necessarily paired *in vivo*; however, in many cases it is likely that affinity selection restores the approximate pairings.[54]

Whereas immunized sources or immune priming by natural infection provides quite useful antibody libraries for some antigens, it is not always possible to acquire such libraries. One strategy that has been developed to rescue the immunological memory of an individual exposed many years ago, but exhibiting no antibody titer, exploits the SCID mouse model. Transfer of human peripheral blood lymphocytes into this mouse with antigen stimulation allows for the resurrection of the extinct response and facile cloning of high-affinity human antibodies from the mouse.[55] Alternatively, since panning has been shown to be capable of selecting rare high-affinity binders from a background of nonbinders, libraries have been created from unimmunized donors.[56–60] Such "naive" libraries should be more diverse than those from hyperimmunized donors. Panning naive single-chain Fv (scFv, Fig. 3–2) libraries with proteins and with small molecules[58] has allowed the isolation of scFv with K_d of approximately 0.1 to 1 mM. From this same library, scFvs with specificity for human self-antigens were isolated.[59] It was shown, however, that appreciable binding was almost exclusively derived from ScFv dimers, and in most cases the affinities of the selected scFvs were too weak to quantify. Since Fab fragments do not dimerize, there should be an inherent advantage in expressing this fragment over an scFv when one seeks to isolate the highest-affinity clone as avidity effects are avoided.

Although antibody libraries cloned from unimmunized donors have been termed "naive," it is likely that bias, for example by plasma cells that produce 1000-fold more mRNA than a resting B-cell, does exist in such libraries. In this case it may be quite difficult to obtain high-affinity antibodies for certain antigens, as antibodies isolated from normal donors will be subjected to *in vivo* editing prior to cloning. As such, these antibody libraries cannot be truly naive. In addition, five of the six CDRs have a limited structural and genetic repertoire.[22,61] It is quite possible that the limit to the "hyper-variability" of antibodies constrains their potential in several areas. In order to more extensively sample the potential and explore the limitations of antibody structure and function, semisynthetic antibodies have been devised. In these molecules, one (or more) of the CDRs has been replaced with a randomized sequence. Unlike antibodies cloned from a donor animal, semisynthetic antibodies can have CDRs of any size with any sequence.

The utility of semisynthesis was first shown in a study to change the specificity of an antibody from tetanus toxoid to fluorescein.[62] A library of approximately 10^7 variants of a tetanus toxoid antibody was created. The variants contained a 16-residue H3 completely randomized by oligonucleotide synthesis (an NNS [where N is any of the four nucleotides and S is either G or C] doping scheme was used to reduce from 64 to 32 the number of possible codons and to eliminate all but one amber stop codon), otherwise they were identical to each other and to the tetanus toxoid binding parent. After four rounds of panning, 46 of 100 clones bound a fluorescein-BSA conjugate by ELISA. Several antibodies were purified and shown to bind fluorescein with good affinity (Fig. 3–7). Several interesting features were noted. Ten clones isolated from a fluorescein elution regimen displayed only three unique sequences. There was also sequence homology among the selected sequences; a Ser-Arg-Pro motif is present in nine of the ten clones. Despite the homology in the amino acid sequences, a variety of codons was used. Thus, selection must have occurred at the protein level, not through nucleotide bias in the "random mixture." This is further demonstrated by the presence of Gly-95 and Asp-101[42] in all ten clones. Both glycine codons in the NNS scheme were used to code for position 95. The conserved Asp has been shown to participate in a structurally important salt bridge with Arg-94 of framework 3 in natural antibodies.[22] Although acid-eluted clones did not show sequence homology, they too bound fluorescein free in solution

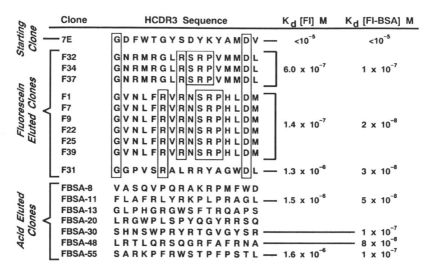

Figure 3–7. Redesign of an antibody based on CDR synthesis and phage display.[62] A human Fab-recognizing tetanus toxoid was used as the template for the production of a semisynthetic library randomized in H3. The library was panned against fluorescein. Separate elution protocols were followed in a parallel panning experiment. The H3 sequence of selected clones is shown, as are corresponding binding constants for fluorescein in solution and a fluorescent protein conjugate.

and as a BSA conjugate with K_ds of $\sim 10^{-8}$ to 10^{-6} M. It should be noted that these affinities are similar to those of antibodies in the secondary response of mice immunized with fluorescein.[63]

Semisynthetic libraries randomized in H3 have also been panned for binders to metal ions and metal oxides,[64] other small molecules,[65] as well as proteins,[66] including receptors.[67] More recently a semisynthetic library was constructed that was randomized in four of the six CDRs.[68] Several features of these semisynthetic antibody studies deserve note. First, unlike CDR grafted (humanized) antibodies, semisynthetics do not have "essential" framework residues. This is a direct result of the panning process whereby binders are *selected* from a nonbinding background. If a *particular* framework residue diminishes the binding of an antibody containing a *particular* CDR, that CDR sequence simply will not be selected. It follows that phage display and selection may provide human antibodies whose CDRs differ from corresponding rodent antibodies by virtue of a different framework context. Second, completeness of a library is, in many cases, not as important as structural diversity.[65] This follows from the hypothesis that a thorough redesign of a binding site may necessitate drastic mutation. The reason that structural diversity often sacrifices completeness derives from a limit to the number of *E. coli* that can be transformed with a foreign plasmid. The most efficient general process is electroporation, which allows up to $\sim 10^9$ trans-formants from 1 microgram plasmid DNA. As a result, current phage systems have the transformation step as the "diversity-limiting" step in the process. The largest libraries yet described approach 10^9 members. The practical outcome is that if a sequence of more than five amino acids is randomized, it is unlikely that each of the possible permutations will be present in the library. On the other hand, perturbing five amino acids out of an entire protein (an Fab is a 50 kDa heterodimer!) may not be enough to generate the unique protein the experimenter desires. To survey, with 99% confidence, a library with five amino acids randomized requires a library size of $\sim 10^8$ whereas six requires greater than 10^9. In an experiment to identify antibodies binding to small molecules,[65] several libraries competed with one another during the panning process. Libraries with HCDR3s of five, 10, and 16 residues were panned. No binders came from the most complete library, the five-residue HCDR3 library. This should be compared to a "winning" library, with both HCDR3 and LCDR3 randomized. Such a library would need approximately 10^{30} members to have every sequence represented, but actually contained only $\sim 10^8$ (or 10^{-20}%) of the possible sequences. In spite of this underrepresentation, several high-affinity binders were isolated from this library (Fig. 3–8). Finally, synthetic libraries (and other mutational strategies discussed later) provide access to sequences not represented in nature. Such sequences may be disallowed for other than structural reasons. For example, position 95 of LCDR3 of murine and human antibodies is predominantly proline. However, no clones selected from a semisynthetic library randomized in LCDR3 panned against small molecules had proline in this position.[65]

Semisynthetic libraries are not limited to total randomization of one (or more) CDR. A more directed approach has recently been used to secure human antibodies to the integrin ligand binding site (Fig. 3–9).[67] Anti-receptor antibodies have received attention recently as potential thera-peutics,[69] yet there has been no general method to make antibodies that can compete with a ligand for binding to its receptor. Antibodies to a receptor may bind either to surface bulge or the receptor cavity. Anti-idiotype Abs are similarly difficult to target. Further, recent structural studies have shown that there can be drastic domain rearrangements when an antibody binds antigen,[7] further exacerbating the search for useful anti-idiotype antibodies. Semisynthetic libraries can be generated that include a minimal ligand in the midst of random sequence. For example, anti-integrin antibodies have been produced[67] by randomizing three residues on either side of an Arg-Gly-Asp motif in H3. Antibodies generated in this fashion were shown to have quite high affinity (K_d ca. 10^{-10} M) and competed with natural ligand (Fig. 3–10). The affinity of these antibodies exceeds any derived from

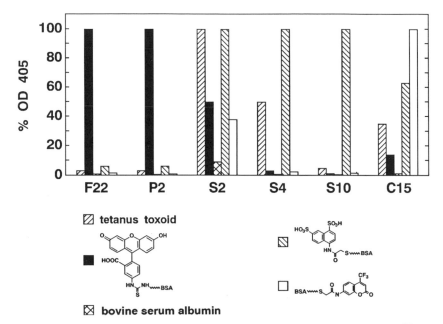

Figure 3–8. Synthetic antibodies with specificity for small molecules.[65] The reactivity profiles of each synthetic antibody as determined by ELISA against several haptens, as well as tetanus toxoid (the original specificity) and BSA as a background control. The antigens are, from left to right, tetanus toxoid, fluorescein, BSA, a sulfonate-containing hapten, and a coumarin derivative. F22 and P2 were selected to bind fluorescein and have dissociation constants of 34 and 80 nM, respectively. S2, S4, and S10 were selected to bind the sulfonate and have dissociation constants of 29, 56, and 37 nM, respectively. C15 was selected to bind the coumarin derivative. F22 is derived from a library where 16 amino acids of the HCDR3 were randomized, while S2 is derived from a library where a total of 20 amino acids from HCDR3 and LCDR3 were randomized.[65]

immunization of mice with integrins. Further, these antibodies could support or block cell adhesion *in vitro* and were potent inhibitors of platelet aggregation. This methodology allows for the rapid refinement and structuring of peptide leads derived from linear libraries. The HCDR3s of these antibodies may represent optimized constrained peptide ligands and thus important lead compounds for the development of small-molecule pharmaceuticals.

Affinity maturation is the *in vivo* process of mutation and selection that is responsible for improving the primary immune response. This process can be approximated *in vitro* to optimize antibodies selected by phage display. One method to improve an antibody is to subject it to PCR mutagenesis to create a library of variants, followed by panning the variant library for new antibodies. This process has been used to increase the affinity of an

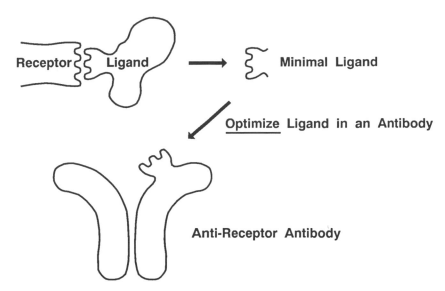

Figure 3–9. The design and selection of human anti-receptor antibodies is aided by a knowledge of residues within the ligand that are involved in binding to the receptor. The first step involves the characterization of this interaction by mapping regions of the ligand involved in binding or by selecting peptides from linear or constrained peptide libraries to bind the receptor. The minimal ligand is then transplanted into a semisynthetic antibody library. The random sequence is provided to optimize the conformational display of the transplanted sequence, which will be dependent on constraints imposed by the antibody. Random sequence also allows for selection of additional contact residues. Selection from a vast library of variants provides the optimal anti-receptor antibody. For integrins that bind RGD-containing peptides, the simple RGD sequence is transplanted.[67]

anti-progesterone antibody 13 to 30-fold[56] and to increase the affinity of an anti-4-hydroxy-5-iodo-3-nitrophenyl acetyl caproic acid antibody 4-fold.[70] As an alternative to random mutagenesis, "semi-random mutagenesis"[71] can be employed. This method takes advantage of information regarding a primary population to determine which residues to mutagenize. Further, a doping scheme for oligonucleotide synthesis is used to create limited amino acid subgroups, such as "hydrophobic" or "neutral," at particular sites. A combination of these two factors can lead to a more efficient search of "sequence space." Additionally, it might be possible to combine individual mutations of independently selected regions that increase binding to create a variant with vastly enhanced affinity.[72,79] We have successfully applied rounds of CDR-directed mutagenesis and selection to prepare antibodies with affinities of greater than 10^{11} M^{-1}. CDR-targeted mutagenesis is advantageous since optimization of these regions is most likely to improve affinity and least likely to create problems of immunogenicity.

Inhibition Constants and Amino Acid Sequences
of RGD-Containing Antibodies

Antibody #	IC$_{50}$ (Molar)			Sequence
	$\alpha_v\beta_3$	$\alpha_{IIB}\beta_3$	$\alpha_v\beta_5$	
4	2.5 x 10^{-10}	2.5 x 10^{-10}	5 x 10^{-7}	TQG - RGD - WRS
7	2.0 x 10^{-10}	5.0 x 10^{-10}	N.I.	TYG - RGD - TRN
8	2.0 x 10^{-10}	3.5 x 10^{-10}	N.I.	PIP - RGD - WRE
9	1.0 x 10^{-10}	1.0 x 10^{-10}	N.I.	SFG - RGD - IRN
10	2.5 x 10^{-10}	2.5 x 10^{-10}	N.I.	TWG - RGD - ERN

Figure 3–10. The ability of the synthetic human antibodies to block vitronectin binding to $\alpha_v\beta_3$ and $\alpha_v\beta_5$ and fibrinogen binding to $\alpha_{IIB}\beta_3$ was determined with a purified receptor binding assay as described. Synthetic antibodies were selected by placing the integrin binding motif, RGD, in the HCDR3 and flanking residues randomized. The sequences flanking the RGD motif of each antibody are shown. N.I. indicates no inhibition at concentrations of up to 5×10^{-7} M. The synthetic antibodies are displaying specificity for the receptor on which they were selected. It should be noted that $\alpha_v\beta_5$ also binds RGD-containing peptides but is not bound by the selected RGD-containing antibodies.[67] The antibodies were selected to bind $\alpha_v\beta_3$ and are demonstrating specificity for the beta subunit.

Chain shuffling has also been used to improve the primary pool of binders. It has been shown that a given heavy chain can combine with a variety of light chains while retaining antigen binding capacity.[73] In one case, an antibody to 2-phenyloxazol-5-one was reshuffled in two stages to isolate an antibody with a 3-fold improvement in its binding affinity. The reshuffled antibody had only one CDR in common with the original antibody.[74]

METHODS

Gene Construction

In order to introduce a random sequence into the CDR of an antibody, a technique known as splicing by overlap extension is employed (see Fig. 3–11)[62,64,65]. This is the most expedient method as no new restriction sites need to be introduced. In this approach the target gene is divided into two pieces with the random sequence to be introduced near one of the ends of the fragments. In the example given here, the random sequence is to be introduced near the 5' end of the 3' half of the gene. Primer pairs for the

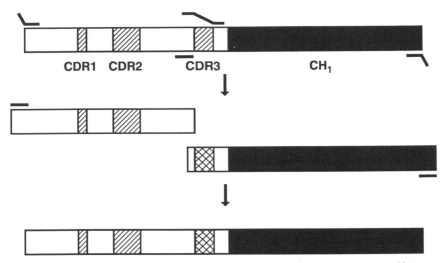

Figure 3–11. Generation of a semisynthetic library randomized in CDR3.[62] Overlap/extension PCR takes place in two stages. First, the CDR1/CDR2 region is amplified, as is the CH1/CDR3 region. The oligos at CDR3 are complementary; the longer one also contains randomized residues. This permits fusion of these two products in a subsequent PCR, while the randomized residues in the oligo are incorporated into CDR3.

5′ portion of the gene are:

    FTX3   5′-G-CAA-TTA-ACC-CTC-ACT-AAA-GGG-3′

    BFR3U   5′-TCT-CGC-ACA-GTA-ATA-CAC-GGC-CGT-3′

This pair will generate a ~ 430 bp fragment including HCDR1 and HCDR2. Primer pairs for the 3′ portion of the gene are:

HCDRD16   5′-GCC-GTG-TAT-TAC-TGT-GCG-AGA-(NNK)$_{14}$-GAC-
                          NNK-TGG-GGC-CAA-GGG-ACC-ACG-GTC-3′

R3B   5′-TTG-ATA-TTC-ACA-AAC-GAA-TGG-3′

This pair will generate a ~ 450 bp fragment including the randomized HCDR3 and the CH1 domain.

The template for both these reactions is a human tetanus toxoid binding Fab in the pComb3 vector in phage display from[47] (Fig. 3–5). The Gene Bank accession numbers are L22156 and L22157 for heavy and light chain respectively. Each of the two PCR reactions is performed with 10 ng of template DNA, 1 μg of each primer, 200 mM dNTPs, 1.5 mM MgCl$_2$, Taq polymerase buffer, and 5 units of Taq polymerase in a final volume of 100 μl. For each reaction type, ten reactions are performed to avoid bias possible in a single reaction. PCR is performed in the Perkin-Elmer 9600 instrument by denaturing for 1 min at 94°C, annealing for 1 min at 50°C, and extending at 72°C for 2 min for 35 cycles. Each PCR product is purified on a 1.5% agarose gel and isolated by electroelution. Following electroelution, the DNA is precipitated by adding 1 μl glycogen, 0.1 volume NaOAc, 2.2 volumes EtOH. The solution is incubated at −80°C for 30 min, centrifuged at 15,000 rpm for 15 min and the pellet is dried in speedvac. For the fusion reaction 1 μg of each purified product is combined and fused in the absence of primers for 35 cycles as above. This fusion product is then amplified using 10 μl of the reaction as template for another PCR reaction where the primers FTX3 and R3B are used. This reaction is performed in ten separate 100 μl reactions where the PCR conditions are identical to those given above. The resulting ∼880 bp product is purified on a 1% agarose gel and isolated as above.

The PCR product, 10 μg, is then digested with 170 U SpeI and 700 U XhoI for 3 hrs. The enzymes liberate a 50 and 150 pb fragment, respectively. The reaction is precipitated and the product purified on a 1.2% agarose gel followed by electroelution. The vector, pComb3 containing a human kappa chain, is prepared by digesting 50 μg DNA with 150 U SpeI and 450 U XhoI for 3 hrs and purified on a 0.7% agarose gel and isolated as above.

Library Construction

Ligation efficiency and background ligation should be tested on a small scale prior to use in library construction.[47,75] Ligate 250 ng vector with 50 ng insert for 2 hrs at room temperature or overnight. For a background control, ligate the vector itself. These ligations should be performed in a 20 μl total volume with 1 U of ligase. Electroplate 1 μl of the ligation mix as described below into 40 μl of cells, incubate, and plate aliquots. The background ligation should be ∼10% of the ligation with insert. The cells used for library construction should be able to produce at least 10^9 transformants per μg pUC19 DNA on electoporation.

1.   For the ligation of a library, ligate 1400 mg of vector containing an H or L chain with 450 ng of the cut PCR product with 40 μl of 5× ligase buffer and 10 μl of ligase in a total volume of 200 μl. Incubate at room temperature overnight. Heat kill ligation mix for 10 min at 70°C. Precipitate as above. Drain pellet and rinse with 70% ethanol and allow to drain inverted on a paper towel. Dry briefly in a Speedvac and resuspend in 15 μl water. Brief heating at 50 to 70°C followed by vortexing may be required to fully resuspend DNA. Place tube on ice for 10 min with electroporation cuvettes. Thaw, on ice, 300 μl of electro-competent cells per ligation. Add cells to tube containing DNA, mix, and set for 1 min. Transfer to electroporation cuvette and electroporate.

2. To electroporate, pulse at 2.5 kV, 0.2 cm cuvette, 25 μFD and 200 Ω.
3. Flush the electroporation cuvette *immediately*, first with 1 ml and then with 2 ml of room temperature SOC (use up to 5 ml SOC to flush cell) and incubate immediately in this SOC for *1 hr* at 37°C in a shaker (250 rpm).
4. Add prewarmed (37°C) Superbroth (SB) (20 μg/ml carbenicillin [carb] and 10 μg/ml tetracycline [tet]) to a total culture volume of 10 ml and immediately titer transformants by plating 100 μl, 10 μl, and 1 μl for test ligation and 10, 1, and 0.1 μl for library ligation on LB plates (100 *μg* → ml carb).
5. Incubate the 10 ml culture for 1 hr at 37°C on a shaker (300 rpm in all remaining steps).
6. Add carb to a final concentration of 50 μg/ml and incubate for an additional hr at 37°C.
7. Add the 10 ml culture to 100 ml of SB containing 50 μg/ml carb and 10 μg/ml tet and incubate 1 hr at 37°C on a shaker.
8. Add helper phage BCS M13 (total of 10^{12} pfu) usually ~1 ml and incubate 2 hr, 37°C.
9. Add kanamycin to 70 μg/ml and incubate on a shaker overnight at 30 to 37°C.
10. Spin cells down at 4000 rpm (JA10 rotor), 4°C for 15 min. Transfer the supernatant to a clean bottle and add 4% (w/v) PEG-8000 and 3% (w/v) NaCl. Place on shaker for 5 min to dissolve.
11. Incubate solution on ice for 30 min.
12. Spin down phage precipitate (9,000 rpm (JA10 rotor) for 30 min at 4°C) pour off and bleach the supernatant. Allow the bottle to drain for 10 min on a paper towel to remove as much PEG solution as possible.
13. Resuspend pellet in 2 ml TBS/1% BSA, centrifuge for 5 min at 14,000 rpm (Eppendorf microcentrifuge) and store the supernatant at 4°C (for long-term storage add 0.02% NaN_3). Packaged phagemid preparations should be used for panning only if they have been freshly prepared, as proteases present in trace levels cleave Fabs from the phage surface. Stored preparations should be reamplified prior to use in the panning protocol, i.e., go to step 9 below.
14. Prepare the phagemid DNA from cell pellet for storage.

Move to step 5 of the Panning procedure and continue.

Panning

A panning experiment is made up of several rounds of recognition and replication (Fig. 3–6).[47,64,75] During each round, specific binding clones are selected and amplified. These clones predominate after four or five rounds.

Recognition

1. Coat ELISA plate (Costar 3690) wells overnight at 4°C with 25 μl of antigen (1 to 0.1 mg/well, 1 mg generally works best) solution in coating buffer (0.1 M bicarbonate pH 8.6, optimal coating buffer may vary with antigen). Coating may also be performed at 37°C for 1 hr. Alternatively, sepharose or some other solid support may be used for the immobilization of the panigen.
2. Shake the fluid out of the wells and wash twice with deionized H_2O.
3. Block by filling the wells completely with 3% BSA in either PBS or TBS and seal the tray, making it airtight, or incubate in a humidified container.
4. Incubate for 1 hr at 37°C, then shake out the fluid as above.
5. Add 50 μl phage suspension to each well (total of about 10^{12} pfu).
6. Incubate for 2 hrs at 37°C after sealing the plate (or placing in a humidified container).
7. Remove the phage, fill well with TBS/0.5% Tween 20 (TBST), and pipette vigorously up and down. Wait 5 min, then remove TBST. In the first round, wash in this fashion once, in later rounds wash 5 times.
8. Elute the phage with 50 μl of elution buffer (0.1 M HCl (adjusted with glycine to pH 2.2)/BSA 1 mg/ml) per well. Incubate for 10 min at room temperature then pipette up and down vigorously. Remove the eluate and neutralize with 3 μl of 2 M Tris. base per 50 μl of elution buffer used. It is also possible to elute with soluble panigen.

Replication

(Reamplification of stored phage begins here.)

9. Infect 2 ml (per panning well) of fresh XL-1 blue ($OD_{600} = 0.5$ to 1; grown with 10 μg/ml tet) with the eluted phage or for reamplification infect 10 ml *E. coli* with 20 μl of phage precipitate and incubate at room temperature for 15 min.
10. Add 10 ml of 37°C prewarmed SB (20 μg/ml carb and 10 μg/ml tet) and immediately titer the eluted phage by plating 10 and 1 μl on LB plates/carb. Incubate the 12 ml culture for 1 hr at 37°C on a shaker.
11. To this culture adjust the carb concentration to 50 μg/ml and incubate for 1 hr at 37°C on a shaker.
12. Add helper phage VCS M13 (total of 10^{12} pfu), usually ~1 ml.

13. Transfer to a 100 ml SB culture with the same antibiotics and incubate on the shaker for 2 hr at 37°C.
14. Add kanamycin to 70 µg/ml and incubate on a shaker overnight at 30 to 37°C.
15. Spin cells down at 4000 rpm (JA10 rotor), 4°C for 15 min. Transfer the supernatant to a clean bottle and add 4% (w/v) PEG-8000 and 3% (w/v) NaCl. Place on shaker for 5 min to dissolve.
16. Precipitate phage on ice for 30 min.
17. Spin down precipitate (9,000 rpm (JA10 rotor) for 20 min at 40°C) and discard the supernatant. Allow the bottle to drain for 10 min on a paper towel to remove as much PEG solution as possible.
18. Resuspend pellet 2 ml TBS/1% BSA, centrifuge for 5 min at 14,000 rpm (Eppendorf microcentrifuge 5415C) and store the supernatant at 4°C (for long-term storage add 0.02% NaN_3). Packaged phagemid preparations should be used for panning only if they have been freshly prepared, as proteases present in trace levels cleave Fabs from the phage surface. Stored preps should be reamplified prior to use in the panning protocol, i.e., go to step 9 above.
19. For subsequent rounds reapply 50 µl of the above packaged phagemid preparation to antigen-coated wells (step 1) and continue from step 5 as above.
20. Titer phase suspension by infecting 50 µl volumes of XL-1 blue cells (OD_{600} = 0.5 to 1) with 1 µl of 10^{-3}, 10^{-6}, and 10^{-8} dilutions of phage suspension for 15 min at room temperature and plating on LB/carb plates (incubate overnight at 37°C).

Production of Soluble Fab

1. Collect the cell pellet from the last round of panning and isolate double-stranded DNA.[47,75]
2. Digest ~5 µg of DNA for 3 hrs with 15 U SpeI and 9 U NheI (Fig. 3–5). Precipitate and load on a 0.6% gel. Isolate the ~4.7 kb band and elute or GeneClean the fragment. Quantitate the recovered DNA.
3. Self-ligate 200 ng of this fragment in 20 µl total volume. Ligate for at least 2 hr at room temperature. Transform 1 µl of this ligation mix into 4 µl of competent cells as described previously. Grow and plate as described for the test ligation, but add an additional plating of a 0.1 µl aliquot of cells.
4. The next day pick a number of colonies for testing on ELISA. Inoculate a single colony into 10 ml of SB containing 20 mM $MgCl_2$ and 50 µg/ml carb. Grow at 37°C for 6 hr and induce by adding IPTG to a final concentration of 1 mM. Incubate overnight at 30°C.
5. Recover cells by centrifugation for 15 min at 1,500 g. Both supernatant

and cell pellet will contain Fab. Generally the concentration of Fab obtained by lysing the cell pellet will be greater than that found in the supernatant.

6. Resuspend the cell pellet in 1 ml of PBS. Lyse by freezing in a dry ice–ethanol bath for 5 min followed by thawing in a 37°C water bath. Repeat this process three times.

7. Clones may now be examined by simple ELISA. Purified Fab may be prepared as previously described.[51]

Points to Consider

Gene Construction: The primers provided produce an HCDR3 of 16 amino acids in length with the penultimate position fixed as Asp. This will provide a structurally diverse library for the isolation of new specificities. For the improvement of affinity of a known clone against the same antigen, six or fewer residues should be targeted for randomization. Note that the random primer region is flanked by 21 bases of perfect match to provide good priming while minimizing biases which may occur with short flanking regions.

Library Construction: Competent cells should generally be prepared with efficiencies of $>3 \times 10^9$ transformants/µg pUC. We suggest using a large volume of cells, 300 µl, to reduce the risk of sparking during transformation.

Panning: Charting the input/output ratio of phage gives an indication of selection for higher affinity clones. As each individual's washing technique and each antigen will generate a different background, high levels (1000-fold) of enrichment should not be expected in all cases. A 10-fold enrichment above binding to the blocking solution generally indicates success.

Production of Soluble Fab: It is not necessary to remove the gIII fusion in order to do a simple ELISA. There is sufficient proteolysis of the fusion to generate Fab for screening. However, for purification and larger-scale production, the fusion should be removed. We recommend inducing cultures at 30°C; however, each clone will have a different optimal temperature for induction.

CONCLUSION

Today, the successful cloning of an antigen-specific antibody need not be the final product. It is evident that a combination of panning and optimization is a general method to produce high-affinity antibodies of a specificity defined by the experimenter. The utility of the method is most obvious for antibodies that are difficult or impossible to obtain through traditional methods. Phage display has already given us myriad human monoclonal antibodies which may be useful for passive immunotherapy. Though we have limited the discussion to the positive selection of clones, it is obvious that negative selections may be performed by the inclusion of competitor in the selection step. Further developments in the field will include technology to access libraries whose diversity is not limited by the transformation efficiency of *E. coli*, for example, by *in vivo* recombination. Also, since selection schemes can be devised for events other than binding,[76] phage display promises to provide a new arena for the isolation and optimization of catalytic antibodies.[77] The creation and improvement of antibodies by the introduction of synthetic segments may one day empower machines that evolve antibodies to suit the experimenter's every requirement.[78]

Acknowledgments

We would like to acknowledge the considerable support and enthusiasm of Richard A. Lerner. Jonathan S. Rosenblum is a National Foundation Predoctoral Fellow. Carlos F. Barbas, III, is supported in part by a Scholar Award from the American Foundation for AIDS Research and an Investigator Award from the Cancer Research Institute.

REFERENCES

1. Wilson, I.A., and Stanfield, R.L. 1993. Antibody-antigen interactions. *Curr. Opt. Struct. Biol.* 3:113.
2. Wu, T.T., and Kabat E.A. 1970. An analysis of the sequences of the variable regions of Bence Jones and myeloma light chains and their implications for antibody complementarity. *J. Exp. Med.* 132:211.
3. Davies, D.R., and Padlan, E.A. 1990. Antibody-antigen complexes. *Annu. Rev. Biochem.* 59:439.
4. Wilson, I.A., Stanfield, R.L., Rini, J.M., Arevalo, J.H., Schulze-Gahmen, U., Fremont, D.H., Stura, E.A. 1991. Structural aspects of antibodies and antibody-antigen complexes. In *1991 Catalytic Antibodies*. Chichester: Wiley. [Ciba Foundation Symposium 159.]
5. Herron, J.N., He, X.M., Ballard, D.W., Blier, P.R., Pace, P.E., Bothwell, A.L.M., Voss, E.W. Jr., Edmundson, A.B. 1991. An auto antibody to single-stranded DNA: comparison of the three-dimensional structures of the unliganded Fab and a desoxynucleotide-Fab complex. *Proteins* 11:159.
6. Herron, J.N., He, X.M., Mason, M.L., Voss, E.W. Jr., Edmundson, A.B. 1989. Three-dimensional structure of a fluorescein-Fab complex crystallized in 2-methyl-2,4-pentanediol. *Proteins* 5:271.
7. Rini, J.M., Schulze-Gahmen, U., and Wilson, I.A. 1992. Structural evidence for induced fit as a mechanism for antibody-antigen recognition. *Science* 255:959–65.

8. Wu, T.T., Johnson, G., and Kabat, E.A. 1993. Length distribution of CDRH3 in antibodies. *Proteins* 16:1.
9. Sanz, I. 1991. Multiple mechanisms participate in the generation of diversity of human H-chain CDR3 regions. *J. Immunol.* 147:1720.
10. Kohler, G., and Milstein, C. 1975. Continuous cultures of fused cells secreting antibody of predefined specificity. *Nature* 256:495.
11. Shawler, D.L., Bartholomew, R.M., Smith, L.M., Dilman, R.O. 1985. Human immune response to multiple injections of murine monoclonal IgG. *J. Immun.* 135:1530.
12. Miller, R.A., Oseroff, A.R., Stratte, P.T., Levy, R. 1983. Monoclonal antibody therapeutic trials in seven patients with T-cell lymphoma. *Blood* 62:988.
13. Shroff, R.W., Foon, K.A., Beatty, S.M., Oldham, R.K., Morgan, A.C. Jr. 1985. Human anti-murine immunoglobulin responses in patients receiving monoclonal antibody therapy. *Cancer Res.* 45:879.
14. Jones, P.T., Dear, P.H., Foote, J., Neuberger, M.S., and Winter, G. 1986. Replacing the complementarity-determining regions in a human antibody with those from a mouse. *Nature* 321:522.
15. Verhoeyen, M., Milstein, C., and Winter, G. 1988. Reshaping human antibodies: grafting an antilysozyme activity. *Science* 23:1534.
16. Foote, J., and Winter, G. 1992. Antibody framework residues affecting the conformation of the hypervariable loops. *J. Mol. Biol.* 234:487.
17. Glaser, S.M., Vasquez, M., Payne, P.W., and Schneider, W.P. 1992. Dissection of the combining site in a humanized anti-Tac antibody. *J. Immunol.* 149:2607.
18. Brown, Jr., P.S., Parenteau, G.L., Dirbas, F.M., Garsia, R.J., Goldman, C.K., Bukowski, M.A., Junghans, R.P., Queen, C., Hakimi, J., Benjamin, W.R., Clark, R.D., and Waldmann, T.A. 1991. Anti-Tac-H, a humanized antibody to the interleukin 2 receptors, prolongs primate cardiac allograft survival. *Proc. Natl. Acad. Sci. USA* 88:2663.
19. Riechmann, L., Clark, M., Waldmann, H., and Winter, G. 1988, Reshaping human antibodies for therapy. *Nature* 332:323.
20. Carter, P., Presta, L., Gorman, C.M., Ridgway, J.B.B., Henner, D., Wong, W.L.T., Rowland, A.M., Kotts, C., Carver, M.E., and Shepard, H.M. 1992. Humanization of an anti-p185[HER2] antibody for human cancer therapy. *Proc. Natl. Acad. Sci. USA* 89:4285.
21. Queen, C., Schneider, W.P., Selick, H.E., Payne, P.W., Landolfi, N.F., Duncan, J.F., Avadalovic, N.M., Levitt, M., Junghans, R.P., and Waldmann, T.A. 1989. A humanized antibody that binds to the interleukin 2 receptor. *Proc. Natl. Acad. Sci. USA* 86:10029.
22. Chothia, C., and Lesk, A. 1987. Canonical structures for the hypervariable regions of immunoglobulins. *J. Mol. Biol.* 196:901.
23. Padlan, E.A., Silverton, E.W., Sheriff, S., Cohen, G.H., Smith-Gill, S.J., and Davies, D.R. 1989. Structure of an antibody-antigen complex: crystal structure of the HyHEL-10 Fab-lysozyme complex. *Proc. Natl. Acad. Sci. USA* 86:5938.
24. Bruck, C., Co, M.S., Slaoui, M., Gaulton, G.N., Smith, T., Fields, B.N., Mullins, J.I., Greene, M.I. 1986. Nucleic acid sequence of an internal image-bearing monoclonal anti-idiotype and its comparison to the sequence of the external antigen. *Proc. Natl. Acad. Sci. USA* 83:6578.

25. Shattil, S.J., Hoxie, J.A., Cunningham, M., and Brass, L.F. 1985. Changes in the platelet member glycoprotein IIb-IIIa complex during platelet aggregation. *J. Biol. Chem.* 260:11107.
26. Taub, R., Gould, R.J., Garsky, V.M., Ciccarone, T.M., Hoxie, J., Friedman, P.A., and Shattil, S.J. 1989. A monoclonal antibody against the platelet fibrinogen receptor contains a sequence that mimics a receptor recognition domain in fibrinogen. *J. Biol. Chem.* 264:259.
27. Taub, R., Hsu, J-.C., Garsky, V.M., Hill, B.L., Erlanger, B.F., and Kohn, L.D. 1992. Peptide sequences from the hypervariable regions of two monoclonal anti-idiotypic antibodies against the thyrotrotin (TSH) receptor are similar to TSH and inhibit TSH-increased cAMP production in FRTL-5 thyroid cells. *J. Biol. Chem.* 267:5977.
28. Levi, M., Sallberg, M., Ruden, U., Herlyn, D., Maruyama, H., Wigzell, H., Marks, J., and Wahren, B. 1993. A complementarity-determining region synthetic peptide acts as a miniantibody and neutralizes human immunodeficiency virus type I *in vitro*. *Proc. Natl. Acad. Sci. USA* 90:4374.
29. Skerra, A., and Pluckthun, A. 1988. Assembly of a functional immunoglobulin Fv fragment in *Escherichia coli*. *Science* 240:1038.
30. Better, M., Chang, C.P., Robinson, R., and Horwitz, R. 1988. *Escherichia coli* secretion of an active chimeric antibody fragment. *Science* 240:1041.
31. Bhat, T.N., Bentley, G.A., Fischmann, T.O., Boulot, G., Poljak, R.J., 1990. Small rearrangements in structures of Fv and Fab fragments of antibody D1.3 on antigen binding. *Nature* 347:483.
32. Rasched, I., and Overer, E. 1986. Ff coliphages: structural and functional relationships. *Microbiol. Rev.* 50:401.
33. Model, P., and Russel, M. 1988. Filamentous bacteriophage. In *The Bacteriophages*, Vol. 2. Valdner, R., ed. Plenum.
34. Russel, M. 1991. Filamentous phage assembly. *Molecular Biology* 5:1607–13.
35. Vieira, J., and Messing, J. 1987. Production of single-stranded plasmid DNA. *M. Enz.* 153:3.
36. Smith, G.P. 1985. Filamentous fusion phage: novel expression vectors that display cloned antigens on the virion surface. *Science* 228:1315.
37. Parmley, S.F., and Smith, G.P. 1988. Antibody-selectable filamentous fd phage vectors: affinity purification of target genes. *Gene* 73:305.
38. Scott, J.K., and Smith, G.P. 1990. Searching for peptide ligands with an epitope library. *Science* 249:386.
39. Devlin, J.J., Panganiban, L.C., and Devlin, P.E. 1990. Random peptide libraries. a source of specific protein binding molecules. *Science* 249:404.
40. Huse, W.D., Sastry, L., Iverson, S.A., Kang, A.S., Alting-Mees, M., Burton, D.R., Benkovic, S.J., and Lerner, R.A. 1989. Generation of a large combinatorial library of the immunoglobulin repertoire in phage lambda. *Science* 246:1275.
41. Saiki, R.K., Scharf, S., Faloona, F., Mullis, K.B., Horn, G.T., Erlich, H.A., and Arnheim, N. 1985. Enzymatic amplification of β-globin genomic sequences and restriction site analysis for diagnosis of sickle cell anemia. *Science* 230:1350.
42. Kabat, E.A., Wu, T.T., Perry, H.M., Gottesman, K.S., Foeller, C. 1991. *Sequences of Proteins of Immunological Interest*. Fifth edition. NIH.

43. Orlandi, R., Gussow, D.H., Jones, P.T., and Winter, G. 1989. Cloning immuno-globulin variable domains for expression by the polymerase chain reaction. *Proc. Natl. Acad. Sci. USA* 86:3833.

44. Sastry, L., Alting-Mees, M., Huse, W.D., Short, J.M., Sorge, J.A., Hay, B.N., Janda, K.D., Benkovic, S.J., and Lerner, R.A. 1989. Cloning of the immuno-logical repertoire in *Escherichia coli* for generation of monoclonal catalytic antibodies: construction of a heavy chain variable region-specific cDNA library. *Proc. Natl. Acad. Sci. USA* 86:5728.

45. Persson, M.A.A., Caothien, R.H., and Burton, D.R. 1991. Generation of diverse high-affinity human monoclonal antibodies by repertoire cloning. *Proc. Natl. Acad. Sci. USA* 88:2432.

46. Burton, D.R. 1991. Human and mouse monoclonal antibodies by repertoire cloning. *Trends in Biotechnol.* 9:169.

47. Barbas III, C.F., Kang, A.S., Lerner, R.A., and Benkovic, S.J. 1991. Assembly of combinatorial libraries on phage surfaces: the gene III site. *Proc. Natl. Acad. Sci. USA* 88:7978.

48. Zebedee, S.L., Barbas III, C.F., Hom, Y.-L., Cathoien, R.H., Graff, R., DeGraw, J., Pyati, J., LaPolla, R., Burton, D.R., Lerner, R.A., and Thornton, G.B. 1992. Human combinatorial antibody libraries to hepatitis B surface antigen. *Proc. Natl. Acad. Sci. USA* 89:3175.

49. Williamson, R.A., Burioni, R., Sanna, P.P., Partridge, L.J., Barbas III, C.F., and Burton, D.R. 1993. Human monoclonal antibodies against a plethora of viral pathogens from single combinatorial libraries. *Proc. Natl. Acad. Sci. USA* 90:4141.

50. Burton, D.R., Barbas III, C.F., Persson, M.A.A., Koenig, S., Chanock, R.M., and Lerner, R.A. 1991. A large array of human monoclonal antibodies to type 1 human immunodeficiency virus from combinatorial libraries of asymptomatic seropositive individuals. *Proc. Natl. Acad. Sci. USA* 88:10134.

51. Barbas III, C.F., Bjorling, E., Chiodi, F., Dunlop, N., Cababa, D., Jones, T.M., Zebedee, S.L., Persson, M.A.A., Nara, P.L., Norrby, E., and Burton, D.R. 1992. Recombinant human Fab fragments neutralize human type 1 immunodeficiency virus *in vitro*. *Proc. Natl. Acad. Sci. USA* 89:9339–43.

52. Barbas III, C.F., Collet, T.A., Amberg, W., Roben, P., Binley, J.M., Hoekstra, D., Cababa, D., Jones, T.M., Williamson, R.A., Pilkington, G.R., Haigwood, N.L., Cabezas, E., Satterthwait, A.C., Sanz, I., and Burton, D.R. 1993. Molecular profile of an antibody response to HIV-1 as probed by combinatorial libraries. *J. Biol. Chem.* 230:812.

53. Barbas III, C.F., Crowe, Jr., J.E., Canaba, D., Jones, T.M., Zebedee, S.L., Murphy, B.E., Chanock, R.M., and Burton, D.R. 1992. Human monoclonal Fab fragments derived from a combinatorial library bind to respiratory syncytial virus F glycoprotein and neutralize infectivity. *Proc. Natl. Acad. Sci. USA* 89:10164–68.

54. Burton, D.R., Barbas III, C.F. 1992. Antibodies from libraries. *Nature* 359:782.

55. Duchosal, M.A., Eming, S.A., Fischer, P., Leturcq, D., Barbas III, C.F., McConahey, P.H., Caothien, R.H., Thornton, G.B., Dixon, F.J., and Burton, D.R. 1992. Immunization of hu-PBL-SCID mice and the rescue of human monoclonal Fab fragments through combinatorial libraries. *Nature* 355:258–62.

56. Gram, H., Marconi, L.-A., Barbas III, C.F., Collet, T.A., Lerner, R.A., and Kang, A.S. 1992. *In vitro* selection and affinity maturation of antibodies from a naive combinatorial immunoglobulin library. *Proc. Natl. Acad. Sci. USA* 89:3576.
57. Kang, A.S., Barbas, C.F., Janda, K.D., Benkovic, S.J., and Lerner, R.A. 1991. Linkage of recognition and replication functions by assembling combinatorial antibody Fab libraries along phage surfaces. *Proc. Natl. Acad. Sci. USA* 88:4363.
58. Marks, J.D., Hoogenboom, H.R., Bonnert, T.P., McCafferty, J., Griffiths, A.D., and Winter, G. 1991. By-passing immunisation human antibodies from V-gene libraries displayed on phage. *J. Mol. Biol.* 222:581.
59. Griffiths, A.D., Malmqvist, M., Marks, J.D., Bye, J.M., Embleton, M.J., McCafferty, J., Baier, M., Holliger, K.P., Gorick, B.D., Hughes-Jones, N.C., Hoogenboom, H.R., and Winter, G. 1993. Human anti-self antibodies with high specificity from phage display libraries. *EMBO J.* 12:725.
60. Chothia, C., Lesk, A.M., Gherardi, E., Tomlinson, I.M., Walter, G., Marks, J.D., Llewelyn, M.B., and Winter, G. 1992. Structural repertoire of the human V_H Segments. *J. Mol. Biol.* 227:799.
61. Kabat, E.A., and Wu, T.T. 1991. Identical V region amino acid sequences and segments of sequences in antibodies of different specificities. Relative contributions of VH and VL genes, minigenes, and complementarity-determining regions to binding of antibody-combining sites. *J. Immunol.* 147:1709.
62. Barbas III, C.F., Bain, J.D., Hoekstra, D.M., and Lerner, R.A. 1992. Semisynthetic combinatorial antibody libraries: a chemical solution to the diversity problem. *Proc. Natl. Acad. Sci. USA* 89:4457.
63. Kranz, D.M., Ballard, D.W., and Voss Jr., E.W. 1983. Expression of defined idiotypes throughout the BALB/c anti-fluorescyl antibody response: affinity and idiotype analysis of heterogeneous antibodies. *Mol. Immunol.* 20:1313.
64. Barbas III, C.F., Rosenblum, J.S., and Lerner, R.A. 1993. Direct selection of antibodies that coordinate metals from semisynthetic combinatorial libraries. *Proc. Natl. Acad. Sci. USA* 90:6385–89.
65. Barbas III, C.F., Amberg, W., Simoncsits, A., Jones, T.M., and Lerner, R.A. 1993. Selection of human anti-hapten antibodies from semisynthetic libraries. *Gene.* 137:57.
66. Hoogenboom, H.R., and Winter, G. 1992. By-passing immunisation human antibodies from synthetic repertoires of germline V_H gene segments rearranged *in vitro*. *J. Mol. Biol.* 227:381.
67. Barbas III, C.F., Languino, L.R., and Smith, J.W. High-affinity self-reactive human antibodies by design and selection: targeting the integrin ligand binding site. *Proc. Natl. Acad. Sci. USA* 90:10003.
68. Garrard, L.J., and Henner, D.J. 1993. Selection of an anti-IGF-1 Fab from a Fab phage library created by mutagenesis of multiple CDR loops. *Gene* 128:103.
69. Taub, R., and Greene, M.I. 1992. Functional validation of ligand mimicry by anti-receptor antibodies: structural and therapeutic implications. *Biochemistry* 31:7431.
70. Hawkins, R.E., Russell, S.J., and Winter, G. 1992. Selection of phage antibodies by binding affinity mimicking affinity maturation. *J. Mol. Biol.* 226:889.

71. Arkin, A.P., and Youvan, D.C. 1992. Optimizing nucleotide mixtures to encode specific subsets of amino acids for semi-random mutagenesis. *Bio/Technology* 10:297.

72. Wells, J.A. 1990. Additivity of mutational effects in proteins. *Biochemistry* 29:8509.

73. Collet, T.A., Roben, P., O'Kennedy, R., Barbas III, C.F., Burton, D.R., and Lerner, R.A. 1992. A binary plasmid system for shuffling combinatorial antibody libraries. *Proc. Natl. Acad. Sci. USA* 89:10026.

74. Marks, J.D., Griffiths, A.D., Malmqvist, M., Clackson, T.P., Bye, J.M., and Winter, G. 1992. By-passing immunisation: building high affinity human antibodies by chain shuffling. *Bio/Technology* 10:779.

75. Barbas III, C.F., and Lerner, R.A. 1991. Combinatorial immunoglobulin libraries on the surface of phage (Phabs): rapid selection of antigen-specific Fabs. *METHODS: A Companion to Methods in Enzymology* 2:119.

76. Matthews, D.J., Wells, J.A. 1993. Substrate phage: selection of protease substrates by monovalent phage display. *Science* 260:1113.

77. Lerner, R.A., Benkovic, S.J., and Schultz, P.G. 1991. At the crossroads of chemistry and immunology: catalytic antibodies. *Science* 252:659.

78. Lerner, R.A., Kang, A.S., Bain, J.D., Burton, D.R., and Barbas III, C.F. 1992. Antibodies without immunization. *Science* 258:1313.

79. Barbas III, C.F., Hu, D., Dunlop, N., Sawyer, L., Cabala, D., Hendry, R.M., Nara, P.L., and Burton, D.R. 1992. In vitro evolution of a neutralizing human antibody to human immunodeficiency virus type 1 to enhance affinity and broaden strain cross-reactivity. *Proc. Natl. Acad. Sci. USA* 91:3809.

CHAPTER 4

Engineering the Antibody Combining Site by Codon-Based Mutagenesis in a Filamentous Phage Display System

Scott Glaser, Karin Kristensson, Todd Chilton, and William Huse

Codon-based mutagenesis, as applied to antibody engineering, is a general method for rapidly and efficiently increasing antibody affinity and for altering specificity towards an antigen. Antibody binding properties can be modified up to 100-fold without devising special screening procedures or requiring detailed antibody structural information.[1] Randomized sequence libraries constructed within the antibody variable region by codon-based mutagenesis and screened by panning, filter lift, or enzyme-linked immunoassay (ELISA) yield functionally modified antibodies at a frequency of three orders of magnitude greater as compared to traditional mutagenesis techniques.

Recently, we have applied codon-based mutagenesis to optimize the affinity of an anti-glycolipid Fab towards antigen 15-fold over that of the parent molecule.[2] Substitution of residues within the combining site of a chimeric antibody directed against a tumor-associated protein have also resulted in a 10-fold increase in antibody affinity (T. Stinchcombe, unpublished results). In one particular example, codon-based mutagenesis was used to decrease nonspecific binding of a murine anti-hapten antibody to a crossreactive epitope by at least an order of magnitude with a concomitant modest increase in binding affinity toward the target antigen.

Two technological advancements have greatly facilitated the successful engineering of antibody genes: (1) the development of rapid methods of cloning and manipulating antibody gene fragments by way of the polymerase chain reaction (PCR),[3–6] and; (2) the efficient expression of soluble, functionally active antibody fragments in bacteria.[7–10] Combining these technologies provides a powerful approach for the design, construction, modification, and screening of genetically engineered antibodies.[11–17]

We have previously described the construction and use of an M13 filamentous phage vector system for the cloning, expression, and mutagenesis of antibody gene fragments.[16] Optimization of the antibody combining site with the goal of improving affinity or redirecting specificity to antigen is readily accomplished in this vector system by mutagenizing one or more of the six complementarity determining regions (CDRs) by codon-based mutagenesis.[1] Codon-based mutagenesis replaces entire codons rather than individual nucleotides, thus allowing production of antibody sequences that may be unlikely to occur using other conventional *in vitro* mutagenesis techniques. Large antibody sequence libraries generated by codon-based mutagenesis can then be expressed in *E. coli* and rapidly screened for novel antibody binding properties.

Prior to initiating a combining site optimization, the antibody genes coding for the desired specificity must be isolated. Polyclonal sets of antibody genes derived from immune tissue can be inserted into the M13 vectors allowing for the rapid construction of combinatorial libraries. Alternatively, single antibody genes derived from hybridomas or Epstein-Barr virus (EBV) immortalized B cells can be cloned into these vectors. Typically, antibody genes are cloned by PCR amplification of cDNA prepared from immune tissue extracts using oligonucleotide primers containing restriction sites found infrequently within immunoglobulin genes.[17] The double-stranded product is digested with restriction enzymes and ligated into a double-stranded replicative form M13 vector containing compatible sites. The heavy-chain library is prepared in a vector that can express the heavy chain as a V_H–C_{H1} fusion product to either the pVIII or pIII viral coat protein.[12–16] The light-chain library is prepared in a similar vector that expresses the light chain as a V_L–C_κ or V_L–C_λ chain. Combinatorial association of the heavy-chain and light-chain populations results in a library that directs the coexpression and assembly of antibody V_H and V_L sequences.

Because the packaged phage genome is single-stranded DNA, subsequent modification of the antibody genes by oligonucleotide-directed mutagenesis is convenient and highly efficient. This vector system can produce soluble Fab and can also synthesize and display Fab fragments on the phage surface via the pVIII fusion product. Antibody fragments displayed on the phage surface permit enrichment of unique antibody specificities by chromato-graphy methods. In addition, the secretion of soluble, functionally active Fab fragments into the bacterial periplasmic space permit rapid biochemical analysis of isolated antibody specificities. Thus, the M13 vector system provides all the features of gene cloning, modification, and expression in a single vector system.

In this chapter, we describe the genetic optimization of the antibody combining site using codon-based mutageneis for the purpose of enhancing affinity or redirecting binding specificity toward antigen. In addition, we describe an improved method for constructing antibody expression libraries in our M13 filamentous phage vector.

CONSTRUCTION OF ANTIBODY EXPRESSION LIBRARIES IN AN M13 PALINDROMIC LOOP VECTOR BY HYBRIDIZATION MUTAGENESIS

Though a single vector system designed for both cloning and expression of antibody fragments has many apparent advantages, there are limitations in cloning PCR amplified antibody genes into double-stranded vectors. Despite designing PCR primers that contain infrequently found antibody V region restriction sites, the presence of these restriction sites within antibody V regions does occur and results in the cloning of incomplete antibody sequences and the concomitant loss of sequence diversity from the antibody repertoire. Moreover, we have found that in some cases substitution of amino terminal V region amino acid residues, a consequence of cloning into double-stranded vectors, results in the loss of antigen binding activity (T. Stinchcombe, personal communication).

A solution to this problem is to clone antibody sequences without the use of restriction endonucleases. Replacement of antibody V regions with homologous V region sequences by mutagenesis has been recently described using M13 cloning vectors.[18] Ideally, it would be desirable to introduce both V_H and V_L antibody chains into a single expression vector because the random association of V_H with V_L gene sequence populations by hybridiza-tion would form the basis of constructing a combinatorial antibody library. This eliminates the conventional steps required for constructing combinator-ial libraries in a two-vector-system amplification of individual V_H and V_L libraries, preparation of replicative form M13 DNA, restriction digestion, and crossing of the V_H and V_L gene sequence libraries. Unfortunately, the efficiency of simultaneously introducing both antibody V_H and V_L gene sequences by hybridization can be highly variable.

To address these concerns we have designed an M13 V_H and V_L coexpres-sion vector for cloning novel antibody V region specificities by hybridization

A.

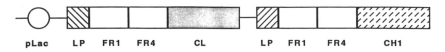

pLac   LP   FR1   FR4   CL   LP   FR1   FR4   CH1

B.

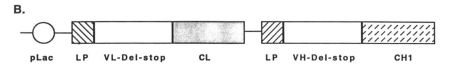

pLac   LP   VL-Del-stop   CL   LP   VH-Del-stop   CH1

Figure 4–1. Schematic representation of the palindromic loop M13 vector used for cloning antibody V_L and V_H genes by hybridization mutagenesis. Each palindromic loop contains an Eco RI restriction site indicated by the solid rectangle. The 3′ ends of the single-stranded V_V and V_C PCR products anneal to the vector at the regions coding for the bacterial leader peptides (L.P.) and the 5′ ends anneal to sequences coding for light-chain constant and heavy-chain C_{H1} regions, respectively. The solid circle indicates the lac promoter and the direction of transcription.

mutagenesis that has the unique feature of allowing selection of recombinant phage that coexpress both V_H and V_L antibody chains. The vector contains two palindromic loops located at the annealing sites for the V_H and V_L region genes, each loop containing an Eco RI restriction site (Figure 4–1). V_H and V_L gene sequences are first amplified using the polymerase chain reaction. The amplified V_H and V_L gene sequences are then reamplified using asymmetric polymerase chain reaction and the single-stranded products simultaneously annealed to the single-stranded palindromic looped vector. The second strand is synthesized *in vitro* by T4 DNA polymerase in the presence of all four deoxynucleotide triphosphates and the nicked molecule ligated. The double-stranded product is then digested with Eco RI restriction endonuclease, cleaving the palindromic loop sequences. Vector that has not incorporated both V_H and V_L gene sequences is converted to linear DNA via the restriction sites within the palindromic loops and fails to form infectious M13 phage upon transfection into an *E. coli* host.

The palindromic loop vector can be designed to accept either murine or human antibody gene sequences. The light-chain portion of the vector retains DNA sequence coding for the first 23 amino acid residues of kappa light-chain FR1 and the last residue in FR4 in addition to the sequence coding for the kappa constant domain (Figure 4–1). Similarly, the heavy-chain portion consists of DNA coding for the first 30 amino acid residues

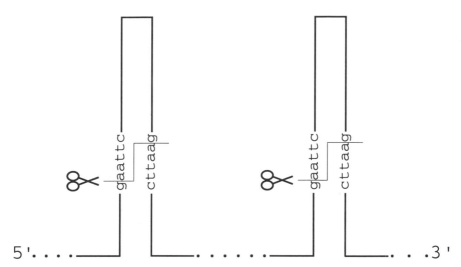

Figure 4–2. Nucleotide sequence of the palindromic loops occurring at the light-chain and heavy-chain variable region annealing sites, respectively. The Eco RI restriction sites are enclosed in the open rectangles.

of FR1, the last residue in FR4, and the heavy-chain CH1 domain. The exact nucleotide sequence of the FR1 and FR4 portions is not critical because these sequences will ultimately be replaced by the donor antibody sequences. Palindromic loops containing the Eco RI restriction site are positioned between the FRI and FR4 regions in both the V_H and V_L domains (Figure 4–2). The vectors can also be designed to express Fabs as any IgG isotype. The successful introduction of full-length antibody V region genes into this vector results in a recombinant molecule identical in format compared to the antibody Fab product resulting from crossed heavy- and light-chain-containing vectors previously described.[16] Secretion of antibody Fab fragments can be monitored by immunochemical detection of Fab molecules immobilized onto nitrocellulose filters or by ELISA.[16]

Generally, we have found that a minimum of 18 nucleotides is sufficient to anneal the PCR amplified DNA to the palindromic loop expression vector. The forward PCR primer sequences are derived from DNA sequences coding for the last 6 carboxy-terminal bacterial leader peptide amino acids and the first 6 amino-terminal variable region amino acids. The reverse PCR primers are complementary to DNA coding for constant domain sequences. The portion of the primers that are used to amplify heavy- and light-chain genes can be derived from Kabat.[19] Conditions for the first round of PCR amplification and the subsequent asymmetric PCR amplification reactions are described in Protocol 1.[3, 5, 20, 21]

Protocol 1. Asymmetric OCR Amplification of Antibody V$_H$ and V$_L$ Genes

Total RNA is prepared from immune tissue or antibody-secreting cells by the guanidinium thiocyanate-phenol/chloroform method described by Chomczynski and Sacchi.[22] First-strand cDNA is synthesized and the cDNA–RNA heteroduplexes are first used as templates for PCR amplification.

1. 5.0 µl 10X PCR buffer (10X PCR buffer = 0.67 M Tris pH 9.2, 16.6 mM NH$_4$SO$_4$, 20 mM MgCl$_2$)
    5.0 µl 2.0 mM dNTPs
    5.0 µl cDNA
    1.0 µl 0.5 M β-mercaptoethanol
    50 pmoles forward PCR primer
    50 pmoles reverse PCR primer
    0.25 µl Taq DNA Polymerase (Boehringer-Mannheim, Indianapolis, IN)
    Sterile H$_2$O to 50.0 µl total volume

    Overlay reaction with 50 µl mineral oil. Conditions for amplification are: denature at 94°C for 2 min, followed by 2 cycles of denaturation of 94°C for 1 min, annealing at 50°C for 1 min, and DNA synthesis at 72°C for 1 min. This is followed by 40 cycles of denaturation at 94°C for 1 min, annealing at 55°C for 1 min, DNA synthesis at 72°C for 1 min, and a final extension at 72°C for 10 min.

2. To a new PCR reaction tube add:
    10.0 µl 10X PCR buffer (10X PCR buffer = 0.67 M Tris pH 9.2, 16.6 mM NH$_4$SO$_4$, 20 mM MgCl$_2$)
    10.0 µl 2 mM dNTPs
    2.0 µl unpurified, double-stranded PCR product
    1.0 nmol reverse PCR primer
    2.0 µl 0.5 M β-mercaptoethanol
    0.25 µl Taq polymerase
    Sterile H$_2$O to 100 µl total reaction volume

    Overlay reaction with 50 µl mineral oil. Conditions for amplification are: denature at 95°C for 1 min, followed by 20 cycles of: denaturation at 94°C for 20 seconds, annealing at 55°C for 30 seconds, and DNA synthesis at 72°C for 30 seconds. This is followed by a final DNA synthesis at 72°C for 7 min. The mixture is then cooled and held at 4°C.

3. Analyze 5 µl of asymmetric PCR product on a 2% agarose gel prepared with 1X Tris-acetate EDTA buffer.[23]

4. Purify the asymmetric V_H and V_L PCR products from a 2% low-melt agarose gel prepared with 1X Tris-acetate EDTA buffer with the Geneclean™ kit (San Diego, CA) according to the manufacturer's protocol.

The 5' ends of the purified V_H and V_L PCR gene fragments are then phosphorylated (Protocol 2) essentially as described by Zoller and Smith.[24] Alternatively, the reverse PCR primer can be phosphorylated prior to asymmetric PCR amplification.

Protocol 2. Phosphorylation of Asymmetric V_H and V_L PCR Products

1. 5 µl (2.0 µg) heavy-chain or light-chain asymmetric PCR product
    2 µl 10X kinase buffer
    1 µl 10 mM ATP
    11 µl sterile water
    1 µl T4 polynucleotide kinase (Boehringer-Mannheim, Indianapolis, IN)
    20 µl total volume. Incubate 37°C, 45 min.
2. Heat inactivate the reaction at 65°C for 10 min.
3. Add 1 to 2 µl 20 mg/ml glycogen (Boehringer-Mannheim, Indianapolis, IN), mix well. Add $\frac{1}{10}$ volume 3 M Na acetate, pH 5.2. Mix and add 2 volumes ethanol to precipitate DNA.
4. Resuspend phosphorylated ssDNA to 50 ng/µl in sterile water.

Hybridization mutagenesis is performed essentially as described by Kundel[5,6] using reagents prepared as described in the Muta-Gen M13 *in vitro* Mutagenesis Kit (Biorad, Richmond, CA) except that the single-stranded PCR amplified antibody gene products are used in place of oligonucleotides. We find that uridinylated palindromic loop vector DNA template lowers the background of non-Fab producing phage approximately 2 to 3 times beyond that achieved with wild type palindromic loop vector. Uracil substituted single-stranded DNA template is prepared by growing M13 phage in the dut⁻ ung⁻ *E. coli* strain CJ236 (BioRad, Richmond, CA). Protocol 3 describes the construction of V region libraries by hybridization mutagenesis. Following the mutagenesis reaction, a portion of the reaction contents are electroporated into *E. coli* and plated.

Protocol 3. Construction of Antibody V Region Libraries by Hybridization Mutagenesis

1. 2 µl (0.1 pmol) uridinylated ss M13 palindromic looped vector
    2 µl (1 pmol) phosphorylated ss heavy chain PCR product

> 2 μl (1 pmol) phosporylated ss light chain PCR product
> 1 μl 10X anneal buffer
> 3 μl sterile water
> 10 μl total volume.

Heat to 96°C for 4 min in heating block. Remove block and allow to cool slowly to 30°C. Microfuge briefly to collect any condensation.
2. Place annealed reaction on ice.
3. Add 1 μl 10X synthesis buffer, 1 μl T4 DNA ligase, and 1 μl T4 DNA polymerase (Boehringer-Mannheim, Indianapolis, IN). Incubate on ice, 5 min, room temperature, 5 min, and 37°C, 1.5 hours.
4. Ethanol precipitate as described in Protocol 2 step 3 and resuspend pellet in 8 μl sterile water.
5. Add 1 μl 10X Eco RI digestion buffer and 1 μl (10 U/μl) Eco RI restriction endonuclease. Incubate 37°C, 2 hours.
6. Ethanol precipitate as described in Protocol 2 step 3 and resuspend pellet in 10 μl sterile water.
7. Electroporate 2.5 μl into electrocompetent bacteria such as DH10B (Gibco BRL, Gaithersburg, MD) and titer into a lawn of F′ bacterium XL-1 Blue (Strategene, San Diego, CA).

Antigen binding specificities can be enriched by panning or directly analyzed for binding to antigen by preparing nitrocellulose filter lifts of Fab-secreting phage libraries.[11–16] Antigen can be derivatized with alkaline phosphatase, horseradish peroxidase, or radiolabeled with ^{125}I and used to screen filter lifts for antigen binding Fabs.[27] Alternatively, biotinylated antigen can be incubated with filter lifts and detected with either streptavidin-alkaline phosphatase or streptavidin-horseradish peroxidase conjugates. Positive clones are then cored, plaque-purified, and reevaluated. In those cases where soluble antigen is not available, such as membrane-associated antigens, filter lifts are first screened for the presence of intact Fab fragments. Plaques that coexpress both heavy- and light-chain sequences are cored, plaque-purified, and used to infect *E. coli* strain MK30-3 [sup0] from which periplasmic fractions can be prepared.[16] These fractions can be analyzed for antigen binding in an ELISA format.

Once clones exhibiting the desired antigen binding properties have been isolated, the V_H and V_L genes are analyzed by DNA sequencing. This analysis serves two purposes. First, the primary functional screen is not usually a quantitative assay, and PCR-generated errors in the antibody gene sequence may produce an antigen-binding positive clone having altered affinity or specificity to antigen. Second, DNA sequence information is required to design the codon-based oligonucleotides used for engineering the antibody combining site. Generally, it is recommended that a sufficient number of clones be evaluated by DNA sequence analysis to be confident that the clone ultimately selected contains no PCR errors.

ENGINEERING ANTIBODY AFFINITY AND SPECIFICITY BY CODON-BASED MUTAGENESIS

During the natural *in vivo* immune response, antibodies of increasing affinity and specificity for antigen emerge with time in a process termed affinity maturation.[28] This progressive refinement for antigen binding is a result of somatic hypermutation events in the antibody V region genes followed by antigen-driven selection and proliferation of those B lymphocytes expressing the responding antibody on the cell surface.[29,30] Berek and Milstein have shown that these mutations are primarily localized to the six hypervariable regions or CDRs and their flanking sequences.[28] Because sequence variability within the six CDRs is the primary correlate of antigen binding, these regions are reasonable choices for alteration by mutagenesis. Substitution of amino acid residues within the antibody combining site with the goal of enhancing affinity or redirecting binding specificity toward antigen is efficiently accomplished using codon-based mutagenesis techniques.

The fundamental problem then depends on how one evaluates the variant antibodies generated by the mutagenesis procedure. If only a limited number of variant antibodies can be produced and screened, then the native amino acids targeted for mutagenesis must be judiciously identified based upon a prediction of the conformation of the antibody combining site. However, if a large number of variant antibodies can be produced and screened, then the mutagenesis procedure can include a greater number of target amino acids without requiring structural information. The procedure and expression of antibody fragments in *E. coli* using M13 phage vector and the ease with which vast numbers can subsequently be screened permits the creation of large, diverse libraries of antibodies containing many mutations in the antibody combining site.

The first step in constructing an antibody sequence library is collecting DNA sequence information of the entire V_H and V_L regions and identifying all six CDRs by sequence variability.[19,31] If structural information is available, only those CDRs predicted to contact antigen might be selected for mutagenesis. For those CDRs that can be classified as belonging to a particular canonical conformation, key residues can be identified and kept as such.[32] Otherwise, sequence libraries can easily be constructed to include any number of amino acid residues.

CDRs selected for mutagenesis are first modified to contain a small deletion immediately followed by a stop codon.[1] The rationale for this step is to facilitate the identification of template molecules that have successfully incorporated mutagenic oligonucleotides. Reconstitution of the CDR will result in the production of immunoglobulin protein that can easily be detected by immunochemical detection. We routinely mutagenize between five and eight amino acid residues within a single CDR. The method for synthesizing and purifying oligonucleotides to be used for condon-based mutagenesis has been described in detail.[1,33] 1 µg of purified oligonucleotides are phosphorylated as described in Protocol 2 and used to mutagenize single-stranded template (Protocol 4).

Protocol 4. Construction of Codon-Based CDR Sequence Libraries

1.   2 µl (250 ng) uridinylated ss M13 antibody template DNA
     2 µl (100 ng) phosphorylated oligonucleotides
     1 µl 10X anneal buffer
     4 µl sterile water
     10 µl total volume.
     Heat to 70°C for 3 min in heating block. Remove block and allow to cool slowly to 30°C. Microfuge briefly to collect any condensation.
2.   Place annealed reaction on ice.
3.   Add 1 µl 10X synthesis buffer, 1 µl T4 DNA ligase and 1 µl T4 DNA polymerase (Boehringer-Mannheim, Indianapolis, IN). Incubate on ice, 5 min.
4.   Incubate room temperature, 5 min.
5.   Incubate 37°C, 1.5 hr.
6.   Place mutagenesis reaction on ice and dilute 2-fold with sterile water.
7.   Electroporate 1 µl into electrocompetent bacteria such as DH10B and titer onto a lawn of F′ bacterium (XL-1 Blue).

Libraries are then screened for the desired phenotype. If one is screening for clones that have redirected binding specificity such as eliminating cross-reactivity to a nontarget antigen, antibody specificity can be selected by enriching for the novel specificity with immobilized antigen in the presence of excess nontarget antigen. Alternatively, replica filter lifts can be prepared. One filter is incubated with the antigen to which binding is to be maintained, and the replica filter is incubated with the cross-reaction material.

Prior to screening a sequence library for higher-affinity antibodies, it is recommended that the optimum antigen concentration to be used in the filter lift assay be determined. Subsaturating concentrations of antigen should allow higher-affinity antibody clones to be distinguished from parental antibody. Once this concentration is determined, replica lifts of the library are prepared, and one lift is assayed for antigen binding and the second lift is assayed for levels of Fab secreted. Those antibodies that are the best candidates for higher-affinity variants secrete normal amounts of Fab compared to the parent molecule, yet produce a more intense signal in the antigen binding assay. Preparation and screening of filter lifts is described in Protocol 5.

Protocol 5. Filter Lift Preparation and Analysis of Fab Expression Libraries

1. To 200 µl log phase XL-1 Blue bacteria (Strategene, San Diego, CA) add 1000 to 2000 pfu of M13 phage library, 3 µl 1 M isopropyl-β-D-thiogalactopyranoside (Sigma Chemical Co., St. Louis, MO) and 3 ml molten top agar. Mix gently and plate onto prewarmed 100 mm LB plates and incubate at 37°C for 6 to 8 hours.
2. Place a dry 0.45 µ nitrocellulose filter (Schleicher and Schuell, Keene, NH) onto plate and adsorb Fab onto filter, room temperature, from 1 hour to overnight, depending on sensitivity of detecting reagent. As a starting point, we recommend using the longer incubation times until the properties of the reagents and assays are characterized.
3. Asymmetrically mark the filter with a needle to allow alignment of filter with plaques at a later time. Remove filter and incubate with blocking solution for at least 2 hours, room temperature to overnight, 4°C to reduce nonspecific binding of detecting reagent with the filter. The composition and a description of a variety of blocking solutions is described in Harbour and Lane.[27]
4. If replica filters are to be prepared, return the plated phage to 36°C for 2 hours. Remove plate and repeat steps 2 and 3.
5. Wash filters 2 × for 5 minutes each with either PBS + 0.05% Tween 20 or TBS + 0.05% Tween 20.
6. Place filters in a clean, empty culture dish containing 15 to 20 ml blocking solution plus primary reagent (antigen, conjugated antigen, antibody, or conjugated antibody). Incubate for 1 hour at room temperature to overnight at 4°C.
7. Wash filters 3 × for 5 minutes each if secondary reagent is to be added or 5 × for 5 minutes each if immediately followed by developing reagents. If filters are ready for developing, go directly to step 10.
8. Add washed filters to clean, empty culture dish containing 15 to 20 ml blocking solution plus labeled secondary reagent. Incubate 30 minutes to 1 hour, room temperature.
9. Wash filters 5 ×, 5 minutes each.
10. Develop filters with appropriate substrates or expose to X-ray film at −70°C with intensifying screen.[27] Wash developed filters extensively in deionized water. Allow filters to dry completely before aligning to plate.

Candidate clones are plaque-purified and the M13 variant clone is then used to produce soluble Fab in *E. coli*. Periplasmic fractions are isolated and the antibody binding properties are evaluated by ELISA. DNA sequence analysis of those clones exhibiting the desired phenotype reveals which amino acid residues contribute to the new binding properties. We routinely synthesize oligonucleotides encoding the novel amino acid residues and mutagenize the parent antibody molecule to confirm that the new phenotype is attributable solely to those residues modified by the codon-based mutagenesis procedure.

Several strategies for optimizing the antibody combining site can be envisioned.[34] One strategy is to screen individually mutagenized CDR libraries, identify clones with the altered phenotype, and then combine the individually mutated CDRs into a single antibody. This approach assumes that each of the amino acid substitutions independently influences antigen binding and that the combined effect will be at least a minimal improvement over the precursor clones. A second strategy is to simultaneously mutagenize two or more CDRs. Extreme diversity can subsequently be generated by recombining a mutagenized light-chain CDR library with a heavy-chain CDR library. A third approach is to optimize the antibody by the stepwise accumulation of CDR mutations. A library mutagenized in a single CDR is screened and variant antibodies are isolated. These clones can then be used as templates for a subsequent round of modification by targeting the remaining CDRs. Reiterative rounds of mutagenesis and screening result in an optimized antibody built upon a previously improved genetic background.

CONCLUSION

Mutagenesis cloning of antibody V region genes can be used to construct combinatorial libraries derived from a variety of tissue sources such as spleen, peripheral blood lymphocytes, bone marrow lymphocytes, and from clonal populations of antibody-secreting cells such as hybridomas and EBV transfected cell lines. Once these antibody genes have been cloned and sequenced, mutagenesis techniques can be used to engineer new properties into the antibody combining site. We have described how codon-based mutagenesis can be used to rapidly and efficiently genetically engineer cloned antibody genes to produce antibody variants with higher affinity or altered specificity toward antigen. The ease with which the sequence of the mutations can be determined also provides a rapid means of elucidating antibody structure/function relationships.

Acknowledgements

We wish to thank Ragen Bradford for assistance with preparing the manuscript and Drs. Paula Boerner and Jeff Gray for their comments and suggestions.

REFERENCES

1. Glaser, S.M., Yelton, D.E., and Huse, W.D. 1992. Antibody engineering by codon-based mutagenesis in a filamentous phage vector system. *J. Immunol.* 149:3903.
2. Glaser, S.M., Huse, W.D., and Yelton, D.E. Manuscript in preparation.
3. Saiki, R.K., Gelfand, D.H., Stoffel, S., Scharf, S.J., Higuchi, R., Horn, G.T., Mullis, K.B., and Erlich, H.A. 1988. Primer-directed enzymatic amplification of DNA with a thermostable DNA polymerase. *Science* 239:487.
4. Orlandi, R., Güssow, D.H., Jones, P.T., and Winter, G. 1989. Cloning immuno-globulin variable domains for expression by the polymerase chain reaction. *Proc. Natl. Acad. Sci. USA* 86:3833.
5. Sastry, L., Alting-Mees, M., Huse, W.D., Short, J.M., Sorge, J.A., Hay, B.N, Janda, K.D., Benkovic, S.J., and Lerner, R.A. 1989. Cloning of the immuno-globulin repertoire in *Escherichia coli* for generation of monoclonal catalytic antibodies: Construction of a heavy chain variable region-specific cDNA library. *Proc. Natl. Acad. Sci. USA* 86:5728.
6. Larrick, J.W., Danielsson, L., Brenner, C.A., Abrahamson, M., Fry, K.E., and Borrebaeck, C.A.K. 1989. Rapid cloning of rearranged immunoglobulin genes from human hybridoma cells using mixed primers and polymerase chain reaction. *Biochem. Biophys. Res. Commun.* 160:1250.
7. Cabilly, S., Riggs, A.D., Pande, H., Shively, J.E., Holmes, W.E., Rey, M., Perry, L.J., Wetzel, R., and Heyneker, H.L. 1984. Generation of antibody activity from immunoglobulin polypeptide chains produced in *Escherichia coli. Proc. Natl. Acad. Sci. USA* 81:3273.
8. Boss, A., Kenten, J.H., Wood, C.R., and Emtage, J.S. 1984. Assembly of functional antibodies from immunoglobulin heavy and light chains synthesized in *E. coli. Nucleic Acids Res.* 12:3791.
9. Skerra, A., and Plückthun, A. 1988. Assembly of a functional immunoglobulin F_V fragment in *Escherichia coli. Science* 240:1038.
10. Better, M., Chang, C.P., Robinson, R.R., and Horwitz, A.H. 1988. *Escherichia coli* secretion of an active chimeric antibody fragment. *Science* 240:1041.
11. Huse, W.D., Sastry, L., Iverson, S.A., Kang, A.S., Alting-Mees, M., Burton, D.R., Benkovic, S.J., and Lerner, R.A. 1989. Generation of a large combinatorial library of the immunoglobulin repertoire in phage lambda. *Science* 246:1275.
12. McCafferty, J., Griffiths, A.D., Winter, G., and Chiswell, D.J. 1990. Phage antibodies: filamentous phage displaying antibody variable domains. *Nature* 348:552.

13. Huse, W. D. 1991. Combinatorial antibody expression libraries in filamentous phage. In *Antibody Engineering: A Practical Guide*, C.A.K. Borrebaeck, ed. W.H. Freeman and Company, New York, pp. 103–20.
14. Kang, A.S., Barbas, C.F., Janda, K.D., Benkovic, S.J., and Lerner, R.A. 1991. Linkage of recognition and replication functions by assembling combinatorial antibody Fab libraries along phage surfaces. *Proc. Natl. Acad. Sci. USA* 88:4363.
15. Barbas, C.F., Kang, A.S., Lerner, R.A., and Benkovic, S.J. 1991. Assembly of combinatorial antibody libraries on phage surfaces: The gene III site. *Proc. Natl. Acad. Sci. USA* 88:7978.
16. Huse, W.D., Stinchcombe, T.J., Glaser, S.M., Starr, L., MacLean, M., Hellström, K.E., Hellström, I., and Yelton, D.E. 1992. Application of a filamentous phage pVIII fusion protein system suitable for efficient production, screening, and mutagenesis of F(ab) antibody fragments. *J. Immunol.* 149:3914.
17. Chaudhary, V.K., Batra, J.K., Gallo, M.G., Willingham, M.C., FitzGerald, D.J., and Pastan, I. 1990. A rapid method of cloning functional variable-region antibody genes in *Escherichia coli* as single-chain immunotoxins. *Proc. Natl. Acad. Sci. USA* 87:1066.
18. Near, R. 1992. Gene conversion of immunoglobulin variable regions in mutagenesis cassettes by replacement PCR mutagenesis. *Biotechniques* 12:88.
19. Kabat. E.A., Wu, T.T., Perry, H.M., Gottesman, K.S., and Foeller, C. 1991. *Sequences of Proteins of Immunological Interest.* 5th ed. United States Department of Health and Human Services, Washington, D.C.
20. McCabe, P.C. 1990. Production of single-stranded DNA by asymmetric PCR. In *PCR Protocols: A Guide to Methods and Applications.* M.A. Innis, D.H. Gelfand, J.J. Sninsky, and T.J. White, eds. Academic Press, Inc., San Diego, CA, pp. 76–83.
21. Gyllensten, U.B., and Erlich, H.A. 1988. Generation of single-stranded DNA by the polymerase chain reaction and its application to direct sequencing of the *HLA-DQA* locus. *Proc. Natl. Acad. Sci. USA* 85:7652.
22. Chomczynski, P., and Sacchi, N. 1987. Single-step method of RNA isolation by acid guanidinium thiocyanate-phenol-chloroform extraction. *Analyt. Biochem.* 162:156.
23. Maniatis, T., Fritsch, E.F., and Sambrook, J. 1989. *Molecular Cloning: A Laboratory Manual.* 2nd ed. Cold Spring Harbor Laboratory, Cold Spring Harbor, NY.
24. Zoller, M.J., and Smith, M. 1984. Oligonucleotide-directed mutagenesis: a simple method using two oligonucleotide primers and a single-stranded DNA template. *DNA* 3:479.
25. Kunkel, T.A. 1985. Rapid and efficient site-specific mutagenesis without phenotypic selection. *Proc. Natl. Acad. Sci. USA* 82:488.
26. Kunkel, T.A., Roberts, J., and Zakour, R. 1987. Rapid and efficient site-specific mutagenesis without phenotypic selection. *Methods Enzymol.* 154:367.
27. Harlow, E., and Lane, D. eds. 1988. *Antibodies: A Laboratory Manual.* Cold Spring Harbor Laboratory, New York.
28. Berek, C., and Milstein, C. 1987. Mutation drift and repertoire shift in the maturation of the immune response. *Immunol. Rev.* 96:23.
29. Allen, D., Cumano, A., Dildrop, R., Kocks, C., Rajewsky, K., Rajewsky, N.,

Roes, J., Sablitzky, F., and Siekevitz, M. 1987. Timing, genetic requirements and functional consequences of somatic hypermutation during B-cell development. *Immunol. Rev.* 96:5.

30. Berek, C., and Milstein, C. 1988. The dynamic nature of the antibody repertoire. *Immunol. Rev.* 105:5.
31. Wu, T.T., and Kabat, E.A. 1970. An analysis of the sequences of the variable regions of Bence Jones proteins and myeloma light chains and their implications for antibody complementarity. *J. Exp. Med.* 132:211.
32. Chothia, C., Lesk, A.M., Tramontano, A., Levitt, M., Smith-Gill, S.J., Air, G., Sheriff, S., Padlan, E.A., Davies, D., Tulip, W.R., Colman, P.M., Spinelli, S., Alzari, P.M., and Poljak, R.J. 1989. Conformation of immunoglobulin hypervariable regions. *Nature* 342:877.
33. Coligan, J.E., Kruisbeek, A.M., Margulies, D.H., Shevach, E.M., and Strober, W. eds. *Current Protocols in Immunology.* Greene Publishing Associates and Wiley-Interscience, New York. Submitted.
34. Huse, W.D., Yelton, D.E., and Glaser, S.M. 1993. Increased antibody affinity and specificity by codon-based mutagenesis. *Intern. Rev. Immunol.* 10:129.

CHAPTER 5

Human Metaphoric Antibodies from a Phagemid Library

Bob Shopes

If antibodies are to be used as therapeutic drugs they must be both safe and effective. Human antibodies do not induce an immune response and are considered generally safe. However, specific human antibodies are difficult to obtain upon demand.[1,2] Monoclonal antibodies from mice, while relatively easy to identify, generally induce an immune response in a significant portion of the intended patient population.[3,4] So, effective mouse monoclonal antibodies exist, but cannot be used, and safe human antibodies of the same effectiveness have not been located. To address this problem, antibody engineers sought to construct artificial antibodies by grafting the six complementary determining regions of an existing mouse antibody onto a human scaffold of framework regions and constant domains.[5-8] This molecular surgery, commonly known as humanization, often requires careful fine-tuning by alternating rounds of computer modeling and genetic modification to retain the specificity and affinity of the original mouse antibody. Humanization has been proved useful but is laborious, and the resulting antibody still contains approximately 50 to 60 residues of mouse antibody sequence. This portion of foreign protein on the surface of humanized antibodies may raise a human immune response, but only limited clinical information on genetically engineered antibodies is available.[9] In view of these circumstances, a reliable method of identifying fully human antibodies would be of considerable value.

Within the past few years many have endeavored to directly clone human antibodies using molecular biology techniques. Initially, lambda phage library cloning methods were used to identify mouse[10] and human antibodies of predetermined specificities.[11] Later, M13 phagemid, and phage, cloning systems were developed to facilitate the cloning of antibodies. The origin of these M13 systems can be traced to the expression of peptides on the surface of this filamentous phage by Smith and coworkers.[12,13] Scott and Smith[13] constructed a hexamer peptide M13 library and elucidated an epitope by sequencing individual clones after several rounds of selective enrichment. Smith's group had modified the M13 genome to display peptides, and this M13 phage display system was extended by Winter and colleagues to the display of single-chain antibody binding fragments.[14–18] Other M13 phagemid display cloning systems based on cloning with *E. coli* compatible plasmids or M13 phage vectors have been used.[19–22]

Recently, we developed a cloning vector to combine the ease of creating a library with a lambda phage vector with the power of M13 phagemid display library screening.[23–25] A portion of gene III from M13, which codes for the carboxyl terminal half (from a.a. 198 to a.a. 406) of the minor coat protein gpIII, was inserted into the cloning region of a lambda vector, ImmunoZAP L,[10,11] to create the SurfZAP cloning vector (Fig. 5–1). In this vector, upstream of gene III, is a portion of a *pel*B periplasmic secretion

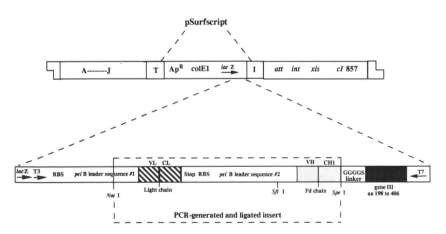

Figure 5–1. Depiction of the SurfZAP lambda cloning vector. The internal pSurf-script plasmid and the location of the cloning region are indicated by the offset dotted lines.

leader sequence and cloning sites for inserts. Inserts are ligated into high-efficiency restriction digested lambda arms. The insert/vector is packaged into lambda phage heads and used to infect *E. coli*. This is a very efficient system to create a library of genes. When an insert is cloned into this vector, in the proper frame, a fusion protein is produced and secreted into the periplasmic space of the infected *E. coli*. The fusion protein is composed of the amino acid sequence encoded by the cloned insert ("foreign domain") linked by a -(GlyGlyGlySer)- flexible spacer to a portion of gpIII. The membrane anchor domain of gpIII becomes imbedded in the periplasmic membrane, and the foreign domain penetrates into the periplasmic space.

In the SurfZAP cloning system the internal pSurfscript plasmid (see Fig. 5–1) can be excised from the bacteriophage lambda vector by co-infection with M13 helper phage. The plasmid, containing the foreign gene or insert, is encapsulated by M13 derived proteins to form a pro-phage. The pro-phage is extruded through the periplasmic membrane and acquires 3 to 5 copies of wild-type gpIII and, usually, one copy of "foreign domain"–gpIII fusion protein. This M13 phagemid bears the binding activity conferred by the display of the fusion protein and contains the DNA insert that codes for that activity. Thus, genotype and phenotype are physically linked in a phagemid display library. The excised phagemid is capable of being selectively propagated as a plasmid since the wild-type gpIII allows for transformation of a cell by infection and the plasmid bears an antibiotic selection marker. The physical linkage of phenotype and genotype, coupled with the biologic survival of the plasmid, sets the stage for the enrichment of specific clones by the application of selection pressure. With a phagemid display library it is possible to enrich for, and identify, clones of interest on the basis of a binding activity given judicious use of selection pressure.

Phagemid display libraries can be enriched for binding clones by bio-panning on polystyrene wells or dishes[13–22] or, by our preferred method, biochromatography.[24,25] In biochromatography, an amplified M13 phagemid display library is loaded onto a column composed of ligand (e.g., an antigen) covalently coupled to a resin. For either biopanning or biochromatography the nonspecific phage are removed by copious washing and the bound phage are eluted by disrupting the "foreign domain"–ligand interaction. The elution of phage or phagemid particles can be accomplished with low or high pH,[13,14,19,23] chaotropic agents,[26] or free ligand (or antigen).[14,19,23,24] The eluted phage constitute an enriched library. The enriched library can be amplified and another biochromatographic cycle performed, if desired. Typically, enrichments of ten-fold to several thousand-fold are observed, but this depends greatly on the particular enrichment system. Isolated single clones can be obtained from the enriched library by infecting *E. coli* and selecting colonies that bear the antibiotic-resistant marker carried by the plasmid. When sufficient enrichment has occurred, usually in three to six cycles, the binding specificity of isolated clones can be determined by an appropriate binding assay, for example an ELISA. In addition, the sequence of genes encoding the binding domains can be obtained readily from the plasmid contained within the phage.

In this chapter the identification of a human "metaphoric" antibody will serve as an example to supply details of the construction and enrichment of a phagemid display library. The metaphoric process transforms mouse antibodies to fully human antibodies and is an attractive alternative to humanization[25] (see Fig. 5–2). The basis of the metaphoric process is the promiscuity of antibodies. Promiscuity refers to the ability of any given heavy chain to combine with many different light chains (or vice versa) to create an effective binding site for an antigen. This promiscuity has been observed in "chain shuffling" where a cloned chain is paired with a library of matching chains. In chain shuffling it has usually been found that many different combinations produce useful binding interactions.[15,18,23] Chain shuffling has been previously done within the boundaries of a single species, for example, a human heavy chain with a library of human light chains, for the modification of affinity or specificity. The metaphoric process is chain reshuffling across species borders.

In the metaphoric process, the light chains of an existing mouse monoclonal antibody is used to identify active human heavy chains (see Fig. 5 2). (Alternatively, one could begin with a mouse heavy chain and a library of human light chains and proceed in an analogous manner). An antibody library is constructed, in which each member has a different human heavy chain but the same mouse light chain. Screening this library yields Fab clones that comprise a human heavy chain and a mouse light chain and bind to the desired antigen. One (or more) selected human heavy chain(s) is then recombined with a library of human light chains to form a metaphoric library. From this human–human library Fabs are selected with desired affinity for the cognate antigen. The resulting metaphoric antibodies are fully human. The cloned human sequence of the metaphoric antibody could be expressed as Fab–phagemid or as Fab[27–28] and used for diagnostics. Alternatively, the human V-genes could be subcloned to appropriate mammalian expression vectors for expression of full-length human antibody for therapeutic use.[29,30]

Figure 5–2. Schematic representation of the metaphoric process to transform a mouse antibody to a fully human antibody. This pathway is referred to as "Method A" in the text. An alternative pathway, "Method B," is not shown. The genes are noted as: V = variable domain, C = constant domain, L = light chain, H = heavy chain, m = mouse, and h = human.

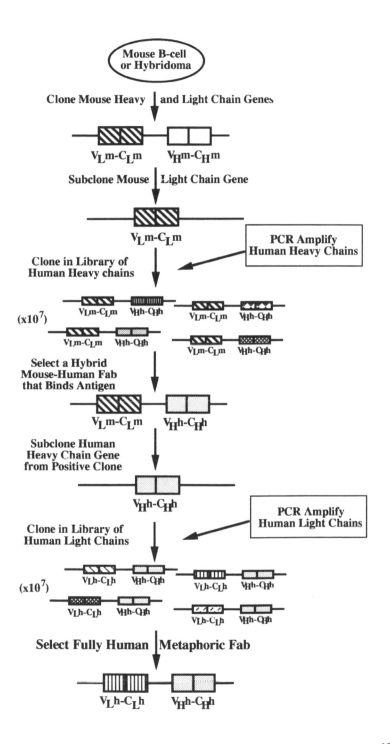

METHODS

Isolation of the Mouse Antibody Genes

The metaphoric process requires cloned mouse genes that code for an antibody that binds the antigen of interest. We have previously reported the cloning of mouse antibody genes using the SurfZAP system,[24] and others have reported the cloning of mouse antibody genes by a variety of methods.[10,31–35] We recommend that the mouse genes, if not originally cloned in SurfZAP, be subcloned in this, or a compatible, system. This generally would involve the amplification of the mouse genes with immuno-globulin-specific primers that add sequence for cloning and expression. A number of the protocols given in this chapter may serve as a useful guide. After cloning it should be determined if the original pair of mouse antibody genes are expressed as active Fab in *E. coli* and if Fab are displayed on the M13 phagemid. In the SurfZAP system the Fd portion of the heavy chain is fused to the carboxyl terminus of gpIII. The light chain is produced with its own leader sequence from the same promoter. The light chain pairs with the Fd–gpIII fusion protein in the periplasmic space and a disulfide bond forms a covalent link. By cloning and expressing the mouse genes in SurfZAP, one can check that each of these processes works for that particular mouse Fab. Here, we assume, for simplicity, that the mouse genes have been cloned in SurfZAP and are adequately expressed.

Construction of a Hybrid Mouse–Human Antibody Lambda Library

The metaphoric process, as described in Fig. 5–2, starts by combining a mouse light chain with a variety of human heavy chains (Method A). Alternatively, one could choose to start by combining a mouse heavy chain with a variety of human light chains (Method B). It is difficult to determine *a priori* whether Method A or Method B will yield the greatest probability of success. Without detailed prior knowledge the choice is somewhat arbitrary. A hint on how to proceed may be found from previous intraspecies chain-shuffling experiments; it is easier to find a matching light chain than a matching heavy chain. This raises the possibility that the preponderance of antigenic determination lies with the heavy chain and that the light chain modulates this interaction. Based on this premise we chose to proceed with Method A, which forces the search for a human heavy chain first. If a human heavy chain is found that binds antigen when paired with a mouse light chain, then a human light chain replacement should be relatively easy to identify. Following the schematic in Fig. 5–2 for our "metaphoric" example, we initially create a library combining a mouse light chain with a variety of human heavy chains.

For our metaphoric example we cloned the mouse genes and expressed Fab on the surface of phagemid. Further subcloning of the mouse light-chain gene is necessary to avoid any contamination of the library with mouse heavy-chain genes (see Protocol 1, for an example). Any mouse heavy-chain gene contamination results in false positives by recreating the original mouse Fab. After subcloning the mouse light chain, a single preparation of plasmid yields adequate material to construct the human–mouse library.

Human heavy-chain genes are amplified by PCR (see Protocol 2, for example). The source of initial genetic material is important in determining the quality of the final library. When starting with mRNA isolated from tissue or peripheral blood lymphocytes, cDNA may be made by first-strand synthesis using an oligo-dT primer and reverse transcriptase. The choice of primers is also an important factor in determining the scope of the resulting library. The primers should be designed to capture a wide range of human immunoglobulin genes (for example, see Table 5–1). The primers we have designed also incorporate restriction sites, leader sequences, and stop codons. To limit plasmid DNA recombination, the nuclei acid sequences of the two *pel*B leaders differ, but for expression purposes the amino acid sequences are identical.

To prepare an antibody library, the light-chain gene and the heavy-chain gene are simultaneously ligated into a cloning vector (see Fig. 5–3 and Protocol 3). In the SurfZAP cloning system we first ligate the light-chain gene and the heavy-chain gene together via a common *Sfi*I restriction site. The *Sfi*I sites containing matching nonpalindromic five-base-pair centers

Table 5–1. Human heavy-chain primers

Upstream; specific for human V_H families
I   5'-TCGCGGCCCAACCGGCCATGGCCCAGGTGCAGCTGGTGCAG-3'
II   5'-TCGCGGCCCAACCGGCCATGCCCAGGTCAACTAAGGGAG-3'
III 5'-TCGCGGCCCAACCGGCCATGGCCCAGGTGCAGCTGGTGGAG-3'
IV 5'-TCGCGGCCCAACCGGCCATGGCCCAGGTGCAGCTGCAGGAGTCG-3'
V   5'-TCGCGGCCCAACCGGCCAATGGCCCAGGTGCAGCTGGTGCAG-3'
VI 5'-TCGCGGCCCAACCGGCCATGGCCCAGGTACAGCTGCAGCACTCA-3'
Downstream; specific for C_H1 of human IgG1 5'-AGCATCACTAGTACAAGATTTGGGCTC-3'

Note: Primers used for the amplification of human heavy chain genes. The 3' end of the "upstream" primers are homologous to the sense strand of first few codons of six V_H families.[39] The 5' end of the "upstream" primers incorporate an *Sfi*I restriction site (underlined) and encode the 3' half of the *pel*B leader sequence #2. The 3' end of the "downstream" primer is homologous to the antisense strand of the last few codons of the C_H1 domain gene of human IgG1. A *Spe*I restriction site (underlined) is incorporated in the "downstream" primer.

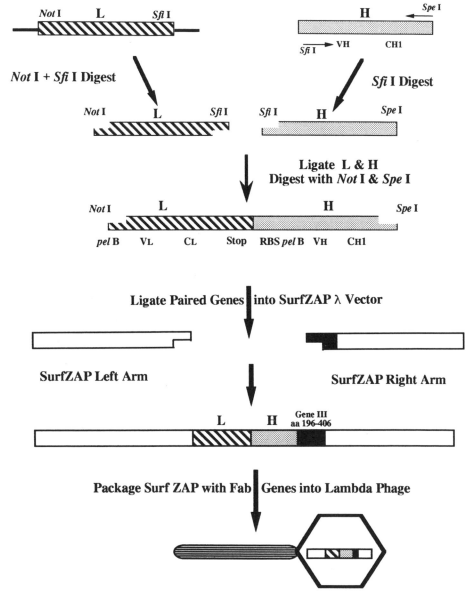

Plasmid with Mouse Light Chain Genes

Not I   L   Sfi I

PCR Amplify Human Heavy Chain Genes

Spe I
H
Sfi I → VH   CH1

Not I + *Sfi* I Digest

Sfi I Digest

Not I   L   Sfi I

Sfi I   H   Spe I

Ligate L & H Digest with *Not* I & *Spe* I

Not I   L   H   Spe I

pel B   VL   CL   Stop   RBS *pel* B   VH   CH1

Ligate Paired Genes into SurfZAP λ Vector

SurfZAP Left Arm

SurfZAP Right Arm

L   H   Gene III aa 196-406

Package Surf ZAP with Fab Genes into Lambda Phage

Figure 5–3. Creation of mouse-human SurfZAP lambda library. The mouse light-chain genes are depicted by a striped box and the human heavy-chain genes are depicted by a gray box. Details of the steps to prepare a library can be found in the text and in Protocol 3.

that allow for directional ligation. The paired genes (the insert) are then digested with *Not*I and *Spe*I restriction enzymes and ligated into prepared lambda vector arms. The insert and vector arms form concatamers that are recognized and packaged in ≈ 40 kb units by lambda derived proteins. Each ≈ 40 kb unit contains a right arm, a left arm, and an insert and is packaged into a lambda phage particle. Ligating of 35 ng of antibody insert into 1 μg of vector arms could, theoretically, yield a library with $\geq 10^9$ clones. Primary SurfZAP antibody libraries of $\approx 10^8$ members have been constructed on occasion, but primary libraries of 10^7 clones are more typical.

The primary library should be amplified to preserve clonal representation. Fresh, early-log phase *E. coli* ($OD_{600\,nm} \approx 0.2$) are desirable for library amplification. Late-log or stationary cultures ($OD_{600\,nm} \approx 1.5$) may have a high percentage of nonviable cell that can act as unproductive sinks for phage. The amplified library should be checked for expression of light chain by an immunoassay (Protocol 4). The percentage of clones that express light chains is one of the best indicators of the quality of the library.

In the metaphoric example the mouse antibody clones used were previously cloned in SurfZAP and bound tetanus toxoid.[24] The mouse light-chain genes were subcloned into a plasmid as in Protocol 1. The starting material for PCR amplification of the human genes was an existing lambda library of human heavy-chain Fd IgG1 sequences obtained from cDNA of peripheral blood lymphocytes of a normal donor.[11] Excellent amplification was observed with all the human V_H family primers of Table 5–1, except for the family II primer, using Protocol 2. The mouse light-chain gene, derived from the plasmid constructed in Protocol 1, was ligated to the human heavy-chain PCR product and a primary library was constructed following Protocol 3 (see also Fig. 5–3). This SurfZAP mouse–human antibody primary library contained 2×10^7 clones. This primary library was amplified following Step 8 of Protocol 3 and the resulting 50 ml of supernatant, containing the amplified mouse–human antibody primary library, had a titer of 10^{10} plaque-forming units (pfu)/ml. An immunoassay for the expression of mouse light chain (Protocol 4) revealed that 95% of the clones expressed a mouse kappa chain.

Mass Excision of the Lambda Library

The SurfZAP lambda library can be converted to an M13 phagemid library by co-infection with an M13 helper phage by excision (see Fig. 5–4). Proteins encoded by the helper phage recognize the I and T sequences imbedded in the SurfZAP lambda vector (see Fig. 5–2). The DNA between T and I is cut, rejoined, and replicated *in vivo* to form a single-stranded circular plasmid (pSurfscript). The helper phase also provides proteins to package pSurfscript as a phagemid.

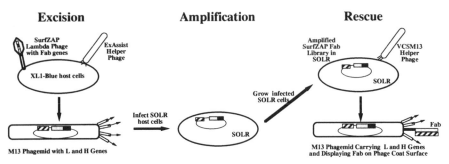

Figure 5–4. Schematic representation of the conversion from an antibody lambda library to an M13 phagemid display library. The mass excision of the amplified lambda library with ExAssist helper phage and XL1-Blue cells is detailed in Protocol 5. The resulting phagemid library is amplified in SOLR cells as in Protocol 6. The phagemid display library is made by rescuing the plasmid from the amplified transformed SOLR cells with VCSM13 helper phage as in Protocol 7. The light-chain genes, and their product, are depicted by a striped box; the heavy-chain genes, and their product, are depicted by a gray box; and Gene III is shown as a black box. Wild-type gpIII is shown as a box with an arrow.

An excess of lambda phage, at least 100-fold over the primary library size, are excised to ensure that each member of the primary library is represented. The amplified lambda library is used to infect *E. coli* and grown in the presence of ExAssist helper phage. The pSurfscript plasmid that resides in the SurfZAP lambda vector is excised, replicated, and packaged as an M13 phagemid particle. The phagemid is extruded from the *E. coli* and accumulates in the media. An extensive incubation will skew the representation of the library to those clones with a growth advantage, so it is important not to incubate longer than 3 hours, at 37°C, for a library excision. Lower temperatures allow for longer growth periods (e.g., overnight at 30°C). The XL1-Blue strain is used for mass excision because they are infectable by lambda and M13 phage and are deficient in a number of recombination systems. The ExAssist helper phage was chosen because it has an amber mutation that rules out propagation in a supO host.[36] The phagemid are titered on SOLR host cells, which are supO and resistant to lambda infection, so neither ExAssist or lambda phage propagate in SOLR. The cleared supernatant, containing excised phagemid, can be stored for 2 to 3 months at 4°C, but the titer may decrease by several logs over time.

For the mouse–human metaphoric library, 4×10^9 amplified hybrid library phage (a 200-fold excess over the primary library size) were added to 4×10^{10} fresh, early-log XL1-Blue and excised with 4×10^{10} ExAssist helper phage (see Protocol 5). The excised mouse–human M13 phagemid library had a titer of 10^8 colony-forming units (cfu)/ml.

Amplification of the Phagemid Library

The SurfZAP phagemid library produced by the mass excision may contain just a few copies of some of the primary clones of the lambda library. To ensure the inclusion of these clones in later steps it is recommended that the phagemid library be amplified. We routinely amplify a portion of the excision supernatant that represents at least ten-fold more clones than found in the primary lambda library. This will ensure, to a >90% probability, that each clone is included at least once.

The excised library is amplified by infecting SOLR host cells and propagating the ampicillin resistant (Amp^r) transformed cells. The plasmid delivered by the phagemid do not carry the M13 genome and will not propagate as a phagemid in the absence of viable helper phage. However, the Amp^r SOLR cells that carry the plasmid will multiply. Slowly growing transformed SOLR cells may indicate toxicity of expressed foreign protein. If this is a problem, amplification may be done in buffered media containing 1% (w/v) glucose to inhibit transcription of the insert fusion protein gene from the *lac* promoter. The addition of glucose may also reduce the rate of DNA rearrangements of instable clones. Using Protocol 6, expect an approximately 100-fold amplification. For the mouse–human metaphoric library we recovered 3×10^{11} Amp^r SOLR cells from the amplification, using Protocol 6 starting with 2×10^9 cfu phagemid from the mass excision.

Conversion to an M13 Phagemid Display Library

A display library can be prepared by infection of the amplified, Amp^r SOLR cells with helper phage and rescuing the plasmid as a phagemid (see Fig. 5–4 and Protocol 7). VSCM13 helper phage can replicate in SOLR and are used for this conversion. To maintain the clonal representation of the library, and to minimize degradation of the gpIII fusion protein, the rescue should not be allowed to proceed to the stationary phase (e.g., $OD_{600\,nm} \geq 1.0$). Rescues may be safely conducted overnight at 30°C. Typically, 1% to 10% of the rescued phagemid bear Fab on their surface. The remainder of the phagemid bear only wild-type M13. After this rescue, the amplified phagemid library can be concentrated 100-fold with two sequential polyethylene glycol (PEG) precipitations. The rescued and concentrated phagemid display library can be stored for several days at 4°C with little loss of titer or degradation of displayed fusion protein. For the mouse–human antibody library we obtained 10^{12} phagemids in 100 µl using Protocol 7.

Enrichment of Binding Clones by Biochromatography

While it is straightforward to provide protocols for the creation of a library, it is somewhat more difficult to give guidance on enrichment procedures as the constraints vary from experiment to experiment. The comments and protocols for enrichment given here reflect our experience with metaphoric antibody identification,[25] other attempts to obtain antibodies,[23,24] and unpublished observations. What we can give as general guidance is biased, but hopefully formative. Every phagemid display library enrichment protocol will differ. The design of the experiment will depend on the antigen and the source of the antibody library. The process of enrichment can be divided into three parts:

1. *The binding of phagemids to a target:* It is best to have a high concentration of targets. This will allow for many opportunities for binding. The target should be tightly attached to the matrix, but the target should not be altered or deformed by this attachment.
2. *The desorption of nonspecifically bound phagemids:* The support matrix should be stable and inert. The lower the nonspecific binding of phagemid to the matrix, the better the enrichment will be. A compatible high salt buffer with a blocking agent (i.e., BSA) is recommended for washes.
3. *The recovery of specifically bound phagemids:* Elution of phagemid can be accomplished by changing conditions so that the binding interaction is disfavored. Low pH, high pH, free ligand, chaotropic agents and detergents may serve to elute the bound phagemids. The matrix should be stable under these conditions.

Our group, and others, have found that biopanning, using ligand- or antigen-coated polystyrene (plates or dishes), or even whole cells, can be useful for enrichment of phagemid display libraries. However, we have found column affinity chromatography of phagemid libraries, or biochromatography, to be both efficient and versatile for enriching binding clones. In general, biochromatography is most successful with purified protein antigens or with small ligands. A column format satisfies most of the criteria for a successful enrichment. Many sophisticated chromatographic resins are available and allow for great flexibility in the design of the enrichment experiment. Even a small column (e.g. ≈ 0.1 ml volume) can have a very large binding capacity with a large number of theoretical plates. The column format also allows for copious washing to remove nonspecific phagemid, and the resins are stable under typical washing and elution conditions.

For the mouse–human metaphoric antibody library we enriched for clones that bound to tetanus toxoid in three rounds of biochromatography.[25] The antigen was coupled to the column matrix, CNBr activated Sepharose, as described in Protocol 8. A phagemid display library was prepared (see above and Protocol 7) and biochromatography performed as in Protocol 9. Phage were eluted with a pH 2.2 buffer. The concentration of phagemid in each elutate fraction was determined, by titre as in step 5 of Protocol 4, and

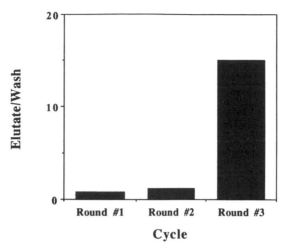

Figure 5–5. A comparison of the number of phagemids per unit volume in the elution fraction and the wash fraction for each cycle of biochromatographic enrichment of the mouse-human phagemid display library.

compared to the concentration of phagemid in the preceding wash fraction (Fig. 5–5). For the first two cycles of biochromatography the concentration of phagemids in the lutate rose only slightly above background. However, for the third round the concentration of phagemid in the elutate was ≈ 12-fold higher than in the preceding wash fraction. This indicated that it was likely that the population of clones had been enriched for binding clones by the selection pressure applied during biochromatography.

Individual specific binding clones were isolated by screening Amp^r colonies for tetanus toxoid binding activity. A dilution of each elutate was used to transform XL1-Blue cells. The resulting Amp^r colonies were replicated and screened with radiolabeled antigen as in Protocol 10. We did not identify any positive clones in screening more than 20,000 members of the original primary library; the frequency of positive clones was less than 0.005%. After three rounds of enrichment approximately 2% of the clones displayed binding to radiolabeled tetanus toxoid. Dozens of positive mouse–human clones could be isolated from a small portion of the third-round elutate. It was important to confirm that the heavy chains recovered were indeed human and not recycled mouse heavy-chain genes or false positives. The heavy-chain genes were sequenced and the C_H1 domain was human IgG1. The variable domain genes were also found to be human. Up to this point in the metaphoric process we have generated a large number of clones that contain one human chain and one mouse chain. The next part is simple: substitute the mouse light chain with human light chain.

Second Half of the Metaphoric Process

The second half of the metaphoric process is shuffling human light chains with the selected human heavy chains and selecting binding clones. First, we subclone a human heavy chain from one of the positive mouse–human clones. This process is similar to the subcloning described in Protocol 1, but we use the *Sfi*I and *Spe*I restriction enzymes to remove the human heavy-chain gene from pSurfscript. Human light-chain genes are PCR amplified using the primers shown in Table 5–2 by Protocol 2. A lambda library is constructed, using Protocol 3, by recombining the isolated human heavy-chain genes with the PCR amplified human light-chain genes. The resulting human–human metaphoric library was converted to an M13 phage display library and enriched by two rounds of biochromatography following Protocols 4 to 8.

For the metaphoric example of this chapter we identified several fully human clones that bound tetanus toxoid. The enrichment of binding clones from the human–human antibody library can be seen in Fig. 5–6. In autoradiographs the positive clones were identified by the dark radioimages. A dozen clones were selected from the screening of the second-round elutate. The binding of radiolabelled antigen was confirmed, and the light chains

Table 5–2. Human light-chain primers

Upstream
5'-GTGCCAGATGTGAGCTCGTGATGACCCAGTCTCCA-3'

*pel*B adapter #1
5'-GAAATCACTCCCAATTAGCGGCCGCTGGATTGTTATTACTCGCTG-
CCCAACCAGCCATGGCC-3'

Downstream
C$_L$ (κ)
5'-TCCTTCTAGATTACTAACACTCTCCCCTGTTGAAGCTCTTTGTGAC-
GGGCGAACTC-3'

*pel*B adapter #2
5'-CATGGCCGGTTGGGCCGCGAGTAATAACAATCCAGCGGCTGCCG-
TAGGCAATAGGTATTTCATTATGACTGTCTCCTTG-3'

Note: Primers used for the amplification of human light chain genes. The 3' end of the "upstream" V$_L$ primer is homologous to the sense strand of the −4 to +8 codons of a consensus human V$_L$ sequence.[39] The *pel*B adapter #1 primer adds a portion of the first *pel*B leader sequence and incorporates an *Not*I restriction site (underlined). The 5' end of the "downstream" C$_L$ primer is homologous to the antisense strand of last codons of human kappa.[39] The *pel*B adapter #2 primer adds the 5' portion of the second *pel*B leader sequence #2 and incorporates an *Sfi*I restriction site (underlined).

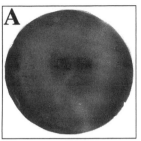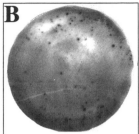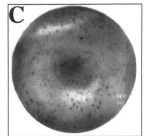

Figure 5–6. Autoradiographs from the screening of ampicillin-resistant colonies with ^{125}I-tetanus toxoid. The positive clones are depicted as dark radioimages. Each screening was done with $\approx$ 10,000 colonies per filter. A. Colonies from human–human metaphoric phagemid library before enrichment. B. Colonies from phagemid library after one round of biochromatography. C. Colonies after second enrichment round.

of these metaphoric clones were found to be human kappa by an immuno-assay.

CONCLUSION

Using the metaphoric process, we were able to transform an antitetanus toxoid mouse antibody clone into a panel of fully human antibody clones that bound the same antigen. This example was a proof of the metaphoric principle and was as well a demonstration of phagemid display library screening. We are continuing to develop this technique to clone metaphoric human antibodies against other targets.

PROTOCOL 1: SUBCLONE MOUSE LIGHT-CHAIN GENE

1. Purify the plasmid containing mouse heavy- and light-chain genes. (If the mouse antibody genes have not been cloned, please see References 31–35 for procedures on PCR cloning of genes from monoclonal cell-lines or clone using SurfZAP.[24] This protocol will assume that the mouse genes have been cloned previously in SurfZAP or pSurf-script.)
2. Digest 4 µg plasmid at 50°C with 20 U *Sfi*I for one hour in buffer. Adjust to 100 mM NaCl and digest with 2 U *Not*I (New England Biolabs) at 37°C for an additional hour.
3. Run the digested DNA on a 1% Sea-Kem (FMC) agarose gel and remove 700 bp band with a sterile scalpel.

4. Remove agarose with Geneclean (Bio101) and precipitate the fragment with two volumes of ethanol. Resuspend in a small volume of water and determine the concentration of DNA.
5. Ligate the fragment with T4 DNA ligase to a *Not*I–*Sfi*I cut vector (e.g., pSurfscript) in a 1:1 insert:vector ratio.
6. Transform competent XL1-Blue *E. coli* with ligation product. Plate on LB-ampicillin agar and grow at 37°C overnight.
7. Pick ampicillin-resistant colonies and grow overnight in 5 ml LB. Min-prep plasmid DNA and map with restriction enzymes. If desired, sequence the insert.

Note: Details of general molecular biology techniques, recipes and buffers used in this and the other Protocols can be found in Refs. 37 and 38. Except where noted all materials are from Stratagene Cloning Systems.

PROTOCOL 2: PCR AMPLIFICATION OF HUMAN HEAVY-CHAIN GENES

1. Add 10 µl 5% Tween-20 (Sigma) to 90 µl of an amplified lambda library containing human heavy chain gene insert. Heat at 95°C for five minutes to disrupt phage heads. (Alternatively, replace the lysed phage with cDNA prepared from lymphocyte mRNA as in Ref. 11.).
2. Mix 10 µl of lysed phage with 200 ng primer for C_H1, 200 ng of one of the V_H primers (see Table 5–1), 10 µl 10X Taq polymerase buffer and 8 µl of 0.25 mM dNTPs. Add to a total volume of 99.5 µl with water and mix. Add 0.5 µl of Taq polymerase (Ampli-Taq, Perkin-Elmer Cetus). Overlay with ≈ 25 µl mineral oil. Prepare positive and negative control samples in parallel.
3. Heat to 92°C for 5 minutes. Cool to 54°C for 5 minutes. Cycle 40 times between 72°C for 2.5 minutes, 92°C for 1.5 minutes, and 54°C for 2 minutes. Finish with 72°C for 10 minutes. Store at 4 to 6°C.
4. Run 10 µl of each PCR reaction on a 1% agarose gel. About 100 to 500 ng of DNA should appear at ≈ 700 bp.

PROTOCOL 3. PREPARATION OF A SURFZAP LAMBDA ANTIBODY LIBRARY

1. Purify the mouse light-chain gene from the plasmid created in Protocol 1 as a *Not*I–*Sfi*I restriction fragment as in steps 2–4 of Protocol 1.
2. Purify the human heavy chain PCR products from step 3 in Protocol 2.

(This procedure is as steps 2–4 of Protocol 1 but omitting the *Not*I digestion.)

3. Combine 500 ng of the mouse light-chain gene product from step 1 with 500 ng of the human heavy-chain PCR products from step 2. Ligate from T4 DNA ligase overnight at 4°C.

4. Heat at 72°C for 20 minutes to kill the ligase. Extract with phenol-chloroform, precipitate with ethanol, dry, and resuspend in 15 µl water. Digest for 1 hour with 10 U *Spe*I at 37°C then add 40 U *Not*I (New England Biolabs) and digest an additional 2 hours at 37°C.

5. Resolve a 1.4 kb band by electrophoresis in 1% SeaKem agarose (FMC) and purify with Geneclean (BIO101). Precipitate with ethanol, dry, and resuspend in 5 µl water. Determine DNA concentration and adjust to 35 ng/µl.

6. Mix 1 µl of SurfZAP vector arms, 1 µl of 1.4 kb insert, 0.5 µl 10X ligase buffer, 0.5 µl 10 mM rATP, 1.5 µl water and 0.5 µl T4 DNA Ligase (0.2 U/µl). For positive and negative controls replace the 1.4 kb insert with test insert and water respectively. Do not exceed 5% glycerol. Incubate at 4°C overnight. Improved results can often be seen when the ligation is extended for 2 to 3 days.

7. Package 1 µl of ligation into phage heads with Gigapack™ II Gold packaging extract according to supplier's instructions. Titer phage. Package rest of ligation if the primary library will be of adequate size (i.e. $> 10^7$ total clones). Combine packaging reactions and titre.

8. As soon as possible amplify the primary lambda library by plating phage at a density of 10^6 plaque-forming units (pfu) per 150 mm LB/agar plate. Grow the primary library plaques at 37°C for 6 hours on an *E. coli* lawn. Harvest the phage by overlaying with 10 ml of SM per plate and gently rock for 12 hours at 4°C. Collect the phage/SM liquid. Add to 2.5% chloroform. Mix for 15 minutes at room temperature. Spin at $3000 \times g$ for 10 minutes. Decant the supernatant to a sterile 50 ml conical tube. The amplified lambda library can be stored at 4°C for several months. (Alternatively, the SurfZAP lambda library may be amplified in liquid culture, but this may alter the representation of clones.)

PROTOCOL 4. MEASURE PERCENTAGE OF CLONES EXPRESSING LIGHT CHAIN

1. Dilute, in SM, a portion of the amplified lambda phage library from step 8 of Protocol 3. Also dilute appropriate negative and positive control samples. Add diluted lambda phage to yield ~100 to 200 pfu to 200 µl XL1-Blue. Incubate for 15 minutes at 37°C. Add 3 ml NZY

top agar (at 50°C) and pour evenly onto 100 mm NZY agar plates. Incubate plates at 37°C until small plaques appear (≈ 6 hours).

2. Saturate nitrocellulose filters (82 mm diameter) with fresh 10 mM IPTG. In the dark, allow the filters to dry and use within an hour.

3. Lay the IPTG-treated nitrocellulose filters on the plates from step 1. Pierce the filters with a needle in several places. Incubate the plates in the dark at room temperature overnight.

4. Carefully remove the nitrocellulose filter and forceps. If the top agar sticks to the nitrocellulose filter, try cooling the plate at 4°C for 2 hours.

5. Immerse the nitrocellulose filters in Tris buffered saline with 0.05% Tween-20 (TBST) and remove any adherent top agar with a gloved finger.

6. Immerse the filters in Tris buffered saline with 1% BSA (TBSB). Agitate gently for 1 hour at room temperature to block.

7. To detect mouse kappa light-chain expression, transfer the blocked filters to a solution of primary antibody. Dilute the rabbit anti-(mouse kappa light chain) antiserum (Cappel) 1:1000 (v:v) in TBSB. Incubate the filters (10 ml/filter) for one hour at room temperature with gentle agitation. Wash the filters 3 times with 10 ml of TBST per filter (5 minutes/wash) to remove unbound antibody.

8. Transfer the filters to a solution of alkaline phosphatase-conjugated goat anti-(rabbit IgG), diluted 1:1000 (v:v) in TBSB. Incubate and wash the filters as in step 7 above.

9. Transfer the filters to a fresh substrate solution (10 ml/filter) containing 330 µg/ml NBT and 165 µg/ml BCIP in pH 9.2 buffered saline. Allow color development to proceed until the positive plaques are clearly discernible above the background (2 to 30 minutes). To halt color development wash the filters thoroughly with water. Dry in the dark.

10. Positive plaques display a darker signal intensity as compared to negative plaques and often exhibit a dark ring on the outer edge of the plaque. True positives always line up with plaques on the master plate.

PROTOCOL 5. MASS EXCISION OF SURFZAP LAMBDA LIBRARY

1. Dilute overnight cultures of XL1-Blue and SOLR host cells to $OD_{600\,nm} = 0.02$ in separate flasks containing 500 ml LB broth. Grow at 37°C while shaking at 250 rpm for ≈ 2 hours to early-log phase ($OD_{600\,nm} \approx 0.2$). Spin at $2500 \times g$ and resuspend the XL1-Blue or SOLR cells in 10 mM MgSO$_4$ to $OD_{600\,nm} = 5.0$ and to $OD_{600\,nm} = 1.0$, respectively.

2. In a 50 ml sterile tube, combine an aliquot of the amplified lambda phage library with XL1-Blue cells at a multiplicity of infection (MOI) of 1:10 (lambda phage:cell) ratio. Assume $OD_{600\,nm}$ of 5.0 equals a cell

concentration of 4×10^9 cells/ml. Use $\approx$100-fold amplified lambda phage over the size of the primary library.

3. Add ExAssist helper phage at a 1:1 (helper phage:cell) ratio. Incubate for 15 minutes at 37°C. Add 20 ml of LB broth. Incubate the culture for 3 hours at 37°C while shaking at 250 rpm.

4. Heat the cultures at 70°C for 20 minutes to lyse the lambda phage particles and the XL1-Blue cells. Spin at 2500 × g for 10 minutes to pellet the cell debris. Transfer the supernatant to a fresh tube and store at 4°C.

5. Determine the titer of ampicillin-resistant colony-forming units (Amp^r cfu) per milliliter. Prepare a dilution series of the supernatant (containing the excised phagemid particles) in TE buffer. Mix 1 μl of each dilution with 0.2 ml of SOLR cells ($OD_{600\,nm} = 1.0$) and incubate at 37°C for 15 minutes. With sterile glass beads or a spreader, plate 0.1 ml of the mixture onto LB plates containing 50 μg/ml ampicillin. Incubate overnight at 37°C. The titer, in cfu/ml, is given as:

$$\frac{\text{Number of colonies} \times \text{dilution factor} \times 2}{\text{Volume of dilution added to cells}}$$

PROTOCOL 6. PHAGEMID LIBRARY AMPLIFICATION

1. Grow fresh, early-log SOLR host cells as in step 1 of Protocol 5 and resuspend the cells in 10 mM MgSO$_4$ to $OD_{600\,nm} = 1.0$ ($\approx 8 \times 10^8$ cells/ml).

2. Mix SOLR cells with a portion of the mass excision supernatant from step 4 of Protocol 5 at a phagemid:cell ratio of 1:2. There must be at least one cell per cfu added, and excess cells should not adversely affect the amplification. Incubate phagemid and cells at 37°C for 15 minutes. For every 10^9 cells, add 100 ml LB broth containing 100 μg/ml carbenicillin (an ampicillin analog) and 50 μg/ml kanamycin. Incubate at 37°C, with shaking at 250 rpm, to mid-log phase ($OD_{600\,nm} \approx 0.5$, in 3 to 4 hr). *Do not amplify overnight.* (Optional: Use 1% glucose in 10 mM HEPES (pH 7.5) buffered LB media.)

4. Spin at 1000 × g for 10 minutes. Discard the supernatant and resuspend the cells in 10 mM MgSO$_4$. The cells should be used as soon as possible for the preparation of recombinant phagemid particles for biopanning.

5. Proceed to Protocol 7, or the amplified library in SOLR cells can be stored in 10 mM MgSO$_4$ at 4°C for 1 to 2. (For long-term storage, 700 μl cells and 300 μl of sterile glycerol to a sterile 1.5 ml tube, mix well by pipette, and freeze to −80°C.)

6. Determine the concentration of Amp^r cells per ml. Spread a dilution series of amplified, transformed SOLR cells on LB/ampicillin plates. Grow overnight and count colonies.

PROTOCOL 7. RESCUE OF AMPLIFIED PHAGEMID LIBRARY

1. If the transformed SOLR cells, containing the amplified phagemid library, have been stored at 4°C or have been grown in the presence of glucose (see Protocol 6), dilute $\approx 10^9$ cells with 10 ml of LB broth (without glucose) containing 100 µg/ml carbenicillin and 50 µg/ml kanamycin. Allow the cells to grow for 1 hour until $OD_{600\,nm} \approx 0.2$ is reached. Pellet the cells at $1000 \times g$ for 10 minutes.
2. Resuspend $\approx 10^9$ SOLR cells, from step 1 above or from step 4 of protocol 6, in 1.0 ml LB. Add 10^{10} VCSM13 helper phage (MOI of 10:1 helper phage to host cells) and incubate at 37°C for 15 minutes. Dilute the cells to $OD_{600\,nm} = 0.1$ with ≈ 10 ml of LB broth containing 100 µg/ml carbenicillin and 50 µg/ml kanamycin (without glucose). Shake the cells at 250 rpm, 30°C, for 20 hours. [Optional: Grow at 37°C for 4 to 8 hours until late phase is reached ($OD_{600\,nm} \approx 1.0$).]
3. Spin at $2500 \times g$ for 10 minutes to pellet the cells. Transfer the supernatant containing the phagemid particles to a fresh centrifuge tube.
4. Add 0.2 ml of 30% PEG (w/v in 2.5 M NaCl) for every 1.0 ml of supernatant. Mix well by inverting the tubes and incubate at room temperature for 15 minutes. Spin at $10,000 \times g$ for 20 minutes. Carefully decant the supernatant and resuspend the pellet in one ml TE buffer. Transfer to a 1.5 ml tube and spin at $10,000 \times g$ for 5 minutes to pellet the residual cell debris. Transfer the supernatant to a fresh microfuge tube.
5. If desired, repeat the PEG precipitation procedure described in the previous step, but resuspend the phage pellet in 100 µl TE buffer. Store at 4°C.
6. Titre the phagemid in XL1-Blue cells as described in step 5 of Protocol 4.

PROTOCOL 8. COUPLING ANTIGEN TO CNBr-SEPHAROSE

1. Three grams of Sepharose CNBr-CL-4B (Pharmacia) in 1 mM HCl and wash three times with 200 mM NaCl.
2. Add 10 ml 0.5 M NaCl, 0.1 M NaHCO$_3$ at room temperature.
3. Add 3 ml purified protein at 1 mg/ml (or the equivalent) and mix for four hours. [We used undiluted purified tetanus toxoid (Connought).]

4. Allow beads to settle and pour off supernatant. Wash three times with Tris-buffered saline. Resuspend in one volume of TBS with 0.1% sodium azide. Store at 4°C.

PROTOCOL 9. BIOCHROMATOGRAPHY OF PHAGE DISPLAY LIBRARY

1. Pour a 0.2 ml column of antigen-Sepharose. A disposable column is convenient, for example, a 10 ml Econo-column (Bio-Rad). Wash with 25 ml high-salt buffer [20 mM Tris, 500 mM NaCl pH 7.5 (TBHS)].
2. Load 100 µl concentrated phagemid ($\approx 10^{12}$ ml) from step 5 of Protocol 7. Collect flow-through and reload twice more. Wash with 200 ml TBHS. Collect last ml fraction as "wash."
3. Elute bound phagemid with 2 ml 0.1 N HCl (to pH 2.2 w/glycine), 0.1% BSA. Collect fractions and neutralize each eluate to pH ≈ 7.5 with small additions of 2 N tris-base.
4. Titer each fraction as in step 5 of Protocol 5.
5. If another round of enrichment is desired, add the eluted phage sample to one ml of XL1-Blue cells ($OD_{600\,nm} = 1.0$). Incubate at 37°C for 15 minutes and add 9 ml LB broth containing 100 mg/ml of carbenicillin. Incubate at 37°C for 1 hour. Add VCSM13 helper phage to the culture at a maximum MOI of 1:1 to 10:1 helper phage to host cells. Incubate at 37°C for 1 hour. Add kanamycin to a final concentration of 50 µg/ml. Continue the incubation at 30°C until late log phase is reached ($OD_{600\,nm} \approx 1.0$ in ≈ 16 to 24 hours).
6. PEG precipitate the phagemid particles as in steps 4–6 in Protocol 7.
7. Repeat the biopanning or screen the amplified phagemid samples for evidence of enrichment (see Protocol 10).

PROTOCOL 10. IDENTIFICATION OF ANTIGEN BINDING CLONES

1. Label 0.1 mg protein with 1 mCi Na^{125}I by the Chloromine-T method. (For the experiment described in the text, tetanus toxoid (Connought) was used.)
2. Add an aliquot of an amplified elution fraction from step 5 of Protocol 9 to 0.4 ml XL1-Blue to yield 1000 to 5000 Amp^r colonies. Incubate for 15 minutes at 37°C. Plate on a 138 nm diameter nitrocellulose filter that has been placed on a large (150 mm diameter) NYZ agar plate with 100 µg/ml carbenicillin. Grow at 37°C for 16 hours or until small colonies can be seen.
3. Remove master filter from plate and place, colony side up, on a piece of adsorbent paper. Align a duplicate filter on top of the master filter. Place

a piece of adsorbent paper to form a sandwich. Apply firm, even pressure with a glass plate. Make orientation marks by puncturing the master and duplicate filters with a needle. Carefully peel the duplicate filter from the master filter. Replace the master filter on its agar plate. Place the duplicate filter on a large NZY agar plate with 100 µg/ml carbenicillin, 1 mM IPTG. Grow the duplicate colonies for 4 hours at 37°C, then overnight at room temperature.

4. Remove duplicate filter from agar plate. Suspend the filter, colony side up, on glass pipettes over a tray of chloroform. Do not let the filters directly touch the chloroform. Cover carefully with aluminum foil. Wait 10 minutes and remove filter. (CAUTION! Perform this step in a well-ventilated fume-hood!)

5. Immerse the nitrocellulose filters in Tris buffered saline with 1% BSA (TBSB) and remove cell debris with a gloved finger. Wash with Tris buffered saline with 0.05% Tween-20 (TBST). Immerse the filters in fresh TBSB. Agitate gently for 1 hour at room temperature to block.

6. Transfer the blocked filters to a solution of 0.1 nm radiolabeled protein in TBSB. Incubate the filters (10 ml/filter) for 1 hour at room temperature with gentle agitation. Wash the filters 3 times with 10 ml of TBST per filter (5 minutes/wash) to remove unbound antigen.

8. Remove the filters from the last wash and air dry. Tape filters to a sheet of paper and cover with plastic wrap. Expose X-ray film overnight with an enhancement screen at −80°C. Develop film. Line up the film to the duplicate filters and replicate the orientation marks in ink on the film. Line up the dark radioimages with colonies on the master plate.

9. Pick the colonies corresponding to positive clones. Repeat this assay to confirm binding activity.

Acknowledgments

I would like to thank Holly Hogrefe, Amy Lovejoy, Jeff Amberg, Martin Gore, and Bev Hay for their dedication which made this work possible.

REFERENCES

1. Larrick, J.W. 1993. Prospects for engineering therapeutic human monoclonal antibodies. In *Therapeutic Proteins.* Kung, A.H.C., Baughman, R.A., and Larrick, J.W., eds.) New York: W.H. Freeman, pp. 9–44.
2. Larrick, J.W. 1989. Potential of monoclonal antibodies as pharmacological agents. *Pharmacological Rev.* 41:4.

3. Boyd, J.E., and James, K. 1989. Human monoclonal antibodies: their potential, problems, and prospects. *Monoclonal Antibodies:* Production and Application 11:1.
4. Dillman, R.O. 1990. Human antimouse and antiglobulin responses to monoclonal antibodies. *Antibody, Immunoconjugates, and Radiopharmaceuticals* 3:1.
5. Jones, P.T., Dear, P.H., Foote, J., Neuberger, M.S., and Winter, G. 1986. Replacing the complementarity-determining regions in a human antibody with those from a mouse. *Nature* 321:522.
6. Queen, C., Schneider, W.P., Selick, H.E., Payne, P.W., Landolfi, N.F. Duncan, J.F., Avdalovic, N.M., Levitt, M., Junghans, R.P., and Waldmann, T.A. 1989. A humanized antibody that binds to the interleukin 2 receptor. *Proc. Natl. Acad. Sci. USA* 86:10029.
7. Tempest, P.R., Bremner, P., Lambert, M., Taylor, G., Furze, J.M., Carr, F.J., and Harris, W.J. 1991. Reshaping a human monoclonal antibody to inhibit human respiratory syncytial virus infection *in vivl. Bio/Technology* 9:266.
8. Carter, P., Presta, L., Gorman, C.M., Ridgway, J.B.B., Henner, D., Wong, W.L.T., Rowland, A.M., Kotts, C., Carver, M.E., and Shepard, H.M. 1992. Humanization of an anti-p185^{HER2} antibody for human cancer therapy. *Proc. Natl. Acad. Sci. USA* 89:4285.
9. LoBuglio, A.F., Liu, T., and Khazaeli, M.B. 1993. Human and chimeric mouse/human monoclonal antibodies. In *Therapeutic Proteins.* Kung, A.H.C., Baughman, R.A., and Larrick, J.W., eds. New York: W.H. Freeman. pp. 45–61.
10. Huse, W.D., Sastry, L., Iverson, S.A., Kang, A.S., Alting-Mees, M., Burton, D.R., Benkovic, S.J., and Lerner, R.A. 1989. Generation of a large combinatorial library of the immunoglobulin repertoire in phage lambda. *Science* 246:1275.
11. Mullinax, R.L., Gross, E.A., Amberg, J.R., Hay, B.N., Hogrefe, H.H., Kubitz, M.M., Greener, A., Alting-Mees, M., Ardourel, D., Short, J.M., Sorge, J.A., and Shopes, B. 1990. Identification of human antibody fragment clones specific for tetanus toxoid in a bacteriophage λ immunoexpression library. *Proc. Natl. Acad. Sci. USA* 87:8095.
12. Parmley, S.E., and Smith, G.P. 1988. Antibody-selectable filamentous fd phage vectors: affinity purification of target genes. *Gene* 73:305.
13. Scott, J.K., and Smith, G.P. 1990. Searching for peptide ligands with an epitope library. *Science* 249:386.
14. McCafferty, J., Griffiths, A.D., Winter, G., and Chiswell, D.J. 1990. Phage antibodies: filamentous phage displaying antibody variable domains. *Nature* 348:552.
15. Hoogenboom, H.R., Griffiths, A.D., Johnson, K.S., Chiswell, D.J., Hudson, P., and Winter, G. 1991. Multi-subunit proteins on the surface of filamentous phage:methodologies for displaying antibody (Fab) heavy and light chains. *Proc. Natl. Acad. Sci. USA* 19:15.
16. Winter, G., and Milstein, C. 1991. Man-made antibodies. *Nature* 349:293.
17. Marks, J.D., Hoogenboom, H.R., Bonnert, T.P., McCafferty, J., Griffiths, A.D., and Winter, G. 1991. By-passing immunization: human antibodies from V-gene libraries displayed on phage. *J. Mol. Biol.* 222.581.
18. Clackson, T., Hoogenboom, H.R., Griffiths, A.D., and Winter, G. 1991. Making antibody fragments using phage display libraries. *Nature* 352:624.

19. Bass, S., Greene, R., and Wells, J.A. 1990. Hormone phage: an enrichment method for variant proteins with altered binding properties. *Proteins* 8:309.

20. Barbas, C.F., Kang, A.S., Lerner, R.A., and Benkovic, S.J. 1991. Assembly of combinatorial antibody libraries on phage surfaces: the gene III site. *Proc. Natl. Acad. Sci. USA* 88:7978.

21. Kang, A.S., Barbas, C.F., Janda, K.D., Benkovic, S.J., and Lerner, R.A. 1991. Linkage of recognition and replication functions by assembling combinatorial antibody Fab libraries along phage surfaces. *Proc. Natl. Acad. Sci. USA* 88:4363.

22. Zebedee, S.L., Barbas, C.F., Hom, Y-L., Caothien, R.H., Graff, R., DeGraw, J., Pyati, J., LaPolla, R., Burton, D., Lerner, R.A., and Thornton, G.B. 1992. Human combinatorial antibody libraries to hepatitis B surface antigen. *Proc. Natl. Acad. Sci. USA* 89:3175.

23. Hogrefe, H.H., Mullinax, R.L., Lovejoy, A.E., Hay, B.N., and Sorge, J.A. 1993. A bacteriophage lambda vector for the cloning and expression of immunoglobulin Fab fragments on the surface of filamentous phage. *Gene* 128:119.

24. Hogrefe, H.H., Amberg, J.R., Hay, B.N., Sorge, J.A., and Shopes, B. 1993. Cloning in a bacteriophage lambda vector for the displaying of binding proteins on filamentous phage. *Gene* 137:93.

25. Shopes, B. Metaphoric antibodies. Manuscript in preparation.

26. Chang, C.N., Landolf, N.F., and Queen, C. 1991. Expression of antibody Fab domains on bacteriophage surfaces. *J. Immunol.* 147:3610.

27. Skerra, A., Pfitzinger, I., and Pluckthun, A. 1991. The functional expression of antibody F_v fragments in *Escherichia Coli*: Improved vectors and a generally applicable purification technique. *Bio/Technology* 9:273.

28. Better, M., Chang, C.P., Robinson, R.R., and Horwitz, A.H. 1988. *Escherichia coli* secretion of an active chimeric antibody fragment. *Science* 240:1041.

29. Gillies, S.D. 1992. Design of expression vectors and mammalian cell systems suitable for engineered antibodies. In *Antibody Engineering: A Practical Guide*, First Edition. C. Borrebaeck, ed. pp. 139–57.

30. Morrison, S.L., and Oi, V.T. 1986. Transfer and expression of immunoglobulin genes. *Ann. Rev. Immunol.* 2:239.

31. Chiang, Y.L., Sheng-Dong, R., Brow, M., and Larrick, J.W. 1989. Direct cDNA cloning of the rearranged immunoglobulin variable region. *BioTechniques* 7:4.

32. Larrick, J.W., Danielsson, L., Brenner, C.A., Wallace, E.F., Abrahamson, M., Fry, K.E., and Borrebaeck, C.A.K. 1989. Polymerase chain reaction using mixed primers: cloning of human monoclonal antibody variable region genes from single hybridoma cells. *Bio/Technology* 7:934.

33. Orlandi, R., Gussow, D.H., Jones, P.T., and Winter, G. 1989. Cloning immuno-globulin variable domains for expression by the polymerase chain reaction. *Proc. Natl. Acad. Sci. USA* 86:3833.

34. Hay, B.N., Sorge, J.A., and Shopes, B. 1992. Bacteriophage cloning and *Escherichia coli* expression of a human IgM Fab. *Human Antibodies and Hybridomas* 3:81.

35. Oi, V.T., Morrison, S.L., Herzenberg, L.A., and Berg, P. 1983. Immunoglobulin gene expression in transformed lymphoid cells. *Proc. Natl. Acad. Sci. USA* 80:825.

36. Hay, B.N., and Short, J.M. 1992. ExAssist Helper Phage. *Strategies* 2:4.

37. Maniatis, T., Fritsch, E.F., and Sambrook, J. 1982. *Molecular Cloning. A Laboratory Manual.* Cold Spring Harbor Laboratory, NY.
38. Ausbel, F.M., Brent, R., Kingston, R.E., Moore, D.D., Smith, J.A., Seidman, J.G., and Struhl, K., eds. 1992. *Current Protocols in Molecular Biology.* Media, Pa.: John Wiley and Sons.
39. Kabat, E.A., Wu, T.T., Reid-Miller, M., Perry, H.M., Gottesman, K.S. eds. 1987. *Sequences of properties of immunological interest.* U.S. Dept. of Public Health, NIH.

CHAPTER 6

Strategies for Humanizing Antibodies

Stephen C. Emery and William J. Harris

The discovery in 1975 of the monoclonal antibody technology by Kohler and Milstein[1] triggered a new era of biotechnology, with respect to the diagnostic industry with monoclonal antibodies now forming the basis of over 50% of the $700 million market. Rodent monoclonal antibodies offer a plentiful and pure source of effective antibody, and very considerable resources were expended during the 1980s to evaluate the *in vivo* human use of these antibodies, particularly aimed at imaging (tumours, blood clots) and therapy (tumors, blood clots, immunosuppression).[2]

While the cumulative data have shown that monoclonals can indeed target and deliver an imaging signal or a therapeutic activity, they have not proved clinically successful except in a very few specialized cases.[3,4] Clinical use is limited by short half life; poor recognition of human effector functions by murine immunoglobulin constant regions, and human anti-mouse immune response (HAMA). The immune response in patients can result in enhanced clearance of the antibody from the serum and block its therapeutic effect. such antibodies can also lead to hypersensitivity reactions with all the resultant complications.

It has proved difficult to make human monoclonal antibodies by hybridoma technologies; however, genetic and protein engineering techniques have now provided the tools to overcome these limitations by allowing the "humanization" of antibodies while still retaining the same binding affinity and specificity of the original rodent monoclonal.[5] Such engineering feats have been facilitated by the modular arrangement of antibody domains such that the Fv (heavy- and light-chain) domains are responsible for antigen binding and the constant domains for effector functions. Each antibody domain is encoded by a different genetic exon, and to build recombinant antibodies such exons are added together. The exons for the variable domains (V genes) can be obtained by polymerase chain reaction (PCR) rescue using specially designed "universal" immunoglobulin primers from cellular mRNA isolates and subsequent cDNA generation. The V genes can be easily linked to those exons encoding the constant domains for the expression of whole antibodies.

The first step toward humanizing rodent monoclonals was to attach a human constant domain to a murine variable domain to generate a chimaeric antibody.[6] Subsequently, a number of mouse–human chimaerics have been constructed and clinical trials carried out. The intravascular half lives of chimaerics have been found to vary considerably from patient to patient (65h–300h) but are disappointing compared to human immunoglobulin, though they are a significant improvement over the murine antibody. *In vitro* and *in vivo* data do show that chimaerics are more effective in sequestering human effector functions of complement fixation and ADCC and HAMA response is also considerably reduced chimaerization (up to 80% reduction).

However, the residual immune reactivity directed against the variable region is still sufficient to block the clinical effectiveness of repeated treatment with the same monoclonal antibody. Since the prime aim of the humanization strategy is to further reduce if not eliminate the remaining HAMA response, a more advanced "humanization" process was developed in which rodent CDRs were transplanted onto a human antibody, thus transferring the antigen binding site and its inherent properties. The construction of such CDR-grafted human antibodies will be discussed in this article.

ANTIBODY STRUCTURE

To aid a better understanding of the concept of humanization, the basic antibody structure is shown in Fig. 6–1. In recent years, there have been great strides in our understanding of the three-dimensional structure of antibodies and, in particular, the structure of the antigen binding site. An important feature to note is that the immunoglobulin molecule consists of an assembly of discrete domains such that the structure of the antigen

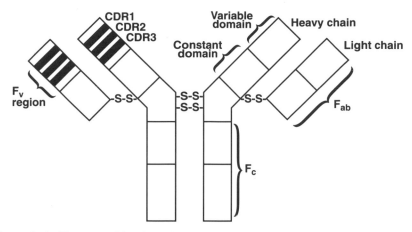

Figure 6–1. Diagram of basic antibody structure.

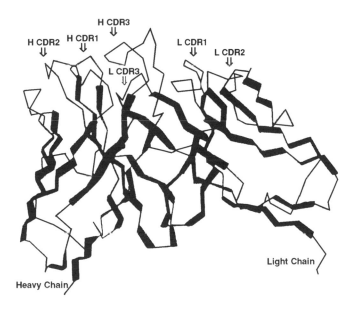

Figure 6–2. Simplified backbone model of the antibody Fv region.

binding site (variable domain, V) is essentially the same in an intact immunoglobulin as in Fab or Fv fragments (Fig. 6–1).

In a more informative model (Fig. 6–2), the Fv variable region constitutes a framework consisting of two domains (heavy and light chains), each of which consists of two antiparallel Beta-sheets arranged in a compressed barrel motif. From these immunoglobulin folds emanate six "hypervariable" loops (three loops each from the heavy- and light-chain domains) which are of particular interest. These loops display high variability in amino acid sequence and size from one antibody to another when compared with the rest of the immunoglobulin framework regions. It is these loops, which are clustered together at one end of the antibody molecule, that contribute the amino acid residues for antigen binding, and hence they are commonly termed the complementary determining regions (CDRs).

HUMANIZED ANTIBODIES

There is general acceptance that the most effective method of achieving humanization is the reshaping technology of Winter and colleagues.[7] This, as previously mentioned, is based upon the transplantation of murine CDRs into human variable-chain frameworks. However, this is slightly over-simplified since additional alterations of individual amino acid residues within the framework may be necessary to recreate the antigen binding site. Thus it is invariably necessary to consider other possible interactions between the Beta-sheet framework and the CDR loops.

A basic outline of this CDR-grafting technique is shown in Fig. 6–3. Detailed technical methods have been described recently.[8,9] The steps, starting with a murine hybridoma cell line, can be summarized as follows.

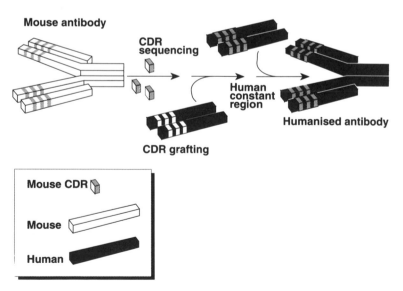

Figure 6–3. Scheme of CDR-grafting for the development of humanized antibodies.

Cloning and Sequencing of Murine Immunoglobulin Variable Region Segments

Amplification of cDNA for the heavy and light variable chains of any murine hybridoma is possible using the polymerase chain reaction (PCR) since there is a high degree of conservation of nucleotide sequence at the 5′ ends of both the VH and VK genes and the 3′ ends of VDJ and VJ rearranged genes (Fig. 6–4).[10] A subset of synthetic oligonucleotide primers with a small level of degeneracy can therefore be used to amplify the immunoglobulin variable genes for all families of murine V genes beginning with RNA extracted from a culture of about 2×10^7 hybridoma cells.

A number of authors have evaluated different primer sets[10–12] that are also designed to include appropriate restriction sites for direct cloning and sequencing in MA3mp18 and M13mp19 by standard methods.[13] It is very common for hybridoma cells to synthesise more than one heavy- and light-chain mRNA even though they do not express functional protein. A number of M13 clones are therefore sequenced to obtain reliable and reproducible information, thereby attempting to negate any misleading PCR errors or spurious sequences. As an additional safeguard, the genes determined to be definitive for the particular antibody are expressed (and tested) as simple chimaeric antibodies to confirm that the deduced sequences do indeed represent the desired antibody.

Preparation of Mammalian Cytoplasmic RNA

Note: Use disposable plasticware when possible; all glassware whould be baked overnight at 180°C prior to use. Use separate reagents for RNA work. Prepare reagents and reactions with DEPC treated water (0.2 ml DEPC/100 ml water) autoclaved. For general solutions in all the following protocols please refer to Sambrook et al.[13] or the specific reference given.

Materials (Favoloro et al. *Methods in Enzymology* 65, 718–749)
  1. Ice-cold PBS (Phosphate-Buffered Saline)
  2. Ice-cold RNA lysis buffer
  3. VRC (vanadyl-ribonucleoside complexes)
  4. 2 × PK buffer

Procedure
  1. Wash cells three times with ice-cold PBS.
  2. Resuspend cells (10^7 to 10^8) in 6 ml ice-cold lysis buffer containing 10 mM VRC and vortex for 10 s.

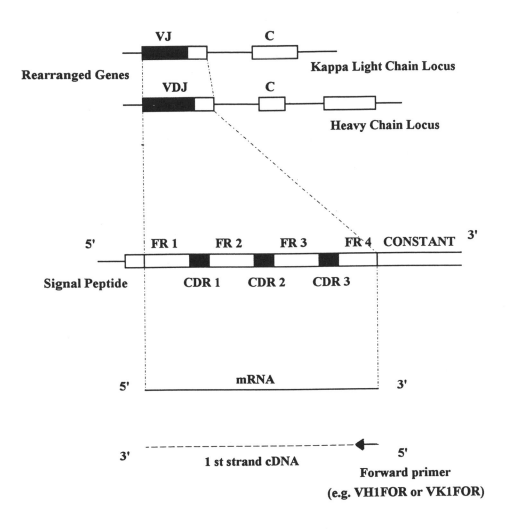

Figure 6–4. Amplification scheme for the rescue of antibody VH and VK genes.

3. Overlay this solution on an equal volume of lysis buffer containing 24% (w/v) sucrose and 1% NP-40. Store on ice for 5 min.
4. Centrifuge at 4000 rpm for 30 min at 4°C.
5. Remove the upper cytoplasmic phase to an equal volume of 2 × PK buffer. Add proteinase K to a final concentration of 200 µg ml^{-1}. Incubate at 37°C for 30 min.
6. Extract the solution with an equal volume of phenol/chloroform.
7. To the aqueous phase add 2.5 vol ethanol and store at −20°C overnight.
8. Collect the RNA by centrifugation (4000 rpm, 30 min), dry in a vacuum desiccator, and dissolve in DEPC water.
9. Measure the RNA concentration by spectrophotometry assuming A_{260} 1 = 40 µg/ml.
10. Run to 2 µg on a 1% agarose gel using TAE as running buffer. Good quality RNA will show sharp ribosomal RNA bands with no evidence of degradation.

Preparation of First-Strand Variable Region cDNA

Materials (Marks et al.[11] for primer sequences)
1. Total cytoplasmic RNA prepared from hybridoma cells.
2. 5X reverse transcriptase buffer (Gibco BRL).
3. Primer: for VK use VK1FOR, VK3FOR, or CKSFOR; for VH use VH1FOR or the CGFOR primer corresponding to the isotype of the antibody; 25 pmol/µl.
4. dNTP stock solution: 2.5 mM each dATP, dCTP, dGTP, dTTP (from Pharmacia 100 mM solutions).
5. Ribonuclease inhibitor (RNAguard, Pharmacia).
6. M-MLV reverse transcriptase (Gibco BRL).

Procedure
1. In a microcentrifuge tube mix:

   5 µg RNA
   10 µl 5X reverse transcriptase buffer
   1 µl forward primer (25 pmol)
   5 µl 2.5 mM dNTPs
   5 µl 10 mM DTT
   0.5 µl RNAguard (Pharmacia)
   DEPC H$_2$O to 50 µl

2. Heat at 70°C for 10 min and then cool slowly to 37°C.
3. Add 100 u (0·5 µl) M-MLV reverse transcriptase and incubate at 37°C for 1 hour.

Primary Amplifications

1. Take 5 µl of the cDNA reaction and add

   5 µl dNTPs (2.5 mM)
   5 µl 10X Enzyme buffer
   1 µl 25 pMole/µl back primer (e.g., VH1back [VH] or VK1back [VK])
   1 µl 25 pMole/µl forward primer (e.g., CG2for [VH] or CK2for [VK])
   0.5 µl AmpliTaq (Perkin Elmer)
   DEPC water to 50 µl

   PCR for 25 cycles at 94°C for 30 s; 50°C for 30 s; 72°C for 60 s ending last cycle with 72°C for 5 min.
2. Check products on a 1.5 to 2% agarose gel.
3. If products are of the expected size, excise the band, purify the DNA, and digest fragment with the relevant enzymes in preparation for subsequent vector cloning.

Cloning Into M13

Materials
1. M13mp18 or M13mp19 RF DNA.
2. Restriction enzyme buffer(s).
3. Restriction enzyme(s).
4. TE-saturated phenol.
5. Chloroform/isoamylalcohol (24:1, v/v).
6. 7.5 M ammonium acetate.
7. Absolute ethanol.
8. 70% ethanol.
9. 5X ligation buffer (Gibco BRL).
10. T4 DNA ligase (Gibco BRL).

Procedure
1. Digestion reaction:

   100 ng M13mp18 or M13mp19 RF DNA
   5 µl 10X buffer (as recommended by the manufacturer)
   5 u each restriction enzyme
   H_2O to 50 µl

   Set up a similar reaction containing the fragment to be cloned. Incubate both reactions at the recommended temperature for 30 min.
2. To each reaction add an equal volume of phenol/chloroform/isoamyl

alcohol (25:24:1, v/v/v), vortex and centrifuge in a microfuge for 3 min.

3. Remove the upper, aqueous phase to a fresh tube and similarly extract with an equal volume of chloroform/isoamyl alcohol (24/1, v/v).

4. Precipitate the DNAs by adding 0.5 vol 7.5 M ammonium acetate and 2 (new) vol of ethanol. Mix and centrifuge in a microfuge for 10 min.

5. Discard the supernatants and wash the pellets with 100 μl 70% ethanol, centrifuging for 2 min and discarding the supernatants.

6. Dry the DNA in a vacuum desiccator and dissolve each DNA in H_2O.

7. Set up the ligation reaction to include the digested M13 RF DNA and fragment in an approximate 1:3 molar ratio. In addition add 3 μl 5X ligation buffer and 0.5 u T4 DNA ligase in a total volume of 15 μl. Incubate at 16°C for 15 min to overnight.

8. Transform into *E. coli* TG1.

Preparation of Single-Stranded M13 DNA

Unlike other coliphages M13 does not lyse its host, rather the phage particles containing ssDNA are continuously secreted by the cells. "Plaques" are seen because infected cells grow 2 to 3 times slower than uninfected cells. M13-infected cells are relatively unstable and have a tendency to lose the F episome, which is required for phage infection. Therefore, plaques should be picked from fresh overnight plates.

Materials
1. Overnight culture of TG1 cells.
2. 20% PEG/2.5 M NaCl.
3. TE (10 mM Tris HCl, pH 8.0; 1 mM EDTA).
4. Phenol/$CHCl_3$.
5. 7.5 M ammonium acetate.
6. Ethanol.

Procedure
1. Inoculate 2 ml LB broth with 20 μl of an overnight culture of cells, e.g. TG1 (it is important to use a fresh overnight culture). Insert an inoculating needle into the centre of the well-isolated plaque and transfer the phage to the broth. Shake at 37°C for 5 to 6 hours. It is important not to incubate longer than 8 hours.

2. Spin 1.5 ml of the culture for 5 min in a microfuge in order to pellet the cells. Remove 1.2 ml supernatant to a fresh tube, being very careful not to pick up any cells. Remove 100 to 200 μl supernatant into a screw-cap 1.5 ml tube for a stock and store at 4°C.

3. The supernatant can be stored overnight (or longer) at 4°C. Recentrifuge before the next step.
4. To supernatant add 300 μl 20% OEG/2.5 M NaCl. Mix well and leave to stand at room temperature for 15 min.
5. Spin in a microfuge for 5 min to pellet the phage. If no phate pellet is visible, then there is no point carrying on. Carefully remove the supernatant and discard it.
6. Recentrifuge for 30 s. Carefully remove all remaining traces of PEG with a pipettor.
7. Resuspend viral pellet in 100 μl TE. Add 50 μl TE-saturated phenol and 50 μl chloroform. Vortex vigorously for 20 s. Centrifuge for 3 min. Remove the aqueous (top) phase to a fresh tube.
8. To the phage DNA solution add 0.5 vol 7.5 M ammonium acetate and 2 vol ethanol. Mix well and leave at room temperature for a few minutes. Centrifuge for 10 min. Carefully remove the supernatant and discard it. Add 200 μl ethanol and centrifuge for 2 min. Carefully pour off the ethanol, removing the last few drops from the inverted mouth of the tube with a tissue. Dry the DNA in a vacuum desiccator.
9. Dissolve the DNA in 30 μl TE. Run 5 μl on an agarose gel. DNA concentration should be approximately 100 μg/ml. 8 μl is sufficient for one sequencing reaction. If the DNA to be cloned is larger than 200 bp, then the insertion can be monitored by gel electrophoresis of DNA from SDS-disrupted cultures. Take 20 μl of supernatant from step 2 and mix with 1 μl 2% SDS and 2.5 μl 10X loading buffer. Run on a 0.7% agarose gel. As a control pick some blue plaques and treat them the same way.

CDR Grafting

Deduced nucleotide sequences are translated into their corresponding amino acid sequences and compared to the consensus of the murine immunoglobulin heavy- and light-chain genes both to identify the family of origin and define the limits of the complementarity determining regions. CDRs were defined by Kabat[14] through comparison with the data base of identified amino acid sequences and delineation of the sequences that were hypervariable from antibody to antibody when compared to the relatively conserved framework regions of each family. Figure 6–5 shows an example of a hybridoma cell whose VH and VK are representatives of the murine VHIA and VKV1 families, respectively.

Six synthetic oligonucleotides are then constructed to encode each murine CDR sequence extended by about 15 nucleotide bases complementary to the flanking regions within the human framework to be used. Oligonucleotide-directed mutagenesis is then used to graft the murine CDRs into the human

(a)

```
              10v                           40v        50v
HCMV37Vh  MRVLILLWLFTAFPGILSQVLLQESGPGLVKPSQSLSLTCTVTGYSITSD
            : LILL L :. .G: S:VQLQESGP:LVKPSQ:LSLTC:VTG SITSD
MuVHIA    MGWSLILLFLVAVATGVHSEVQLQESGPSLVKPSQTLSLTCSVTGDSITSD
              1      60v       70v    2   80v      90v       100v
HCMV37Vh  YAWNWIRQFPGNKLEWMGFISYSGSTSYNPSLESRISVTRDTSKNQFFLQL
          Y WNWIRQFPGNKLEWMG:ISYSGST YNPSL.SRIS:TRDTSKNQFFLQL
MuVHIA    Y-WNWIRQFPGNKLEWMGYISYSGSTYYNPSLKSRISITRDTSKNQFFLQL
              110v            120v  3       130v
HCMV37Vh  YSVTTEDTATYYCANMIT-----TSAYW--YFDVWGAGTTVTVSS
          SVTTEDTATYYCA.:  .     ...Y.  FD WG.GTTVTVSS
MuVHIA    NSVTTEDTATYYCAR-YGYYRGDEEDYYAMAFDYWGQGTTVTVSS
```

(b)

```
              10v                           40v        50v
HCMV37Vk  MDFQVQIFSFLLISASVIMSRGQIVLSQSPAILSASPGEKVTMTCRASSSV
          M  ..Q::::LL:   :     :RGQIVL:QSPAI:SASPGEKVTMTC.ASSSV
MuVkVI    MRVPAQLLGLLLLWLPG--ARGQIVLTQSPAIMSASPGEKVTMTCSASSSV
              1      60v       70v    2    80v      90v       100v
HCMV37Vk  SYMHWYQQTPGSSPKPWIYATSNLASGVPTRFSGSGSGTSYSLTISRVEAE
          SYMHWYQQ.:G:SPK.WIY.TS:LASGVP:RFSGSGSGTSYSLTIS.:EAE
MuVkVI    SYMHWYQQKSGTSPKRWIYDTSKLASGVPARFSGSGSGTSYSLTISSMEAE
              110v   3   120v
HCMV37Vk  DAATYFCQQWSSHPPTFGGGTKLEIKR
          DAATY:CQQWSS:P TFG:GTKLE:KR
MuVkVI    DAATYYCQQWSSNPLTFGAGTKLELKR
```

Figure 6–5. Example comparison of a murine antibody's variable region sequences with their respective murine subgroups.

framework. A variety of mutagenesis methods have been used,[15,16] though it is sometimes necessary to perform a second cycle of mutagenesis to incorporate all six CDRs. An example of methodology currently used is given below.

CDR Grafting

Materials
1. Mutagenic oligonucleotides.
2. S.S. uracil-DNA template.
3. Polynucleotide kinase (PNK) buffer.
4. T4 polynucleotide kinase (BRL).
5. 0.1 M DTT.

6. 0.1 M ATP.
7. 6.25 mM dNTPs.
8. T4 DNA ligase (BRL).
9. T7 DNA polymerase (United States Biochemicals).
10. 7.5 M ammonium acetate.
11. Ethanol.
12. 10X uracil DNA glycosylase buffer.
13. Uracil DNA glycosylase (Boehringer Mannheim).
14. 3N NaOH.
15. Pnenol/CHCl₃.
16. Thermalase (International Biotechnologies Inc.).

Procedure

1. Synthesize mutagenic oligonucleotides encoding CDRs and which are flanked by 15 to 18 perfectly matched nucleotides at both ends.
2. The template ssDNA must contain uracil in place of thymine. Inoculate 20 ml LB broth with 20 µl of an overday culture of RZ1032 (dut-ung-) and 0.5 µl M13 supernatant. Shake at 37°C overnight.
3. Prepare ssDNA.
4. Phosphorylate oligonucleotides:

   10 pmol each oligonucleotide
   5 µl 5X kinase buffer
   water to 24.5 µl
   5 u (0.5 µl) T4 polynucleotide kinase (GIBCO BRL)

   Incubate at 37°C for 1 h.
5. Annealing primers to template:

   ~0.2 pmol (0.5 µg) ssU-DNA
   1 pmol each phosphorylated oligonucleotide
   1 pmol oligo 10
   4 µl 5X annealing buffer
   water to 20 µl

   Heat to 90°C for 30 s, rapidly cool to 70°C and slowly cool to 37°C.
6. Extension and ligation: to the annealed DNA from step 3 add the following:

   2 µl 5X annealing buffer
   2 µl 0.1 M DTT
   0.3 µl 0.1 M ATP
   1 µl 6.25 mM dNTPs
   2.5 u (0.5 µl) T7 DNA polymerase

0.5 u (0.5 µl) T4 DNA ligase
water to 30 µl

Incubate at room temperature for 1 h.

7. Add 15 µl 7.5 M ammonium acetate and 90 µl ethanol. Precipitate at room temperature for a few minutes.
8. Centrifuge for 10 min at 12,000 rpm, wash pellet with 100 µl ethanol, dry, and dissolve in 44 µl water.
9. Add 5 ml 10X uracil glycosylase buffer and 1 µ (1 µl) uracil DNA glycosylase. Incubate at 37°C for at least 1 h.
10. Add 43.3 µl water and 6.7 µl 3N NaOH and incubate at room temperature for 5 min.
11. Neutralize the sample by adding 50 µl 7.5 M ammonium acetate and 300 µl ethanol.
12. Collect the DNA by centrifugation at 10,000 rpm for 10 min, wash pellet with 100 µl ethanol, dry, and dissolve DNA in 20 µl TE.
13. The extended/ligated DNA is amplified by PCR:

    2 µl mutant DNA
    5 µl 10X Thermalase buffer
    5 µl 2.5 mM dNTPs
    25 pmol oligo 10
    25 pmol oligo 11
    water to 49.5 µl
    0.5 µl (1 u) Thermalase

    15 cycles of 94°C, 30 s; 50°C, 30 s; 72°C, 1 min; ending with 5 min at 72°C. Run a 5 µl aliquot on an agarose gel to check for the presence of full-length VH or VK.

14. Extract sample with an equal volume of phenol/CHl$_3$, ethanol precipitate the DNA, wash, dry, and dissolve in 20 µl TE for subsequent digestion with HindIII and BamHI in preparation for cloning into the M13mp19 vector.

Mutagenesis by Overlap/Extension and PCR

This is the method of choice for removing restriction sites or changing a small number of amino acids in VH and VK. The mutagenesis can be accomplished with two PCRs and does not require any DNA fragments to be purified. The starting point is usually VH or VK DNA cloned in M13mp19. (Higuchi et al. (1988). *NAR* 16:7351–67.)

Materials
1. Two overlapping mutagenic oligonucleotides.
2. M13 universal and reverse sequencing primers (oligos 10 and 11).

3. dNTPs.
4. Vent DNA polymerase (New England Biolabs) (IBI).
5. 10X Vent or Thermalase reaction buffers (NEB or IBI).

Procedure
1. Synthesize two oligonucleotides encompassing the sites to be mutated with a 15 to 18 bp clamp at both ends. Oligos must overlap each other by at least 16 bp. The outside primers are usually M13 universal and reverse sequencing primers (oligos 10 and 11).
2. Two PCRs:

   ~0.2 μl supernatant of M13 clone
   25 pmol mutagenic oligo 1 or 2
   25 pmol oligo 10 or 11
   5 μl 2.5 mM dNTPs
   5 μl 10X Vent polymerase buffer (NEB)
   water to 49.5 μl
   0.5 μl (1 u) Vent DNA polymerase (NEB)

   Typically, the samples are subjected to 12 to 15 cycles of 94°C, 30 s; 50°C, 30 s; 75°C, 20 s to 1 min, depending on the length of extension (assume 1 min for every 1 kb).
3. Run 10 μl on an agarose gel to check that the PCRs have worked. If not, lower the annealing temperature by 5°C and/or increase the number of cycles of PCR.
4. Two two products are joined in a second PCR using the outside primers only:

   ~1 μl each DNA from 1st PCRs
   5 μl 10X Vent buffer (NEB)
   2 μl 6.25 mM dNTPs
   25 pmol oligo 10
   25 pmol oligo 11
   water to 49.5 μl
   0.5 μl (1 u) Vent DNA polymerase

   10 to 12 cycles at 94°C, 30 s; 50°C, 30 s; 75°C, 40 to 60 s.
5. Run a 5 μl aliquot on an agarose gel. There should be visible bands corresponding to the DNAs from the first PCRs, but also species of approximately 880 or 680 bp corresponding to full-length VH or VK respectively. If there is no product of the expected size, then lower the annealing temperature by 5°C and/or increase the number of PCR cycles. Sometimes Vent DNA polymerase gives a very smeary product in the second PCR. If so, use Thermalase (IBI) and its reaction buffer in place of Vent DNA polymerase.

6. Usually the DNA from the second PCR can be extracted with phenol/chloroform, ethanol precipitated, digested with HindIII and BamHI, and cloned directly without purification.
7. All DNAs that have been amplified by PCR must be sequenced to ensure the absence of spurious mutations.

Expression of Humanized Antibodies

Humanized V genes are cloned into separate expression vectors of the type depicted in Fig. 6–6. The essential features of such vectors include (a) effective promoter and enhancer elements such as immunoglobulin signals; (b) the heavy- and light-chain constant domains usually in the form of genomic

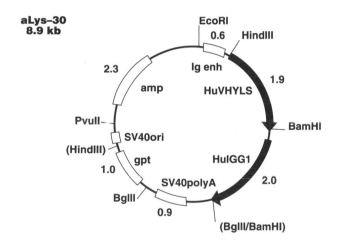

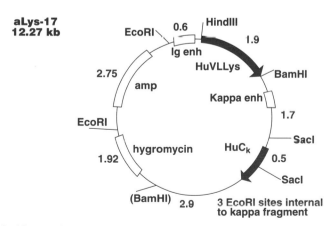

Figure 6–6. Mammalian cell vectors for the expression of humanised V regions.

constructs; (c) antibiotic selection markers. Both vectors are cotransfected into myeloma or CHO cells by electroporation, and while selection can be made for both antibiotic markers, in practice the efficiency of cotransfection is very high and a single antibiotic selection (generally for the heavy-chain expression plasmid) is usually adequate. Resulting clones generally secrete 5 to 15 µg/ml of antibody, as it is unusual for a simple CDR-grafted antibody to have antigen-binding affinity fully restored. Hence, a number of alternate constructs with additional changes in the VK and particularly the VH framework regions are made. Many researchers therefore use a transient expression system to screen rapidly for restoration of binding affinity. Both COS cells and CHO-K1 cells have been used.[17,18]

Transfection of Mammalian Cells

Materials
1. DMEM with glutamine, antibiotics, and 10% FCS (GIBCO BRL).
2. Bio-Rad Gene Pulser and cuvette.
3. Selective DMEM (to 500 ml complete medium add 5 ml 2.5 mg/ml xanthine in 0.5 M NaOH and 160 µl 2.5 mg/ml mycophenolic acid in ethanol.
4. 24-well tissue culture plates.

Procedure
1. DNA to be transfected should be linearized to improve efficiency. For the transfection of expression vectors pSVgptVHHuIgG1 and pSVhygVKHuCK into YB2/0 or NS0 where selection will be for the heavy-chain plasmid, prepare PvuI digests of about 3 and 6 µg of the plasmids respectively. If Sp2/0 are to be transfected, 5 times these amounts will be required. The digested DNA should be ethanol precipitated and dissolved in 50 µl H_2O.
2. Resuspend cells from a semiconfluent 75 cm² flask and collect by centrifugation at 1,000 rpm for 5 min. Discard supernatant.
3. Resuspend cells in 0.5 ml DMEM and transfer to a Gene Pulser cuvette (Bio-Rad). Mix the DNA with the cells by gentle pipetting and leave on ice for 5 min.
4. Insert the cuvette between the electrodes of the Bio-Rad Gene Pulser and apply a single pulse of 170 V, 960 µF.
5. Return cuvette to ice for 20 min.
6. Transfer cell suspension to a 75 cm² flask containing 20 ml DMEM and allow to recover for 1 to 2 days.
7. Harvest cells and resuspend in 36 ml selective DMEM and aliquot 1.5 ml to each well of a 24-well plate.
8. About four days later gently tap the plate and remove most of the

medium from each well to clear away the majority of dead cells. Replace with 1.5 ml fresh selective DMEM per well.

9. Approximately 10 days from the start of selection, colonies should be visible to the naked eye and the well supernatants should be assayed for antibody. On the basis of the level of antibody production and the number and size of the colonies in the well, clones can be chosen for expansion. To resuspend the cells from the designated wells, rub the end of a 1 ml pipette across the surface and transfer the medium to a 25 cm^2 flask containing 4 ml fresh selective DMEM.

10. Cells should be expanded in order to lay down liquid nitrogen stocks and to provide medium for antibody purification.

ELISA for Secretion of Human Antibodies

Materials
1. ELISA coating buffer.
2. PBST (PBS + 0.05% Tween 20).
3. OPD (o-phenylene diamine) tablets.
4. OPD substrate buffer.
5. H$_2$O$_2$.
6. 12.5% sulphuric acid.
7. Goat anti-human IgG antibodies (Sera-Lab).
8. HRPO goat anti-human kappa antibodies (Sera-Lab).
9. ELISA plate (Immulon 1, Dynatech).
10. Tissue culture supernatant.
11. Human IgG, K control antibody.

Procedure
1. Dilute goat anti-human IgG antibodies 1:1000 in coating buffer and add 100 μl to each well. Wrap in clingfilm and leave at 37°C for 1 h or overnight at 4°C.
2. Empty wells and wash 3× with PBST.
3. Add 2 to 20 μl supernatant diluted with PBST to give a volume of 100 μl per well. Include a row of control human IgGK antibody, serially diluted twofold, starting from 100 ng and a row containing PBST only. Incubate at room temperature for 1 h.
4. Empty wells and wash 3× with PBST.
5. Dilute HRPO goat anti-human kappa antibodies 1:2000 in PBST and add 100 μl to each well in a sequential manner. Incubate at room temperature for 1 h.
6. Empty wells and wash 4× with PBST.
7. Just prior to use, dissolve 20 mg OPD tablet in 45 ml water and add 5 ml 10X OPD substrate buffer and 10 μl 30% H$_2$O$_2$. Add 100 μl to

each well and leave at room temperature until the yellow color develops, usually less than 10 min.

8. Stop the reaction in a sequential manner by adding 50 µl 12.5% H_2SO_4. Read the absorbance at 492 nm.

Purification of Reshaped or Murine Antibodies

For purification of reshaped human IgG1 and IgG4 antibodies and murine IgG2a and IgG2b antibodies use protein A agarose. For purification of murine IgG1 antibodies use protein G agarose.

Materials
  1. Culture medium containing secreted antibody.
  2. Protein A or G agarose (Boehringer).
  3. 1 M TrisHCl pH 8.0.
  4. 0.1 M TrisHCl pH 8.0.
  5. 0.01 M TrisHCl pH 8.0.
  6. 0.1 M glycine pH 3.0.
  7. OBS.

Procedure
  1. Remove cells from culture medium by centrifugation at 10,000 rpm for 10 min.
  2. To cell free supernatant add 0.1 volume 1 M TrisHCl pH 8.0 and sodium azide to 0.05%. Add 0.5 to 1 ml protein A agarose beads (which should bind at least 10 mg antibody) and stir gently at room temperature for at least 3 h.
  3. Collect beads by centrifugation at 2,000 rpm for 2 min and remove most of the medium.
  4. Pool protein agarose beads, wash with 0.1 M TrisHCl pH 8.0, collect by centrifugation, and load into an Econo-column (Bio-Rad).
  5. Wash column with 10 ml 0.1 M TrisHCl pH 8.0.
  6. Wash column with 10 ml 0.01 M TrisHCl pH 8.0.
  7. Elute antibody with 4 × 1 ml aliquots 0.1 M glycine pH 3.0 into tubes containing 100 µl 1 M TrisHCl pH 8.0.
  8. Measure the A_{280} of each fraction and pool those fractions containing significant antibody.
  9. Dialyse against PBS.
  10. Sterilize solution by filtering through a 0.22 µm membrane (Costar Spin-X). Take a spectrum of a portion of the antibody and estimate concentration assuming $A_{280}1 = 714$ µg/ml.

RESHAPING TECHNOLOGIES

Simple application of CDR grafting methods, although they provide a good yield of recombinant antibody, often result in the loss of antigen-binding affinity. This is not unexpected since it is clear from structural analysis and three-dimensional computer graphic modeling that a CDR must assume the correct conformation to bind antigen, and the correct conformation is dependent upon interaction between specific residues within the CDR and within the framework. The delineation of a CDR through comparison of linear amino acid sequences is not an accurate reflection of the structural extent to hypervariable loops. Commonly, for example, structural analysis places amino acid residues of VH 27–30 within the first loop (Fig. 6–7) extending from residues 26–32,[19] while by the definition of Kabat,[14] they fall outside CDR-1 designated as 31–35. The current state of the art in computer graphic three-dimensional modeling is inadequate to predict the conformation of CDRs, though they do assume a limited number of canonical forms.[20] However, in the absence of knowledge of the structure of the antigen, it is doubtful whether a single definitive structure could be predicted for such flexible moieties.

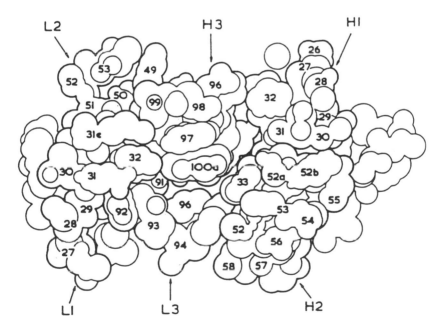

Figure 6–7. Representation of an Fv region model as viewed from above the antigen-binding site (amino acid residues are numbered and CDR position indicated).

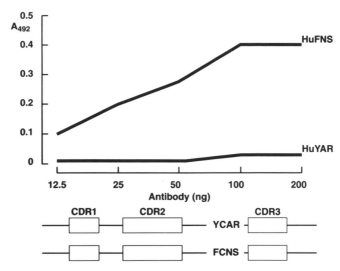

Figure 6–8. Graphic representation illustrating the restoration of antibody binding by residue substitution in the heavy-chain framework of the anti-RSV antibody.[25]

A number of indirect methods have therefore been developed to assist in the prediction of additional changes that need to be introduced into human frameworks to restore antigen-binding affinity. A typical example is shown in Fig. 6–8, which demonstrates that antigen-binding affinity is restored by strategic alterations to framework regions. Atomic coordinates are available for 11 Fab antibody fragments, and 35 different Fabs have been elucidated by X-ray diffraction studies. Padlan[21] has resolved all the residues that impact upon the CDRs in these structures and thereby provided a platform for selection. Similarly, Foote and Winter[22] defined a number of the residues that most commonly impact upon CDRs.

Selection of Human Framework

The original method to restore antigen-binding affinity following CDR grafting was described by Winter.[7] First, the murine monoclonal variable region amino acid sequence is assigned to its family subgroup, and then the primary sequence is compared with the consensus sequence for that subgroup. This aims to identify unusual amino acids which may have been introduced during *in vivo* affinity maturation of the antibody-producing cell and hence may be required for correct orientation of CDRs. These specific residues are then introduced into the human framework. Second, the selected human framework is compared with its subgroup with its subgroup consensus sequence, in this case to remove unusual amino acid residues whose function may be to interact with the endogenous human CDRs (which are being replaced). Finally, the murine and human sequences are compared to eliminate consistent species differences, particularly those which would cause a significant change in size, charge, hydrophobicity, etc. This method has been described in detail[8] and successfully applied to a number of antibodies,[9] initially using the human domains NEWM (VH) and REI (VK) to form a human Fv framework. These two human protein domains were chosen since their structure has been resolved and some direct study of CDR-framework interactions are possible.

A modification of this method has been described by Queen.[23] The linear amino acid sequence of murine variable chains as derived from the hybridoma is compared directly to available human sequences, and the human sequence with the highest homology is taken as the acceptor framework. This is then modified by the inclusion of more commonly found human residues, as described above, to construct a more generic human acceptor sequence. Computer graphic analysis is then used to identify critical residues from the murine sequence for inclusion within the human framework. An advantage to this approach is the increased chance of restoring antigen-binding affinity without the need to test alternative constructions. A number of antibodies have been reshaped by this method.[9]

A further modification of this method termed "veneered" antibodies is based upon selective replacement of murine surface exposed residues in the murine antibody, with human residues occupying the equivalent position in the human antibody selected on the basis of the highest homology.[24]

Reshaping Using Fixed Human Frameworks

Antibody structure can be described as a relatively conserved barrel structure of Beta-pleated sheets upon which CDRs are mounted, and the three-dimensional structures of antibodies are very similar indeed (Fig. 6–2). From structural and functional studies it is known that very different primary amino acid sequences can fold to form the same structural motifs. There is therefore no *a priori* reason why the basic scaffold structure of human framework should possess a linear amino acid sequence that exhibits a high homology with a murine antibody. In most cases it should be possible to use the same human framework to host the CDRs for any rodent monoclonal and, by judicious alteration of particular framework residues, to restore antigen-binding affinity.

In many examples of reshaped human antibodies, we have transferred rodent CDRs into the same human frameworks, NEWM for VH and REI

Table 6–1. Sites of Inclusion of Murine Residues
in Addition to CDRs Transferred into VH
(NEWM)

Kabat Residue	*30*	*71*	*94*
NEWM	STFS	LV	YCAR
RSV-HuFNS	*F*TFS	LV	*F*CN*S*
α-TOX	*FTFS*	LV	YCN*A*
ANTI-JUNIN	*Y*TFS	LV	YCAR
HCMV-37	*YSIT*	LV	YCA*N*

for VK. Typical examples from over 20 different studies are shown in Table 6–1.[25,26] As with homology matching, murine and humanized V region protein structure is analyzed by computer in order to deduce potential molecular interactions. Additionally, we have adopted the approach of introducing minimal modifications into human frameworks through an interactive approach of confirming that any particular murine residue is essential for antigen-binding affinity. This approach generally results in only a few murine framework residues being included. Substitutions commonly occur within the flanking regions surrounding the CDRs (Table 6–1). It is important then not to rely on the traditional definition of CDR regions, nor consider the inclusion of the entire sequence defined by CDR as essential for restoration of antigen binding.

HUMANIZED PRODUCTS

It is likely that over 50 monoclonals have been successfully humanized.[9] Murine, rat, and bovine (unpublished data) monoclonals have been used as starting material. They also represent a broad range of target antigens, including anti-infectious disease and anti-cancer antibodies. In most cases the affinity has been restored to within a few fold of the original antibody. It is clear then that this technology is widely applicable. Although the bulk of antibodies have been reshaped as human IgG1 class, this has been a deliberate choice as the subclass most likely to be clinically effective. IgG4 and IgG2 have also been used. In theory, antibodies of the IgM class could also be made, but efficient expression of recombinant IgM molecules in mammalian cells is more difficult to achieve.

CONCLUSIONS

The technique of CDR grafting was first described in 1987 and has now matured into a major protein engineering technology. Experience over the last 5 years has resulted in the evolution of a number of different methods to convert the technology into reproducible practice. The very evolution of a variety of methodologies, however, has led to the development of humanized antibodies with great variation in the number of rodent amino acid residues retained in the final human antibody. The term *humanization* is used to describe a range of antibodies, some of which have been reshaped to possess only murine CDRs, while others are more accurately described as hyperchimaeric, retaining a significant percentage of murine residues within the human framework. There is great optimism that the major problems associated with the *in vivo* clinical use of rodent monoclonal antibodies have been solved by this technology.

Acknowledgments

It is a pleasure to thank our colleagues in Scotgen Ltd., Frank Carr, Phil Tempest, and Anita Hamilton, for unpublished data and valuable discussions.

REFERENCES

1. Kohler, G., and Milstein, C. 1975. Continuous cultures of fused cells secreting antibody of predefined specificity. *Nature* 256:52–53.
2. Larrick, J.W., and Fry, K.E. 1991. Recombinant antibodies. *Hum. Antibod. Hybridomas* 2:172–89.
3. Borrebaek, C.K., and Larrick, J.W. 1990. *Therapeutic monoclonal antibodies*. New York: Stockton Press.
4. Zuckier, L.S., Rodriguez, L.D., and Scharff, M.D. 1989. Immunologic and pharmacologic concepts of monoclonal antibodies. *Seminars in Nuclear Medicine* 19:166–86.
5. Winter, G., and Milstein, C. 1991. Man-made antibodies. *Nature* 349:293–99.
6. Morrison, S.L., Johnson, M.J., Herzenberg, L.A., and Oi, V.T. 1984. Chimaeric human antibody molecules: mouse antigen-binding domains. *Proc. Nat. Acad. Sci. USA* 82:6851–55.
7. Verhoeyen, M., Milstein, C., and Winter, G. 1988. Reshaping human antibodies: grafting an anti-lysozyme activity. *Science* 239:1098–1104.
8. Gussow, D., and Seeman, G. 1991. Humanization of monoclonal antibodies. *Method. Enzymol.* 203:99–121.

9. Harris, W.J. 1993. Humanising monoclonal antibodies for *in vivo* use. *Animal Cell Biotechnology* 6. In press.

10. Orlandi, R., Gussow, D.H., Jones, P.T., and Winter, G. 1989. Cloning immunoglobulin variable domains for expression by the polymerase chain reaction. *Proc. Nat. Acad. Sci. USA* 96:3833–37.

11. Marks, J.D., Hoogenboom, H.R., Bonnert, T.P., McCafferty, J., Griffiths, A.D., and Winter, G. 1991. By-passing Immunization. *J. Mol. Biol.* 222:581–97.

12. Sastry, L., Alting-Mees, W.D., Huse, J.M., Short, J.M., Sorge, J.A., Hay, B.N., Janda, K.D., Benkovic, S.J., and Lerner, R.A. 1989. Cloning of the immunoglobulin repertoire in *E. coli* for generation of monoclonal catalytic antibodies: Construction of a heavy chain variable region-specific cDNA library. *Proc. Nat. Acad. Sci. USA* 86:5728–2.

13. Sambrook, J., Fritsch, E.F., and Maniatis, T. 1989. *Molecular Cloning. A Laboratory Manual.* 2nd edition, vol. 2. Cold Spring Harbor Laboratory Press.

14. Kabat, E.A., Lu, T.T., Reid-Miller, M., Perry, H.M., and Gottesman, K.S. 1987. *Sequences of Proteins of Immunological Interest.* 4th edition. Washington D.C.: United States Dept. of Health and Human Services.

15. Nakamaya, K., and Eckstein, F. 1986. Inhibition of restriction endonuclease NciI cleavage by phosphothiorate groups and its application to oligonucleotide-directed mutagenesis. *Nuc. Acids. Res.* 14:9679–98.

16. Clackson, T., and Winter, G. 1989. "Sticky feet"-directed mutagenesis and its application to swapping antibody domains. *Nuc. Acid. Res.* 17:10163–71.

17. Boshart, M., Weber, F., Jahan, G., Dorsch-Hasler, K., Fleckenstein, B., and Schaffner, W. 1985. A very strong enhancer is located upstream of an immediate early gene of human cytomegalovirus. *Cell* 41:521–30.

18. Bebbington, R.B. 1991. Expression of antibody genes in nonlymphoid mammalian cells. *METHODS: A Companion to Methods in Enzymology* 2:136–45.

19. Chothia, C., and Lesk, A.M. 1987. Canonical structures for the hypervariable regions of immunoglobulins. *J. Mol. Biol.* 196:901–17.

20. Chothia, C., Lesk, A.M., Tramontano, A. Cevitt, M., Smith-Gill, S.J., Air, G., Sheffiff, S., Padlan, E.A., Davies, D., Tuplip, L.R., Colman, P.M., Spinelli, S., Alzari, P.M., and Poljak, R.J. 1989. Conformations of immunoglobulin hypervariable regions. *Nature* 342:877–83.

21. Padlan, E.A. 1991. A possible procedure for reducing the immunogenicity pf antibody variable domains while preserving their ligand-binding properties. *Mol. Immunol.* 28:489–98.

22. Foote, J., and Winter, G. 1992. Antibody framework residues affecting the conformation of hypervariable loops. *J. Mol. Biol.* 224:487–99.

23. Queen, C., Schneider, W.P., Selick, H.E., Payne, P.W., Landolfi, N.F., Duncan, J.F., Aldalovic, N.M., Levitt, M., Junghans, R.P., and Waldmann, T.A. 1989. A humanised antibody that binds to the interleukin-2 receptor. *Proc. Nat. Acad. Sci. USA* 86:10029–33.

24. Daugherty, B.C., Demartino, J.A., Law, M.F., Kawka, D.W., Singer, I.T., and Mark, G.E. 1991. Polymerase chain reaction, facilitates cloning, CDR-grafting, and rapid expression of a murine monoclonal antibody against CD18. *Nuc. Acid. Res.* 19:2471–76.

25. Tempest, R.R., Bremner, P., Lambert, M., Taylor, G., Furze, J., Carr, F.J., and Harris, W.J. 1991. Reshaping a human monoclonal antibody to inhibit respiratory syncytial virus infection *in vivo*. *Biotechnology* 9:266–71.
26. Harris, W.J., and Tempest, P.R. 1991. Reshaping of human antibodies for the treatment of infectious diseases. *J. Cell. Biochem.* 15E, 136.

CHAPTER 7

Single-Chain Fv Design and Production by Preparative Folding

James S. Huston, Andrew J.T. George,
Mei-Sheng Tai, John E. McCartney, Donald Jin,
David M. Segal, Peter Keck,
and Hermann Oppermann

Although the first single-chain Fv proteins were described only six years ago, in 1988,[1,2] they engendered widespread application of single-chain immuno-technology to Fv regions from antibodies and other members of the immuno-globulin superfamily.[3,4] Engineering of antibody combining regions has shown the single-chain Fv (sFv) to be a readily accessible and stable form of the Fv region. Self-targeted sFv fusion proteins may also be constructed by recombinant DNA methods,[3–11] to yield fusion of an effector protein at either the amino terminus[6–7] or carboxyl terminus[8–10] of an sFv. For example, an sFv-toxin fusion protein exhibited cytotoxicity that was signifi-cantly enhanced over the crosslinked immunotoxin control.[8] Fusion of sFv to the gene III coat protein of fd filamentous phage led to assembled particles termed phage antibodies, which express their genomically encoded sFv on the bacteriophage surface.[9] These are vehicles for refinement of sFv binding sites, as well as for selection of murine or human sFv protein of any desired specificity that may be derived from V gene libraries.[9,12] An alternative to phage antibodies involves fusion of sFv genes to the peptidoglycan associated lipoprotein (PAL), which upon expression in *E. coli* results in sFv-PAL fusion protein decorating the outside of the *E. coli* cell that contains the responsible sFv-PAL plasmid.[13,14] Medical applications of sFv proteins are of considerable promise,[3] based on their properties *in vivo*, where upon intravenous administration of sFv to animal models, the sFv species show a large volume of distribution, rapid clearance, and highly specific localization at intended target sites with little nonspecific binding.[15–17]

At a practical level, the literature indicates that the preferred methods for production of particular single-chain Fv proteins vary widely. Several groups have had success in preparative folding or sFv or sFv fusion proteins expressed as bacterial inclusion bodies.[1–8,15–24] Preparative folding has allowed us to make as much as 110 to 130 mg of active sFv or sFv fusion proteins per liter of fermented cells.[6,18] Cell secretion methods can obviate the need for refolding in some cases,[9,10,12–14] but there are particular sFb proteins that cannot be secreted in active form or do so at unacceptably low levels. It is not uncommon for bacterially secreted sFv protein to form insoluble aggregates in the periplasm, which then necessitates refolding. Cytoplasmic overexpression to form inclusion bodies may be desirable to augment protein yield, or it may be an obligatory route to produce sFv-toxin fusions or other sFv proteins that harm the host cell if made in active form. Compared to secretion, refolding of inclusion body protein can significantly reduce the cost of producing isotopically labeled sFv for nuclear magnetic resonance (NMR) spectroscopy, which is becoming an increasingly useful approach in molecular immunology. Consequently, refolding remains important to both large- and small-scale production of sFv analogs and fusion proteins. This chapter will address sFv design and production, with emphasis on the methodical application of preparative folding methods. The techniques discussed are applicable to sFv constructions that not only utilize antibody variable regions but also corresponding regions from diverse members of the immunoglobulin superfamily.[4,19]

Fv STRUCTURE AND sFv DESIGN

Constructing an sFv involves manipulation of heavy- and light-chain variable regions, which must be connected at the gene level by an oligo-nucleotide sequence that codes for an appropriate peptide linker (Fig. 7–1). Figure 7–1(a) schematically compares the V_H–V_L and V_L–V_H forms of the sFv to IgG and its fragments. Size differences are an obvious distinction between 150 kDa IgG, 50 kDa Fab fragment, 25 kDa Fv fragment, and 26–27 kDa sFv. The Fv and sFv are the only species with intact combining sites that are devoid of constant regions, which can translate into reduced nonspecific binding during *in vivo* use.[3,15–17] The single polypeptide composition of the sFv makes domain association a unimolecular reaction that is independent

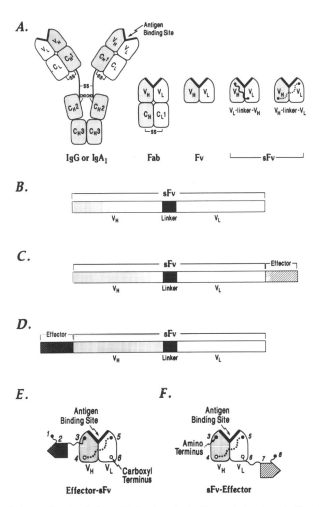

Figure 7–1. Schematic depiction of single-chain Fv proteins in relation to IgG and its fragments. A. IgG antibody: 4 chains connected by disulfides, 150,000 mol wt. Fab fragment: 2 chains connected by a disulfide, 50,000 mol wt. Fv fragment: 2 chains, 25,000 mol wt. sFv protein: 1 chain, about 26,000 mol wt.; the sFv is shown in its two possible permutations, V_L–V_H and V_H–V_L, with the dotted linker on the back face and solid linker on the front, for the given orientation of V regions. B. Single-chain Fv polypeptide chain. The typical sFv protein consists of about 250 amino acids and has a molecular weight of approximately 26,000 to 27,000 (the specific value depends on actual sequences of the V regions and linker segment). C. Polypeptide chain comprising an sFv fusion protein with a protein effector fused to the carboxyl terminus of the sFv. D. Polypeptide chain comprising an sFv fusion protein having a protein effector used to the amino terminus of the sFv. E. Schmeatic drawing of an effector-sFv fusion protein in its native conformation (corresponding to the N-terminal fusion protein of Fig. 1D). F. Schematic (continued).

of protein concentration, in contrast to the bimolecular association reaction of Fv heterodimers, where dilution favors domain dissociation.[11] As indicated in Fig. 7–1(c)–(e), the fusion of an effector domain at either terminus of the polypeptide can yield an sFv fusion protein inherently able to target the effector protein. Although protein crosslinking has long been useful, the commercial preparation of proteins for human use is very costly, and there is significantly advantage in combining all the desired targeting and effector domains in one multifunctional protein, instead of purifying them in separate processes. Furthermore, sFv fusion proteins sometimes show biological activity that is superior to larger crosslinked antibody conjugates.[8]

Linker Considerations. The linker, in its simplest form, must connect the V_H and V_L domains of a given Fv without perturbing interdomain contacts or interfering with domain folding. We favored a linker that would have few side chains yet exhibits reasonable solubility and maximal flexibility. Thus we arrived at the 15-residue $(Gly_4Ser)_3$ peptide,[1] which is the most widely used linker in the sFv literature (Table 7–1). Other successfully used sFv linkers are also in this compilation,[2,13,24–28] which emphasizes that the uncharged flexible linker is distinct from the others, which contain charges as well as glycine and serine. We have avoided charged residues in the linker because they could potentially interact with sidechains of opposite charge in proximity on the Fv surface, which could sometimes alter binding-site conformation.[11] The peptide segments that link lobes or domains in natural proteins of known structure have recently been shown to often be rich in glycine and serine, quite analogous to our uncharged linker.[29] Within native Fv structure, the linker must bridge the distance between the C-terminus of the amino-terminal or first V region and the N-terminus of the second V region. Reference to Table 7–2 shows calculations of the linear distances between these V domain termini based on PDB coordinates[30] for nine Fab crystal structures,[31–39] oriented in both sFv configurations. The V_H–V_L

Figure 7–1 (*continued*). drawing of an sFv-effector fusion protein in its native conformation (corresponding to the C-terminal fusion protein of Fig. 7–1(c). Numbers indicate parts of the proteins in Figs. 7–1(e) and 7–1(f), as follows: **1**, the first residue of the effector domain, which is the N-terminus of the fusion protein; **2**, N-terminal effector protein domain; **between 2 and 3**, spacer sequence that facilitates dual function of effector and sFv; **3**, first amino acid residue of V_H; **4**, last amino acid residue of V_H; **between 4 and 5**, sFv linker segment; **5**, first amino acid residue of V_L; **6**, last amino acid residue of V_L; **between 6 and 7**, spacer sequence that facilitates dual function of effector and sFv; **7**, C-terminal protein effector domain; **8**, the carboxyl terminus of fusion protein. Color coding: light chain and V_L, white; heavy chain and V_H, gray; sFv linker, black; effector protein, such as a toxin or growth factor, striped. Antigen binding site is part of the Fv region, indicated by a V-shaped docking site. [Reproduced with permission from Huston et al., *Intern. Rev. Immunol.* **10**, 195–217, 1993,[3] Harwood Academic Publishers GmbH.]

Table 7-1. List of commonly used single-chain Fv linker sequences

Identity/Linker, Specificity	V Order	Linker Length	Linker Sequence # Uses 1988–92 [Initial Ref]
26-10 sFv/(G$_4$S) linker Anti-digoxin	V$_H$–V$_L$	15	GGGGS GGGGS GGGGS 36 [1]
FB-sFv$^{26-10}$	V$_H$–V$_L$ V$_L$–V$_H$	15 10–25	GGGGS GGGGS GGGGS [6] (GGGGS)$_n$, n = 2–5 [11]
18-2-3 sFv/202 linker Anti-fluorescein	V$_L$–V$_H$	14	EGKSS GSGSE SKST 6 [2]
4-4-20 sFv/212 linker Anti-fluorescein	V$_L$–V$_H$	14	GSTSG SGKSS EKG 10 [24]
3C2 sFv/59 linker Anti-bov. growth hormone	V$_L$–V$_H$	18	KESGS VSSEQ LAQRF SLD 2 [2]
Se155-4 sFv/Switch region linker, Anti-CHO	V$_H$–V$_L$ V$_L$–V$_H$	16 14	LTVSS ALTTP PSVYP L 3 [25] SSPSV TLFPP SSNG
MA-12C5 sFv/K$_{12}$G$_0$ linker Anti-fibrin	V$_L$–V$_H$	10	IKRAG QGSSV 2 [26]
4-4-20/205 linker, Anti-fluorescein	V$_L$–V$_H$	25	SSADD AKKDA AKKDD AKKDD AKKDG 1 [27]
Ox sFv/CBH linker Anti-2-phenyloxazolone	V$_H$–V$_L$	27	PGGNR GTTTT RRPAT TTGSS PGPTQ SH 2 [28]
Humanized D1.3 sFv/tubulin epitope linker Anti-lysozyme	V$_H$–V$_L$	18	GSASA PKLEE GEFSE ARE 3 [13]

Source: Reproduced with permission from Huston et al., 1993, Cell Biophysics 22, 189–224,[5] Humana Press.

Table 7–2. Comparison of bridging distances in Fv regions of known structure

			Euclidean (Linear) Distance Between Linker Ends (Å)		
Structure	Ref.	PDB	V_H–V_L	V_L–V_H	$(V_L$–$V_H)$–$(V_H$–$V_L)$
NEW	[32]	3fab	29.30	39.80	10.50
KOL	[33]	2fb4	33.06	39.18	6.12
McPC 603	[34]	2mcp	34.55	39.10	4.55
D1.3	[31]	1fdl	35.44	43.27	7.83
HyHEL-5	[35]	2hfl	35.28	42.97	7.69
HyHEL-10	[36]	3hfm	31.93	36.55	4.62
J539	[37]	2fbj	34.72	41.44	6.72
R19.9	[38]	1f19	34.74	43.04	8.30
4-4-20	[39]	f4ab	34.27	39.50	5.23

Using the Biosym Insight II program and coordinates from the Brookhaven Protein Data Bank[57] (PDB), from data sets noted above in the PDB column, the Euclidean distance between linker ends was estimated by the following method. The algorithm was used to calculate the distance between the C-terminal α-carboxyl carbon of the first V region and the N-terminal α-amino nitrogen of the second V region; the linker is peptide bonded to each of these positions in sFv proteins. For the V_H–linker–V_L order, these end points are V_H position 128 to V_L position 1 by the numbering scheme in Fig. 7–3(#). For the V_L–linker–V_H orientation, these termini are V_L position 117 and V_H position 1 (Fig. 7–3). For each crystal structure, the quantity $[(V_L$–$V_H)$–$(V_H$–$V_L)]$ is the difference in Euclidean bridging distances for the V_L–linker–V_H and the V_H–linker–V_L constructs; the distance is always longer for V_L–linker–V_H. Minimum linker lengths for any of these possible constructions must be somewhat longer than the distances noted, in order to compensate for curvature of the protein surface.
Source: Reprinted from Huston et al.[4] with permission of the Portland Press.

bridging distances range from 29 Å to about 36 Å, which requires at least 10 residues and preferably 15 to bridge either orientation. Our analysis shows that the V_L–V_H bridging distance is always 5 to 10 Å greater than the corresponding V_H–V_L distance for a given molecule. The computer-generated sFv structures given in Fig. 7–2 were calculated from Fv region coordinates of the D1.3 Fab structure.[31] The V regions have been connected by (Gly$_4$Ser)$_3$, with the V_H–V_L orientation in Fig. 7.2(a) and the V_L–V_H order in Fig. 7–2(b). The necessity of presenting a static linker conformation for a given picture is misleading, however, as it is actually quite mobile and may be considered to occupy a large ensemble of configurations. The flexibility and absence of significant surface contacts for this linker have been confirmed by NMR relaxation studies, which compared isotopically enriched McPC 603 sFv made with the (Gly$_4$Ser)$_3$ linker to Fv heterodimer; McPC 603 sFv was produced by refolding of labeled protein overexpressed as inclusion bodies.[40]

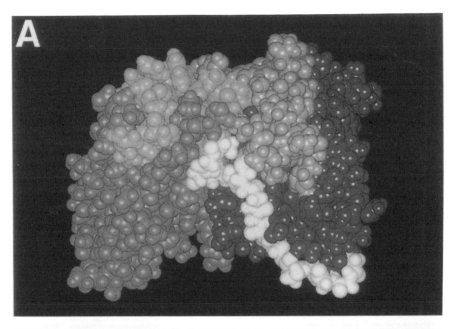

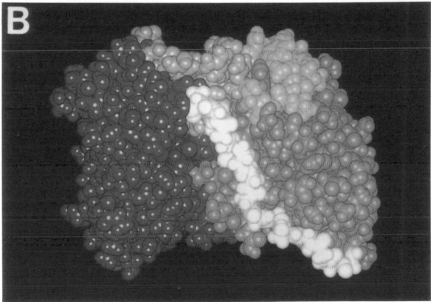

Figure 7–2. Computer-generated models of anti-lysozyme D1.3 single-chain Fv proteins. The displayed D1.3 sFv models have the $(GGGGS)_3$ linker bridging variable regions [11], constructed in (A) as V_H–linker–V_L or in (B) as V_L–linker–V_H; (*continued*)

Alternative Strategies for Expression of sFv Inclusion Bodies

Having chosen sFv inclusion bodies as the starting material for production of active protein, there is still a choice to be made, whether one uses cytoplasmic inclusion bodies that can easily be scaled up for large preparations, or periplasmic inclusion bodies, which can sometimes be simpler to renature. The cytoplasmic and periplasmic environments differ in a number of respects, which result in different properties for sFv inclusion bodies, which may have correspondingly distinct protocols for sFv folding. The cytoplasm maintains a reducing milieu wherein cysteinyl residues cannot form disulfide bonds that are essential to the integrity of native variable domains. The high concentration of recombinant protein causes aggregation of the molecules into inclusion bodies, which facilitates initial collection of sFv protein. However, these inclusion bodies must be fully reduced and denatured to take full advantage of the large quantity of sFv produced, since upon bacterial lysis the protein can rapidly oxidize to randomly form disulfide crosslinks between molecules. For cytoplasmic inclusion bodies, our procedures all start with sFv protein in a fully reduced random coil state,[41] whereas periplasmic inclusion bodies are assumed to contain properly oxidized sFv polypeptides.

An alternative location for the production of sFv is the periplasmic space. This is the area in gram negative bacteria that lies between the inner and outer membranes and contains the bacterial peptidoglycan and various other proteins.[42] This area has a number of features that are similar to the endoplasmic reticulum, and, in a limited sense, expression of proteins into the periplasm mimic the expression of proteins in the endoplasmic reticulum of eukaryotic cells.[43] Thus, in a similar manner to the endoplasmic reticulum, the periplasm has an oxidizing environment, encouraging formation of disulfide bonds. The periplasm, like the endoplasmic reticulum, also contains

---

Figure 7–2 (*continued*). as indicated in Fig. 7-1(a), the linkers of these two forms are on opposite faces of the Fv region, such that the Fv region in (A) would be rotated 180° about a vertical axis to give the orientation shown in (B). The molecular models were based on the X-ray crystallographic coordinates of Fischmann et al.,[31] obtained as data set 1FDL from the Protein Data Bank at the Brookhaven National Laboratory.[30] The molecular graphics models were constructed on a Silicon Graphics 4D/70 GTX Superworkstation (Silicon Graphics, Mountain View, CA), using programs *Insight II* and *Discover* (Biosym, San Diego, USA). The linker conformations shown in this figure are two representations from the ensemble of accessible low-energy conformations calculated in the absence of solvent. The color scheme is the following: framework region of V_H (dark gray), linker (white), framework of V_L (gray), and their corresponding complementarity-determining regions (lightest shapes of gray) at the top of each variable domain. [Reproduced with permission from Huson et al., *Intern. Rev. Immunol.* **10**, 195–217, 1993,[3] Harwood Academic Publishers GmbH.]

proteins such as disulfide isomerases and chaperonins,[44,45] which aid in the refolding of the newly synthesized proteins. In addition, secretion of proteins into the periplasmic space necessitates vectorial transport of the nascent polypeptide through a lipid bilayer and cleavage of a signal peptide from the chain, features that have all been suggested to be involved in the correct folding of native proteins. It should be stressed, however, that there are also many differences between the periplasm and the endoplasmic reticulum.[43] Thus, in prokaryotic cells translation of the RNQ is relatively rapid when compared to the rate of translocation through the membrane, and proteins may be transported into the periplasm after already being synthesized. In eukaryotic cells the lower rate of translation means that the synthesis of the polypeptide chain and its translation into the endoplasmic reticulum can occur simultaneously.[43] A further problem is that the high levels of recombinant proteins produced in some expression systems can overwhelm the refolding pathways utilized by bacteria. This is highlighted by the finding that the transfection of the chaperonins GroES and GroEL can increase the yield of recombinant molecules expressed in *E. coli.* Indeed, such chaperonins have been used to increase the yield of correctly folded Fab fragments expressed in the periplasm using a phage display system.[46] It should also be remembered that the chaperonins and similar molecules in the periplasm are different from those seen in eukaryotic cells, and molecules known to have specialized roles in the refolding of immunoglobulins, such as heavy-chain binding proteins (BiP), are absent from the periplasm. The absence of a glycosylation pathway in the periplasm, while important in the expression of many proteins, is not normally an issue in the production of sFv.

FOLDING OF sFv AND sFv FUSION PROTEINS

The recovery of antigen-binding activity in a single-chain Fv, first achieved for the 26 10 sFv,[1] was based on folding of the cloned protein, and thus it continued the long tradition of studies on antibody refolding. Successful renaturation of antibody binding regions was first accomplished some 30 years ago.[47-51] Recovery of binding activity from Fab fragments was achieved for rabbit Fab fragments that were fully reduced and denatured,[47,48] as well as unreduced and denatured, keeping original disulfide bonds intact.[49,50] These experiments on polyclonal Fab fragments examined statistically large populations of variable regions, and thus offer some general insights that are not possible when examining a single monoclonal antibody fragment. Fab with intact disulfides gave much higher refolding yields[49,50] than did Fab with fully reduced disulfides.[47,48] Part of the advantage in keeping disulfides intact was that Fd (the amino-terminal half of H chains present in Fab) and L chain retained their natural pairing, being crosslinked through an interchain disulfide.[49,50] Renaturation of the reduced and denatured Fab lowered its apparent affinity for hapten, presumably from mispairing of V_H and V_L regions. However, the disulfide-intact material recovered nearly 72% of the antigen-binding activity of native Fab,[50] whereas the corresponding reduced anti-dinitrophenol (DNP) Fab regained about 11% of the active protein that was denatured and refolded.[48] Part of this low yield was due to precipitation or other protein losses during refolding; based on the specific activity of the soluble Fab renatured from 6 M GuHCl/β-mercaptoethanol, anti-ribonuclease Fab recovered 27% of its native activity[47] while the anti-DNP Fab gave a maximum of 24%, unless free hapten was added during renaturation, resulting in 34% recovery.[48] Thus, these early studies suggested that refolding of Fab fragments was clearly more successful if their original disulfide bonds were kept intact; Rowe and Tanford subsequently studied the kinetics of reversible unfolding and folding of a myeloma chain and Fab.[55,56] Refolding studies on MOPC 315 Fv fragments[52–54] extended these observations to a homogeneous antigen-combining fragment, for which recovery of active Fv required obligatory formation of its intrachain disulfide bonds before refolding; contemporary experiments on recombinant MOPC 315 sFv[7,53] confirm that the same refolding requirements apply to the MOPC 315 single-chain Fv, implying that the refolding properties of particular V regions dominate even after they have been conected in an sFv.

Fidelity of sFv Combining Site. The demonstration of specific binding activity in 26-10 sFv[1] was followed by refinements in binding analysis that confirmed full recovery of both affinity and specificity of antigen binding shown by the parent 26-10 monoclonal antibody.[6] Equilibrium binding data for the 26-10 sFv and its amino-terminal fusion with the FB fragment of protein A, FB-sFv[26-10], are compared with those for control 26-10 IgG and Fab (Table 7–3). These results confirm that, within experimental error, all 26-10 species bound digoxin with the same apparent dissociation constant of $K_{d,app} = 4.2 \times 10^{-10}$ M.[6] Inhibition assays were used to measure the relative ability of a panel of related cardiac glycosides to block binding of radioiodinated digoxin to each 26-10 species; each 26-10 species was specifically bound to a microtiter plate coated with goat anti-mouse Fab IgG. The concentration of each glycoside needed for 50% inhibition of maximal binding by ^{125}I-digoxin was determined; each 50% value was normalized by dividing with the unlabeled digoxin concentration at the midpoint of its binding isotherm. These normalized specificity values[1,6,11] were converted to relative dissociation constants, $K_{d,rel}$, by multiplying these normalized constants by the $K_{d,app}$ for digoxin binding.[3] Thus the specificity profile of each 26-10 species is given by the set of dissociation constants in each column pertaining to each cardiac glycoside noted at the left. Each row presents inhibition measurements with a single glycoside, and the $K_{d,rel}$

Table 7–3. Relative dissociation constants of 26-10 binding sites and cardiac glycosides

Cardiac Glycoside	$K_{d,rel} \times 10^{10}$ (M)			
	sFv	*FB-sFv*	*Fab*	*IgG*
Digoxin	4.2	4	4.5	4.2
Digoxigenin	5.5	4.8	3.6	3.8
Digitoxin	10.1	4.4	5.9	4.6
Digitoxigenin	7.6	4.4	5	4.2
Acetylstrophanthidin	11.8	6.8	11.7	4.6
Gitoxin	50	52	50	37
Ouabain	143	160	176	105

$K_{d,app}$ values are given for digoxin complexes with 26-10 species noted above each column. Free digoxin concentration was measured by ultrafiltration and the dissociation constant calculated to be 4.2×10^{-10} M, within experimental error for all 26-10 species.[11] Results for other cardiac glycoside complexes were calculated from specificity assays, wherein affinity purified goat anti-mouse Fab was adsorbed to microtiter plates, followed by the 26-10 species of interest.[1,6,11] $K_{d,rel}$ values were derived from the product of the corresponding digoxin $K_{d,app}$ value and the normalized concentration of glycoside that inhibited to 50% of maximal ^{125}I-digoxin binding. For each type of 26-10 species, normalization involved dividing the concentration of each glycoside at 50% inhibition by the concentration of digoxin at 50% inhibition. *Source:* Modified with permission from Tai et al.[6] (1990) *Biochemistry* **29**, 8024–30, American Chemical Society.

values measured for each 26-10 species are the same within the error of these experiments.[3,6] These studies verified that binding properties of the 26-10 sFv combining site, in both engineered forms, are indistinguishable from those of 26-10 IgG or Fab fragment.

Transition Curves of 26-10 sFv. An interesting perspective on the 26-10 sFv refolding process is given in Fig. 7–3, which presents equilibrium unfolding and refolding data for the 26-10 sFv with its native disulfides intact. Both in urea and in guanidine hydrochloride (GuHCl), transition curves appear to be fully reversible, whereas only about 35% of the 26-10 sFv protein refolds from the fully reduced and denatured polypeptide.[6,18] These transition curves were determined by measuring changes in tryptophan fluorescence of 26-10 sFv as a function of denaturant concentration. Both Trp-37 and Trp-175 are conserved in all V_H and V_L domains, respectively, where they flank the C-terminal side of the first CDR and pack internally against the conserved disulfide bond of each domain;[57] in addition, 26-10 has Trp-105 in the H3 loop, where numbering refers to that of the 26-10 sFv sequence given previously.[1,6,11]

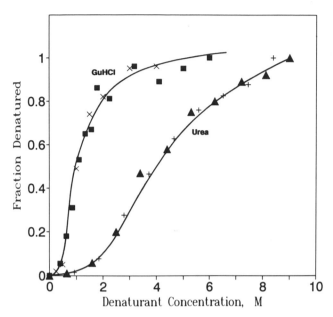

Figure 7–3. Transition curves for the 26-10 sFv in guanidine hydrochloride (GuHCl) and urea. The coincidence of denaturation (× or +) and renaturation points (filled squares or triangles) indicates reversibility of the folding process when the disulfide bonds of the variable regions remain intact. The fraction denatured was based on the tryptophan fluorescence of the 26-10 V regions. The 3 M urea conditions used for refolding 26-10 sFv are thus seen to correspond to mostly native species at equilibrium. Conditions of experiments were the following: a Perkin Elmer LS-5 Spectrofluorometer was used with excitation at 295 nm and emission at 350 nm; samples were incubated at a constant temperature, in all experiments between 22.4 and 22.7°C; tryptophan fluorescence was enhanced by unfolding and quenched by refolding. All stock protein solutions were in PBSA and the 26-10 sFv concentration for measurements was 0.75 μM.

Preparative folding is sometimes the method of choice for production of sFv fusion protein. A convenient strategy for assessing the potential of this approach, as well as to optimize it, is to collect refolding data on each of the separate partners in the fusion. Thus the sFv and its partner would be likely to exhibit distinct transition curves, which would probably be spread more in urea than GuHCl. Hypothetically using Fig. 7–3 as an example of this experiment, assume that the curve labeled "GuHCl" is for the fusion partner, and that labeled "urea" is for the sFv. They overlap somewhat, but the fusion partner is mostly denatured at the midpoint of the sFv transition, where 50% of the sFv has recovered its native state. In this example, the sFv fusion protein could potentially be refolded in stages, first refolding the sFv portion at 3.5 M urea, and then refolding the fusion partner at 1.5 M urea.

General Aspects of sFv Folding. Although general understanding of protein refolding remains a formidable research problem, a number of considerations can facilitate the refolding of sFv polypeptides from inclusion bodies. One feature central to the recovery of active sFv is the pattern of disulfide bonds, which we have previously discussed in detail.[11] The three possible disulfide bonding patterns of fully oxidized sFv monomer are shown in Fig. 7–4. The top panel depicts the properly oxidized sFv polypeptide, which emphasizes the partitioning of each disulfide *within* the V_H or V_L domain of the sFv.[11] Drawings on the left present a schematic comparison of the native and nonnative forms, (a) and (b), each of which involves two disulfides *between* variable regions. The figures on the right are meant to convey the distances between α-carbons of pairs of cysteinyl residues corresponding to native or nonnative bonds in McPC 603 sFv. Clearly, the native sFv protein conformation is compatible only with the normal disulfide bond pattern shown in the top row of figures. Whereas the normal distance between α-carbons is about 6.5 Å, the distances between the incorrect pairs are 24 Å for (a) and 18 or 31 Å for (b), which are physically impossible. Nonetheless, oxidation of reduced and denatured sFv monomers results in the formation of all three disulfide-bonded forms (drawn to scale in middle column of pictures). The nonnative forms that are folded from such a mixture cannot yield native sFv, and their separation from authentic, properly oxidized and refolded forms is critical to use of the disulfide-restricted refolding procedure, which for some sFv proteins is the obligatory method of recovery of activity.[7,15,23]

The refolding properties of sFv proteins are depicted in Fig. 7–5, where spontaneous refolding from the reduced random coil is categorized as dilution refolding (A) or redox refolding (B). Those sFv proteins that do not spontaneously refold are frequently renaturable by the disulfide-restricted refolding method (C), which refers to folding from the crosslinked random coil state. Oxidation of reduced and denatured sFv monomers yields a mixture of the three forms if fully oxidized, or a maximum of ten forms if all partially oxidized forms and the fully reduced form are included. Complete oxidation is desirable to ensure the optimal starting material, since theoretically only a third of the oxidized protein would be capable of yielding native sFv. Given a choice of several sFv proteins against the same target, a spontaneous refolding species should be chosen to make later scale-up reasonable. This category of sFv should also make better fusion proteins, if they need to be refolded. An excellent example of a well-behaved sFv fusion protein is the FB-sFv[26-10], which has been described by Tai et al.,[6] and is shown in Fig. 7–6 through the course of its preparation monitored by sodium dodecyl sulfate (SDS) polyacrylamide gel electrophoresis (PAGE); the corresponding purification is described in Protocol IIIB. The inclusion bodies (lane 2) are highly enriched in the FB-sFv[26-10] polypeptide and minimally contaminated by extraneous proteins. Any DNA present would

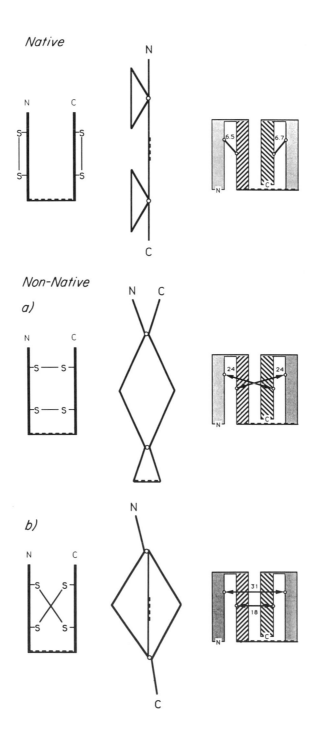

be removed by the DE 52 chromatography step (lane 3). Importantly, the refolding conditions do not yield visible disulfide-crosslinked aggregate upon SDS-PAGE in nonreducing conditions (lane 6). This is illustrative of the simplicity of affinity purification (lane 7), where in this case excess cardiac glycoside (20 mM ouabain) was used to elute the FB-sFv$^{26-10}$ without recourse to acid pH or chaotropic agents.

In optimizing the refolding conditions for a given sFv, the use of SDS-PAGE on nonreduced samples can be very informative. A common indicator of refolding problems is the formation of disulfide-stabilized aggregates, visible on SDS-PAGE as an accumulation of protein at the junction of the stacking and running gel. If the protocol involves redox refolding, this may be due to the suboptimal choice of urea concentration during renaturation, or excessive concentration of protein. For disulfide-restricted refolding protocols, aggregate formation may also stem from a poor choice of redox conditions and insufficient strength or concentration of denaturant during oxidation. Figure 7–7 gives examples of intermediates forming during disulfide-restricted refolding of FB-sFv^{315} in panel (A), 26-10 sFv in panel (B), and FB-sFv$^{26-10}$ in panel (C).[11] As conditions of FB-sFv^{315} oxidation were varied in 6 M GuHCl-Tris buffer solution [in panel (A)], different patterns of oxidized species were obtained (lanes 3–5). Affinity purification on trinitrophenyl-lysine-Sepharose proved that the upper band of the middle doublet was the active FB-sFv^{315}, which was shown to have recovered the native dinitrophenyl-lysine binding affinity of the MOPC 315 IgA. The preferred oxidation conditions for FB-sFv^{315} were 1 mM oxidized and 0.1 mM reduced glutathione,[7] which corresponds to the sample run in lane 4 (Fig. 7–7). Using these conditions (approximately given in Protocol 2), 4% of the refolded protein was active FB-sFv^{315}, which was twice the yield

---

Figure 7–4. Schematic representations of the native and nonnative disulfide bonding patterns of fully oxidized sFv. In all figures the crenulated line corresponds to the linker segment. The left-hand drawings show the linear connectivities of the four cysteinyl residues in forming three possible disulfide bond patterns for the sFv. The native form has one disulfide in each variable domain (vertical line), whereas the nonnative form has two disulfides between variable regions. The middle diagrams are drawn linearly to scale, showing the size and disposition of disulfide-bonded loops in the three forms of oxidized sFv. The right-hand diagrams are drawn to convey the relative orientation of the sFv cysteinyl residues. They are part of either the inner β-sheets (hatched walls) or outer β-sheets (shaded walls). The numbers adjacent to black bars or arrows represent the distance in Angstroms between α-carbons of these residues, based on the crystal structure of McPC 603. The native distance is to be contrasted with the alternative nonnative pairings, which are clearly impossible in the context of a native sFv (also compare Fig. 7–2(a) and (b). [Reproduced with permission from Guston et al., *Methods in Enzymology* **203**, 46–88, 1991,[11] Academic Press, Inc.]

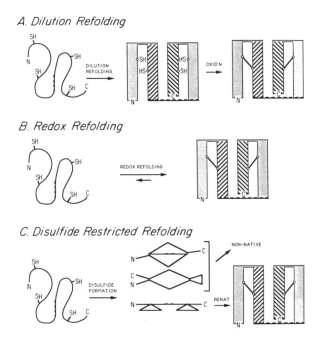

Figure 7–5. Refolding schemes for sFv proteins. The left-hand figures represent the fully reduced and denatured sFv polypeptide, with the crenulated line indicating the linker segment. As noted in the legend to Fig. 7–4, the disulfide-bonding pattern is depicted by the right-hand figures, while the intermediate forms in (C) are noted in the middle schematic diagrams. A. Dilution refolding involves a two-stage refolding process, with renaturation of the reduced protein followed by air oxidation. B. The redox refolding process does not involve isolated intermediates, since refolding and reoxidation occur concurrently in solution. C. Disulfide-restricted refolding leads to three potential forms of the fully oxidized monomeric protein, shown as intermediate species (note 7–4), of which only the bottom species can yield a native sFv. [Reproduced with permission from Huston et al., *Methods in Enzymology* **203**, 46–88, 1991,[11] Academic press, Inc.]

obtained by Cheadle et al. for directly expressed MOPC 315 sFv oxidized in 8 M urea[53] rather than 6 M GuHCl.[7] The advantages of sFv proteins that refold spontaneously from the reduced random coil are readily apparent for 26-10 sFv in panel (B) (lanes 2–5), as these samples are devoid of nonnative forms, which were visible as a lower band on SDS-PAGE when the same 26-10 sFv was made by disulfide-restricted refolding (lanes 6 and 7). The FB-sFv$^{26-10}$ in panel (C) was analyzed on SDS-PAGE prior to affinity purification to reveal the minor nonnative forms in the nonreduced lanes. As discussed previously,[11] the nonnative species are only sometimes distinguishable from the native sFv, but when they are separable on SDS-PAGE, it provides a very helpful measure for comparing products of

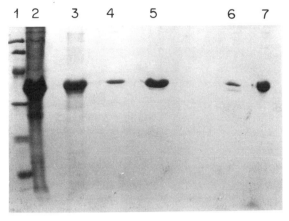

Figure 7–6. SDS-PAGE analysis of FB-sFv 26-10 during purification.[6] SDS-PAGE analysis of FB-sFv at progressive stages of purification. Samples in numbered lanes of the Coomassie blue-stained 15% gel were the following: (1) Standards (Mr $\times$ 10^{-3}), 14.4, 20.1, 29, 43, 67, 94; (2) unpurified inclusion bodies; (3) Whatman DE 52-purified FB-sFv[26-10]; (4) refolded FB-sFv[26-10] mixture [reduced]; (5) affinity-purified FB-sFv[26-10] (reduced); (6) refolded FB-sFv[26-10] mixture (unreduced); (7) affinity-purified FB-sFv (unreduced). As discussed for 26-10 sFv,[1,11] the FB-sFv[26-10] migrated with a higher apparent molecular weight than its calculated molecular weight of 33,106; this is at least partially due to the low residue weight of glycine and serine in the linker sequence. [Reproduced with permission from Tai et al., *Biochemistry* **29**, 8024–30, 1993,[6] American Chemical Society.]

different refolding protocols. The best procedures in our laboratory yield only species with correctly paired disulfide bonds.

Purification of Native sFv Protein

Affinity Chromatography. In some cases the antigen bound by a given sFv is readily availble. In these cases the simplest purification procedure is to use affinity chromatography against the antigen or an appropriate analog, eluting either with soluble antigen or with destabilizing conditions of low pH (e.g., 0.1 M glycine, pH 2.9) or chaotropic agents (e.g., 3.5 M LiCl). The anti-digoxin 26-10 sFv has been isolated by affinity chromatography with immobilized ouabain, which has the advantage of binding to 26-10 species about 100-fold more weakly than digoxin while being comparatively soluble in water. As indicated in the specific methods of Protocol III, this combination of properties greatly simplifies the affinity binding and elution of active 26-10 sFv; the dialysis of eluted 26-10 sFv removes the ouabain used for elution.[1,6,11,18]

Size-Exclusion Chromatography. In the absence of affinity chromatography, the use of size-exclusion chromatography as a final purification step is often quite effective.[8,16,18,19] Although others have used HPLC ion exchange

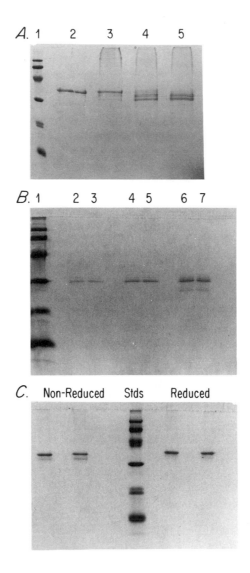

Figure 7–7. SDS-PAGE profiles of oxidized sFv or FB-sFv species. A. Unreduced mixture of randomly oxidized FB-sFv[315] analyzed by SDS-PAGE on 15% gels. Disulfide crosslinked species span a lower M_r range than the reduced chain at about 29 kD. Reduced and denatured FB-sFv[315] (0.2 mg/ml) was oxidized overnight at 20°C in 25 mM Tris, 10 mM EDTA, 6 M GuHCl, pH 9.0, in the presence of different concentrations of reduced and oxidized glutathione. Glutathione was removed by dialysis against 25 mM Tris, 10 mM EDTA, 8 M urea, pH 8.0, followed by renaturation of FB-sFv[315] by dialysis against 166 mM H_3BO_3, 28 mM NaOH, 150 mM NaCl, 6 mM NaN_3, pH 8.2. Samples were analyzed by 15% SDS-PAGE. Oxidized samples were electrophoresced in the absence of reducing agent. Numbered lanes contain the following samples: (1) Standards ($Mr \times 10^{-3}$), 94, 67, 43, 29, 20, 14; (2) FB-sFv[315], reduced; (3) FB-sFv[315] treated in 6 M GuHCl solution with 10 mM

(continued)

chromatography at the final stage of purification,[20] we have found chromatography on Sephadex S-100 HR or Superdex-75 (Pharmacia LKB Biotechnology, Uppsala, Sweden) to be quite effective in specific cases.[16,18,19] While the affinity purification necessarily relies on interactions with active sFv binding sites, size-exclusion chromatography depends only on molecular size and aggregation state. Successful purification by size-exclusion chromatography thus requires that the most compactly folded protein recovers native structure and binding properties, which is often true, but must be tested empirically to determine its validity in any particular case. In studies with 741F8 sFv directed against c-*erb*B-2, we find that a size separation yields protein that is generally at least 95% active, and physicochemical measurements show essentially quantitative binding between a divalent form, 741F8 (sFv')$_2$, and the cloned extracellular domain of c-*erb*B-2.[16,21,22] Aggregates in the void volume tend to be denatured protein, although slow aggregation can occur for some sFv proteins.[18] Generally the aggregated protein is not of interest if the goal is to isolate discrete sFv monomers or dimers. Nonetheless, there are recent examples of tandem Fv dimers and multimers forming from sFv proteins, which suggests the prudence in checking aggregated protein for binding activity.[58,59]

Metal Chelate Affinity Chromatography. In many cases the antigen is not available for the sFv and there may be a low concentration of protein in the refolding solution. In these cases one of the simplest methods of sFv purification is metal chelate affinity chromatography,[60] also termed immobilized metal-chelate affinity chromatography, or IMAC. A sequence of five or six histidines is incorporated into the sFv, usually at its C-terminus, in our case between the end of the last V region and the *myc* peptide, when present.[19] This allows purification of the sFv on Zn^{2+} or Ni^{2+} columns, with elution by either imidazole or by low pH buffer. This method has the advantage that the sFv can be isolated in large amounts and under denaturing conditions, while it has the disadvantage that it is insensitive to whether or not the protein is properly folded. The His$_6$-peptide has been reported to be of relatively low immunogenicity,[61] but it could still be useful for *in vitro* studies even if it should pose problems *in vivo*; if the His$_6$ is problematical, an alternative mode of production can be developed, or if

---

Figure 7–7 (*continued*). oxidized glutathione; (4) FB-sFv[315] treated with 1 mM oxidized and 0.1 mM reduced glutathione, (5) FB-sFv[315] oxidized with 0.1 mM oxidized and 1 mM reduce glutathione. B. Unreduced mixture of 26-10 sFv oxidized by (2 and 3) air oxidation during dilution refolding, (4 and 5) glutathione during redox refolding, or (6 and 7) oxidation with glutathione in 6 M guanidine hydrochloride during disulfide-restricted refolding. C. Redox-refolded samples of oxidized FB-sFv[26-10] prior to affinity purification, run with or without reduction, as noted. [Reproduced with permission from Huston et al., *Methods in Enzymology* **203**, 46–88, 1991,[11] Academic Press, Inc.]

necessary, a proteolytic cleavage site can be inserted between the His_6 and the sFv for final processing. This procedure produces very pure sFv material, as determined by SDS-PAGE visualized with Coomassie blue and silver staining; an analogous approach has been used for purifying clinical-grade vaccine candidates.[61] However, in some cases, especially where the production of sFv is low, a single contaminating bacterial product can be seen at about 30 kDa, probably representing β-lactamase, which unfortunately migrates much the same on SDS-PAGE as a typical sFv. This contaminant is readily distinguished from sFv by Western blotting and probing with the anti-peptide antibody. Antigen-bonding activity of the sFv protein must be quantitated, of course, and inactive protein removed by other methods.

Storage of sFv Protein. We have examples that demonstrate that soluble monomeric sFv appears to be stable at 4°C in PBSA[6,16] or in 0.1 M Tris, 0.4 M arginine, 2 mM EDTA pH 8.0 buffer for several months. If necessary we dialyze the samples into PBS or an appropriate buffer immediately before use, with no apparent loss of activity. Indeed, the sFv appears to be stable in PBS, at a concentration of at least 0.5 mg/ml, for months. However, the solubility and tendency for self-association are known to vary for different sFv proteins, and the optimal storage conditions may likewise be variable.

Extinction Coefficients. The numbers of specific aromatic residues in variable regions are not constant because of the hypervariability of CDR loops. Differences in amounts of tryptophan and tyrosine between different Fv regions lead to a considerable range of extinction coefficients, which for sFv proteins lead to observe absorptivities at 280 nm in the range of 1.4 to 2 per mg/ml (for a 1 cm pathlength), and this range can vary more widely for sFv fusion proteins. Approximate calculations of extinction coefficient can be based on amino acid composition,[62] which is most easily determined as an adjunct of V gene sequencing, or by amino acid analysis; however, the native structure of folded proteins can make such predictions somewhat unreliable, and the empirical determination of extinction coefficients is desirable. One rapid method of measuring absolute concentration to calculate extinction coefficients involves using a Model E analytical ultracentrifuge equipped with a Rayleigh interferometer;[63] another approach for measuring protein concentration is quantitative amino acid analysis.[1]

Classical methods for determining extinction coefficients are still useful, since the availability of pure sFv protein is often not limiting. A normal day weight determination uses 15 to 20 mg of dialyzed protein at a concentration of about 1 mg/ml, and proceeds as follows. The absorption spectrum of the dialyzed sFv protein should be carefully determined from 400 nm to 240 nm, so that Rayleigh scattering corrections can be made, if needed. The mass of solute in aliquots of the sample (5 ml each) and dialysate (10 ml each) are measured at intervals during the drying process by standard methods of gravimetric analysis. The drying oven should be maintained at 107°C and the weighing bottles cooled for a standard time (30 min) in a desiccator before weighing in an analytical balance. As the evaporation is completed and the protein is dried further, the protein mass reaches a minimum but may then start to increase with further drying; the minimum value is used for the protein mass calculations. The extinction coefficient (ml mg^{-1} cm^{-1}) is calculated by dividing absorbance at the wavelength of interest for the stock sFv solution (in a 1 cm cuvette) by the actual mass of protein (mg/ml) in the sample, corrected for the weight of buffer solute present (all averaged from their triplicate determinations).

METHODOLOGY FOR sFv PROTEIN PRODUCTION

Construction of sFv Genes

Cloning of V Region Genes. The cloning of specific variable region genes is done mostly by PCR methods. For design of PCR primers and sequence predictions the sequence atlas of Kabat has been a most valuable resource.[64] PCR primers have been used that anneal to many V_H and V_L genes in segments encoding their FR1 and FR4 framework region sequences.[65] At the 5' end of V_H or V_L one may use primers for FR1, whereas at the 3' end one should initially use PCR primers for the constant regions. After the nucleotide sequence has been determined, one may engineer appropriate restriction sites in a second PCR reaction. The convenience of this procedure compensates for later effort needed to correct errors caused by using consensus primers. Nucleotides that generate erroneous amino-terminal residues for V_H or V_L should be corrected.[11] The authentic sequences may be determined by N-terminal sequencing of the H and L chains of the parent monoclonal antibody, or through the use of additional PCR primers that anneal within flanking signal sequences,[11] by anchored PCR,[66] or by cDNA library cloning.

Assembly of sFv Genes. The structural conservation and sequence homology of framework regions in variable domains invite general schemes for the assembly of single-chain Fv proteins. One may use PCR with overlapping primers for sFv assembly, but this will require relatively long primers. Instead, we have cloned and sequenced the linker in a pUC plasmid, adding V_H and V_L in two additional cloning steps. The following is one possible approach that allows easy assembly of both configurations, V_H–V_L and V_L–V_H, due to the design of suitable restriction sites at the boundaries of V_H and V_L by silent mutagenesis.

1. At the N-terminus of V_H or V_L we have placed an NcoI site, encoding methionine, required for direct expression depending on the configuration of the sFv (CC*ATG*GAG or CC*ATG*GAT).

2. At the C-terminal end of murine and human V_H, one typically finds two serine residues. An SacI site can be embedded in these codons (by silent mutagenesis) for the purpose of connecting with the single-chain linker. A stop codon placed behind these serines is useful if V_H is the second element in the sFv and it may generate an Afl-II site (GAGCTCT*TAA*G).

3. At the C-terminal boundary of V_L in murine and human kappa light chains, one often finds a Lys–Arg dipeptide sequence. For a V_H–V_L orientation, one can engineer an HpaI site overlapping with the arginine codon, which can supply the stop codon (CGT*TAA*C). For a V_L–V_H orientation, one can fuse the HpaI site of V_L to an SacI site blunted by treatment with T4 polymerase at the 5′ boundary of the linker and thereby convert the stop codon of this V_L into a serine residue (CGT*TCC*).

4. The synthetic oligonucleotide encoding the $(Gly_4Ser)_3$ single-chain linker contains an upstream SacI site and a downstrean NcoI site (CCATGG) (Fig. 7–8). This NcoI site can be used to attach V_H or V_L downstream of the single-chain linker. The methionine encoded by the NcoI site adds one C-terminal residue to the linker, which should be removed, preferably by site-directed mutagenesis.[67] Ausubel et al.[68] is a good source for molecular biology methodology used in these manipulations.

```
      VH-3'end                                                    VL-5'end
---GTGAGCTCCTAA                                              CCATGGATATC---
---ValSerSerstop                                               MetAspIle---
      SacI                                                    NcoI EcoRV


   GAGCTCCGGAGGTGGCGGGTCT GGTGGTGGAGGCAGC GGTGGTGGGGGATCCATGG
    SerSerGlyGlyGlyGlySer GlyGlyGlyGlySer GlyGlyGlyGlySerMet
   SacI                                                       NcoI
         BspEI                                             BamHI

      VL-3'end                                                    VH-5'end
---AAACGTTCC                                                 CCATGGAGGTG---
---LysArgSer                                                   MetGluVal---
      HpaI/SacI fused                                        NcoI

      VL-3'end
---AAACGTTAAC
---LysArgstop
      HpaI
```

Figure 7–8. A scheme of restriction sites for the ends of V_H and V_L is shown that allows construction of either sFv orientation, with V_H–V_L above the single-chain linker and V_L–V_H below it. Nucleotide sequences at the ends of V_H and V_L and separately of the $(Gly_4$-$Ser)_3$ linker are underlined. The corresponding amino acid residues are shown in bold print. The SacI and NcoI sites flanking the linker are shaded. The methionine residue, useful for direct expression and at the end of the linker, is shown in italics. After the sFv assembly this codon can be removed from the linker by site directed mutagenesis.

Cytoplasmic Expression of sFv Inclusion Bodies

Direct Expression. Direct expression of sFv in *E. coli* is most readily accomplished using the T7 system developed by Studier and Moffatt,[69] which is commercially available as pET-3d and related vectors (Novagen, Inc., Madison, WI). In this system the sFv gene is inserted into the expression vector between a transcriptional promoter and terminator derived from T7 bacteriophage. The plasmid is then transfected into a specially engineered *E. coli* host that contains the gene for T7 RNA polymerase, genomically integrated and under transcriptional control of the *lac* promoter. Induction of the *lac* promoter during fermentation by the addition of IPTG results in the expression of T7 RNA polymerase, which interacts with the T7 promoter to produce high levels of the mRNA encoding sFv. The translation of this foreign mRNA results, in most cases, in misfolded polypeptides which build up in cytoplasmic inclusion bodies.

The overproduction of sFv protein to form inclusion bodies can be readily monitored under the phase contrast microscope. At 1,000-fold magnification, using an oil immersion lens, intracellular inclusion bodies are visible as refractile bodies near the poles of cells which may contain one or more particles of varying size. This microscopic examination is instantaneous and may be used to monitor the status of a fermentation without need for SDS-PAGE until the fermentation has been completed. Large inclusion bodies are desirable, since their size facilitates isolation by low-speed centrifugation.

The T7 system performs reliably with most gene constructs, but the final expression level will depend on the protein. This may be due to various factors, such as sFv protein breakdown, limitations on the rate of synthesis caused by particular mRNA structure, or feedback effects of the gene product. Protein stability may be enhanced in some host strains deficient in cytoplasmic proteases, as is the case with *E. coli* BL21 (DE3), which is a host for T7 expression. However, the initial cloning of the expression plasmid should be done in a different strain, such as JM101, where the T7 promoter is silent due to the absence of the T7 polymerase gene.

Fusion Protein Expression. In most cases, sFv or sFv fusion protein can be obtained by direct expression using the T7 system. Higher yields of inclusion bodies may often be obtained by attaching a bacterial leader upstream from the sFv gene. The leader protein may also be useful for affinity chromatography and may be modified, for example, to contain a His$_6$-peptide for purification by metal chelate affinity chromatography. Unfortunately, isolation of sFv from fusion proteins requires a chemical or enzymatic cleavage step to remove the leader.[1] The presence of some leader proteins, such as the 58 residues FB, need not compromise antigen binding activity of sFv, as demonstrated for FB-sFv[26-10].[6] FB-sFv proteins will usually express at high levels, while the FB often moderates sFv solubility problems and facilitates affinity purification by IgG-Sepharose chromatography.[11] FB-sFv genes have been produced in combination with the *tac* promoter in JM101 containing the *lac*[lq] repressor. Fusion of only the first nine residues of the FB (ADNKFNKDP) may be used to elevate expression levels of a given sFv,[16] although without the Fc-binding properties of intact FB fusions. Another useful leader has been the 60 residue modified trp LE leader (MLE),[1] which is used in combination with the *trp* promoter;[1] 14 residue truncated form of the MLE, termed the SLE (MKAIFVLKGSL-DRDP), also promotes high expression levels.[3,15]

Vectors and Bacterial Strains. For direct expression, one may obtain various forms of the pET-3d vector, which contains the T7 promoter from commercial sources (Novagen, Inc., Madison, WI). Bacterial strains such as BL21 (DE3) are kept as frozen stocks in LB with 7% glycerol at $-70°C$; competent cells are made by the $CaCl_2$ method. Transformed colonies appear quickly and the JM101 strain can be infected by M13 phage. Hence, this strain is useful for preparation of templates from phagemid vectors. JM101 contains the *lac*[lq] repressor which permits expression with the *tac* promoter.

The vectors are derived from pBR322, pUC, or phagemid. The drug resistance marker for routine work is β-lactamase, which confers ampicillin resistance. For fermentations at larger scales, the tetracycline marker is preferred to avoid contamination with β-lactamase, which migrates on SDS-PAGE similarly to sFv. Modifications of the expression vector have involved, for example, cloning of the T7 promoter and terminator into pBR322 and replacing part of the β-lactamase gene.

Small-Scale Fermentation. For production of useable quantities of sFv protein, it is convenient to grow cultures of 1 or 2 liters in shaker flasks maintained in a shaker incubator at 37°C. Induction of protein expression should be done at a cell density of approximately 1 OD, by adding IPTG to a concentration of 1 mM. Inclusion bodies develop after about 2 hours and increase in size over the next few hours; in some cases one may lower the temperature to 30°C after the induction to improve the level of inclusion body formation.

Refolding Protocols for Cytoplasmic Inclusion Bodies

The following protocols are intended as a general starting points for refolding of directly expressed sFv. For any given sFv there is a progression of refolding procedures that one may test, given a reliable activity assay, which is an invaluable adjunct to this approach. In the case of cytoplasmically expressed protein, any oxidation of disulfides within the inclusion body tends to randomly crosslink polypeptides, and thus all procedures start with complete reduction and denaturation of the inclusion bodies. This contrasts with the vectorial process of periplasmic secretion, discussed in later sections, where proper disulfides should be formed during secretion and hopefully retained in periplasmic inclusion bodies that may form as the primary end product.

I. *General Protocol for Redox Refolding of sFv from Reduced and Denatured State.* The following procedure applies to sFv proteins that refold by dilution or redox-refolding methods[11] from the fully reduced random coil state. This is based in part on procedures developed for 26-10 sFv made in the T7 expression system; after redox refolding and affinity chromatography, a 1 liter fermentation typically yields 130 mg of active 26-10 sFv protein, which is soluble and well-behaved in its solution properties at neutral pH.[18]

1. The inclusion bodies should be washed twice by resuspension in 25 mM Tris-HCl, pH 8, and 10 mM EDTA followed by brief sonication and centrifugation at 14,300 g, 4°C for 30 min.

2. The pellet is dissolved in 6 M GuHCl, 25 mM Tris HCl, pH 8.7, 10 mM EDTA, 10 mM DTT, and incubated at room temperature overnight, and recentrifuged before refolding the reduced and denatured protein. Inasmuch as 6 M GuHCl is 50.2% GuHCl by weight, a convenient way of correcting for the weight of water in the inclusion body pellet is to add additional solid ultrapure guanidine (Heico, Inc., Delaware Water Gap, PA) equal to the weight of the inclusion body pellet being dissolved. Thus the final guanidine concentration will be accurately maintained at 6 M, regardless of the properties of the inclusion bodies, such as the percentage of sFv protein in the pellet.

3. The reduced and denatured sFv solution may be refolded by dilution into 3 M urea, 25 mM Tris-HCl, pH 8.1, 1 mM oxidized glutathione, 0.1 mM reduced glutathione to a final protein concentration of 0.01 to 0.1 mg/ml (occasionally much higher concentrations are workable). The refolding solution should be incubated in the cold for 16 to 18 hrs.

4. The 3 M urea-redox buffer may be removed by dialysis against PBS (50 mM potassium phosphate, 0.15 M NaCl, pH 7) for two days in the cold.

5. Isolation of the active protein may be conveniently accomplished by affinity chromatography, if applicable, or by size exclusion chromatography on Sephacryl S-100 HR or similar resins. In the latter case, aggregated protein will elute in void volume, while active sFv should be the dominant part of the material eluting at the position of monomer or dimer, depending on the self-association properties of the given sFv.

II. *General Protocol for Disulfide-Restricted Refolding of sFv from Reduced and Denatured State.* The following procedure applies to sFv proteins that do not refold spontaneously from the fully reduced random coil state, but only after reoxidation of disulfide bonds. This procedure is based in part on that described by McCartney et al. for MOPC 315 sFv.[7] To maximize the amount of correct starting material for disulfide-restricted refolding of sFv proteins, iterative experiments are usually needed to find reasonably optimal conditions. Testing both urea and GuHCl as denaturants is desirable, as is checking a range of denaturant concentrations and oxidation conditions. A similar disulfide-restricted folding method has been described for recombinant Fab fragments,[70] and Kurucz et al. used this method to recover an active sFv analogue to the 2B4 T cell receptor.[23] The following protocol should thus be considered only a reasonable starting point.

1. Preparation of inclusion bodies. Inclusion bodies should be washed and solubilized as in Protocol 1, steps 1 and 2 above. The inclusion bodies should be washed twice by resuspension in 25 mM Tris-HCl, pH 8, and 10 mM EDTA followed by brief sonication and centrifugation at 14,300 g, 4°C for 30 min.

2. The pellet is dissolved in 6 M, GuHCl, 25 mM Tris HCl, pH 8.7, 10 mM EDTA, 10 mM DTT, incubated at room temperature overnight, and recentrifuged before dilution and oxidation.

3. The concentrated inclusion body solution should be adjusted to 0.05 mM DTT in 6 M GuHCl solution before oxidation. This may be accomplished by dialysis against several changes of appropriate small volumes of buffer or by 200-fold dilution of the material from II-3 into 6 M GuHCl, 25 mM Tris-HCl, pH 8, 10 mM EDTA. Sulfhydryl oxidation is accomplished by making the solution 1.05 mM in oxidized glutathione to yield a final redox buffer of 1 mM oxidized glutathione and 0. 1mM reduced glutathione. The sFv protein (at 0.1 to 1 mg/ml, again depending on optimization) should be oxidized overnight at 20°C.

4. Glutathione should be removed by dialysis against 8 M urea, 25 mM Tris-HCl, pH 8, 10 mM EDTA, before denaturation by dialysis against or dilution into aqueous solution, such as PBSA (0.15 M NaCl, 0.05 M potassium phosphate, pH 7.0, 0.03% NaN_3) or BSB (166 mM H_3BO_3, 28 mM NaOH, 150 mM NaCl, 6 mM NaN_3, pH 8.2).

5. Active protein may be monitored by an appropriate assay, such as an ELISA or dot blot procedure, and for preparative purposes isolated by an appropriate purification method, as discussed above under *Purification of Native sFv Protein.*

6. Enrichment of active sFv by limited proteolysis of refolded protein mixture. Pepsin digestion has been used to enrich active sFv core protein from FB-sFv315,[7] and this procedure should be applicable to other sFv proteins, although possibly with a different choice of protease. Thus as one form of this general approach, the refolded sFv

mixture at 0.1 to 1 mg/ml should be digested with 2.5% pepsin (5 µg/ml) for a 0.2 mg/ml sFv solution) in 50 mM sodium acetate, 50 mM NaCl, pH 4.6, for 1.5 hr at 37°C (obviously, aliquots should be examined as a function of time). The reaction is stopped by addition of 0.2 volumes of 1.5 M Tris, pH 8.8, followed by dialysis against BSB. SDS-PAGE and activity assays should be used to monitor the effect of limited proteolytic enrichment of native sFv protein.[7]

III. *Specific Protocols for Refolding 26-10 sFv and FB-sFv²⁶⁻¹⁰ from Reduced and Denatured States.* The following purification protocols and procedures are reproduced and modified from Huston et al.[11] with permission. These particular examples specifically involve various protocols for the preparation of 26-10 single-chain Fv species. These procedures all share a common starting point, in that the recombinant protein initially was derived from *E. coli* inclusion bodies.[1,6,11] The cell paste for small-scale preparation was suspended in 25 mM Tris-HCl, pH 8, and 10 mM EDTA, treated overnight with 0.1 mg lysozyme per ml solution, sonicated at a high setting for three 5-minute periods in the cold, and spun in a preparative centrifuge at 11,300 × *g* for 30 minutes. The pellet was then washed with a buffer containing 3 m urea, 25 mM Tris-HCl, pH 8, and 10 mM EDTA; after recentrifugation, the protein pellet served as the starting material for each of the following protocols, or it was stored at −20°C until use. For large-scale preparation of inclusion bodies, cells are concentrated by ultrafiltration, lysed with a laboratory homogenizer (model 15MR, APV Gaulin, Inc.), and inclusion bodies are collected by centrifugation. The following are examples of sFv refolding from the fully reduced random coil state (reproduced and modified from Huston et al.,[11] with permission).

A. Preparation of 26-10 sFv by redox refolding. This differs from Procedure I in being designed specifically for acid-cleaved 26-10 sFV fusion protein, stored as an ethanol precipitate. When applied, for example, to directly expressed sFv protein, this protocol offers the advantage of refolding relatively pure sFv protein after ion exchange chromatography, which removes DNA and may improve the final yield of native sFv.

1. Transfer of sFv protein into urea buffer for ion exchange chromatography. Dissolve the directly expressed sFv (or ethanol-precipitated sFv cleavage mixture, as in early studies) in 6 M GuHCl, 0.2 M Tris-HCl, pH 8.2, with addition of solid GuHCl equal to the pellet weight; adjust pH if needed before adding 2-mercaptoethanol to 0.1 M. Dialyze this solution into starting buffer for ion exchange chromatography consisting of 6 M urea, 2.5 mM Tris-HCl, pH 7.5, 1 mM EDTA, 5 mM dithiothreitol.

2. Removal of uncleaved fusion protein from 26-10 sFv. Chromatograph

solution on a column of Whatman DE 52 equilibrated in starting buffer III(A-1). SDS-PAGE is used to monitor purity of the flow-through protein, and leading fractions that contain pure 26-10 sFv are pooled.

3. Renaturation of 26-10 sFv. Starting with the DE 52 pool of 26-10 sFv in 6 M urea buffer III(A-2), dilute to a concentration below 0.1 mg/ml with 3 M urea, 25 mM Tris-HCl, pH 8, 10 mM EDTA, 1 mM oxidized glutathione, 0.1 mM reduced glutathione, and incubate at room temperature for 16 hours.

4. Removal of redox couple from 26-10 sFv. Dialyze in the cold against 3 M urea, 25 mM Tris-HCl, pH 8, 10 mM EDTA.

5. Removal of denaturant from 26-10 sFv. Dialyze in the cold for two days against 0.01 M sodium acetate, pH 5.5, 0.25 M urea.

6. Affinity isolation of 26-10 sFv. Follow step III(C-6) for ouabain-amine-Sepharose affinity chromatography, adding 0.25 M urea to all buffers.

B. Purification of FB-sFv$^{26-10}$ fusion protein (Fig. 7–6).

1. Solubilization and dialysis of inclusion bodies. Dissolve 6 M GuHCl, 0.2 M Tris-HCl, pH 8.2, 0.1 M 2-mercaptoethanol, and incubate at room temperature for 1.5 hours. Dialyze the denatured and reduced inclusion body protein into 6 M urea, 2.5 mM Tris-HCl, pH 8, 1 mM EDTA, 5 mM dithiothreitol.

2. Purification of FB-sFv$^{26-10}$ in denaturant. Chromatograph inclusion body protein on a column of Whatman DE 52. Pool flow-through fractions that are free of contaminants according to SDS-PAGE analysis.

3. Renaturation of FB-sFv$^{26-10}$ by redox refolding procedure. Follow step III(A-3).

4. Removal of denaturant and redox couple. The 3 M urea buffer and glutathione redox couple are removed in one step by dialysis for two days against 50 mM potassium phosphate, pH 7, 0.25 M urea.

5. Affinity isolation of FB-sFv$^{26-10}$. Follow step III(C-6), except that all column buffers contain PBSA instead of acetate buffer, and the final dialysis to remove ouabain is against PBSA.

C. Purification of 26-10 sFv by acid cleavage of MLE-sFv$^{26-10}$ fusion protein.[1,11]

1. Solubilization of fusion protein in acid cleavage buffer (6 M GuHCl, 10% acetic acid, pH 2.5). Dissolve MLE-sFv$^{26-10}$ inclusion bodies in 6.7 M GuHCl, and then dissolve solid GuHCl equal to the weight of the inclusion body pellet. Add glacial acetic acid to 10% of total volume and adjust to pII 2.5 with concentrated IICl.

2. Cleavage of the unique Asp-Pro bond at the junction of leader and 26-10 sFv. Incubate the solution from III(C-1) at 37°C for 96 hours. Stop reaction by adding 9 volumes of cold ethanol, storing at -20°C for several hours, followed by centrifugation to yield a pellet of precipitated 26-10 sFv and uncleaved fusion protein. The cleavage mixture may also be stored as a pellet at -20°C until needed.

3. Transfer of cleaved protein into urea buffer for ion exchange chromatography. Dissolve the precipitated cleavage mixture in 6 M GuHCl, 0.2 M Tris-HCl, pH 8.2, with addition of solid GuHCl equal to the pellet weight; adjust pH if needed before adding 2-mercapoethanol to 0.1 M. Dialyze this solution into starting buffer for ion exchange chromatography consisting of 6 M urea, 2.5 mM Tris-HCl, pH 7.5, 1 mM EDTA, 5 mM dithiothreitol.

4. Removal of uncleaved fusion protein from 26-10 sFv. Chromatograph solution on a column of Whatman DE 52 equilibrated in starting buffer III(C-3). SDS-PAGE is used to monitor purity of the flow-through protein, and leading fractions that contain pure 26-10 sFv are pooled.

5. Renaturation of 26-10 sFv by the dilution refolding procedure. Dilute DE 52 pool of 26-10 sFv at least 100-fold into 0.01 M sodium acetate, pH 5.5, to a concentration below 10 µg/ml, and dialyze at 4°C. The two intrachain disulfides reform by air oxidation of the four cysteinyl residues in 26-10 sFv.

6. Affinity purification of the anti-digoxin 26-10 sFv from the mixture of refolded species. The sample from step III(C-5) is loaded onto a column containing ouabain-amine-Sepharose 4B, and the column is washed successively with 0.1 M sodium acetate, pH 5.5, followed by two column volumes of 1 M NaCl, 0.01 M sodium acetate, pH 5.5, and then 0.01 M sodium acetate once again to remove salt. Finally, the active protein is displaced from the resin by 20 mM ouabain in 0.01 M sodium acetate, pH 5.5. Absorbance measurements at 280 nm indicate which column fractions contain active protein. However, the spectra of the protein and ouabain overlap, and ouabain must be removed by exhaustive dialysis against 0.01 M sodium acetate, pH 5.5, 0.25 M urea, in order to accurately quantitate the protein yield. If the protein is to be kept at 4°C for long periods and is not intended for animal use, sodium azide is added to a concentration of 0.03%. Alternatively, the protein may be lypholized and stored at -20°C.

D. Purification of 26-10 sFv from FB-sFv$^{26\text{-}10}$ or FB-FB-sFv$^{26\text{-}10}$

1. Cleavage of fusion protein. Since crude inclusion bodies will not dissolve in acetic acid, the purified or at least refolded fusion protein

offers the starting material for this procedure. The purified and refolded fusion protein, prepared according to steps III(B-1) to III(B-5), is the starting material for this procedure. The active fusion protein in PBSA is mixed with 0.11 volumes of glacial acetic acid to make a 10% solution, and the solution is incubated at 37°C for 96 hours. Cleavage in 10% acetic acid results in acid hydrolysis of the Asp-Pro bond at leader-sFv junction; the fragment B has two acid labile bonds between residues FB-36 to FB-38 (Asp-Asp-Pro), which are not cleaved in 10% acetic acid but are partially hydrolyzed in 6 M GuHCl, 10% acetic acid, pH 2.5. The cleavage mixture is ethanol precipitated as noted under III(C-2), and following centrifugation the pelleted material can be stored at −20°C until use.

2. Renaturation of proteins in cleavage mixture. Since the intradomain sFv disulfides were preformed, this step would also apply to periplasmic sFv, insofar as this refolding protocol takes advantage of the correct disulfides being present. The pellet is dissolved in 6 M GuHCl, 25 mM Tris-HCl, pH 8, 10 mM EDTA, dialyzed into 6 M urea, 25 mM Tris-HCl, pH 8, 10 mM EDTA. The dialysate is diluted to 3 M urea, 25 mM Tris-HCl, pH 8, 10 mM EDTA, followed by dialysis against the 3 M urea buffer in the cold for two days. Alternatively, renaturation can be performed on the fully reduced random coil, but one loses the advantage afforded by intact disulfides.

3. Removal of uncleaved fusion protein. The remaining intact fusion protein is removed from the dialyzed cleavage mixture by affinity chromatography on human IgG-Sepharose in 3 M urea, 25 mM Tris-HCl, pH 8, 10 mM EDTA. Under these conditions aggregation is negligible, but the FB or FB-FB effector domains bind to the immobilized immunoglobulin. The flow-through fractions contain the purified sFv. The column can be successfully regenerated after elution with 0.2 M glycine, pH 2.9, followed by extensive washing with PBSA.

4. Removal of denaturant from 26-10 sFv. Dialyze in the cold for two days against 0.01 M sodium acetate, pH 5.5, 0.25 M urea.

5. Affinity isolation of 26-10 sFv. Follow step III(C-6) for ouabain-amine-Sepharose affinity chromatography, adding 0.25 M urea to all buffers.

Periplasmic sFv Inclusion Body Methodology

Numerous recombinant forms of immunoglobulin fragments or domains have been produced in active form using periplasmic expression systems, including Fab fragments,[71] F(ab′)₂ dimers,[72] Fv fragments,[73] and single-chain Fv analogues of antibodies[60] and T cell receptors,[75] as well as isolated V_H domains.[76] The requisite for such a system is the use of a signal peptide at the amino terminus of the protein that directs its synthesis into the periplasm. A number of such N-terminal sequences have been used for the expression of sFv, including *pel*B (derived from the pectate lysase gene), the Omp (outer member protein) family of sequences, and leader sequences derived from periplasmic proteins such as alkaline phosphatase and the 5' end of the gene III of filamentous phage.[9] It is not clear, at present, whether any of these leader sequences are better than the others. Recombinant sFv have also been made as fusion proteins with staphylococcal protein A[77] and with bacterial alkaline phosphate,[78] the leader of which directs secretion to the periplasm. Following insertion into the periplasm, the signal sequence is removed by proteases. The sequences recognized by the proteases lie within the signal sequences, and the N-terminal sequence of the mature protein is not critical.[79] One advantage of producing proteins by secretion is that the N-terminal amino acid of the mature protein need not be a methionine. Correct processing of the sFv can be checked by N-terminal amino acid sequencing of the isolated molecule.

Expression of sFv in periplasmic expression systems has been used in two major settings. The first of these, which is outside the scope of this review, is to make the sFv as a fusion protein with the gene III or gene VIII products of filamentous phage (reviewed by George[80]). The synthesis of the fusion proteins is directed into the periplasmic space, so when phage particles are assembled and extruded through the cell membranes, they incorporate the sFv fusion proteins as part of their viral coat.[9] The expression of refolded, active antibody fragments on the surface of the phage allows for affinity selection of novel antibodies. The second major use of periplasmic expression systems has been to make recombinant antibody for experimental use. The simplicity of the system has led to its adoption by a number of groups, who have used a range of leader sequences and promoters to direct synthesis of the sFv to the periplasm. Some groups have reported that sFv made in the periplasm can be isolated from the bacterial culture medium in high yield, presumably due to leakage through the outer membrane of the cells.[76,81] However, this is not a universal finding, as we have shown that, using the same vector and conditions, some sFv proteins can be isolated from the culture supernatant while others cannot.[19] The appearance of sFv in the culture supernatant appears, therefore, to be dependent on the individual antibody variable domains.

The low or undetectable levels of sFv secretion into the culture supernatant have meant that a number of groups have developed ways of isolating antibody from the cells. The simplest way to do this is to solubilize the bacterial periplasm, either by hypo-osmotic shock or by lysozyme digestion, leaving the bacteria intact.[74] If the sFv is soluble, this allows the molecule to be isolated from the supernatant, but in many cases it has proven impossible to obtain sFv fragments by this means, as the protein appears to be insoluble. The insolubility of an sFv may arise either because the secretory pathways of the cell are overloaded, or because high local concentrations of sFv in the periplasm lead to aggregation and precipitation. Indeed, the soluble sFv is frequently isolated after relatively short induction periods (45 minutes),[74] possibly preventing build-up of high concentrations of protein.

If it does not prove possible to obtain solution sFv from either the culture supernatant or the periplasm, then the sFv may need to be recovered from insoluble periplasmic protein by solubilizing and refolding the sFv protein. A number of protocols have been described for cytoplasmic inclusion bodies, and we give Protocol IV that we have found useful in our laboratory for periplasmic sFv inclusion bodies,[19] which others have approached with large-scale techniques that are somewhat more complex.[20] All these protocols contain three fundamental steps: first, solubilization of the protein, normally using a strong denaturing agent such as 6 M GuHCl or a weaker one such as 8 M urea; second, refolding of the protein; and finally, isolation of active material. The refolding step is simpler than from traditional inclusion bodies, as it does not involve reduction of the molecule followed by the formation of disulfide bonds, as these have already formed in the periplasm.

Vectors and Methods for Expression of Periplasmic Inclusion Bodies

We have used standard techniques to construct the sFv. They have been cloned into one of two vectors, either pSW1 V_H poly TAG1 or pHEN1 (both kind gifts of Dr. G. Winter, Laboratory of Molecular Biology, Cambridge University);[76,81] these particular vectors or related ones may be obtained commercially from Cambridge Antibody Technology (Cambridge, UK), and phage antibody kits are also available from Pharmacia-LKB (Uppsala, Sweden). The sFv genes have been cloned downstream of DNA encoding the PelB leader sequence, and immediately upstream of DNA that encodes a short peptide tag derived from c-*myc*. An antibody (*myc*1 9E10.2—available from the ATCC) has been raised against the peptide, which can be used to monitor the production and purification of the sFv by Western or dot blotting. Both vectors carry the β-lactamase gene conferring ampicillin resistance and use the *lacZ* promoter for the production of recombinant protein. We do not have direct experience with other vectors using the refolding scheme to that we will describe; however, others who have used our method report success with different vectors (R. Spooner and P. Savage, personal communications), and we have no reason to believe that this methodology could not be applied to any periplasmic expression system.

We have used JM101 cells for the cloning and expression of sFv in the pSW1 V_H poly TAG1 vector.[19] The pHEN 1 vector, which is designed to be used in a phage display system, allows the construction of 5'-sFv-*myc*-[amb]-gene III-3' in a provisionally open reading frame, with an amber stop codon interrupting the sequence between the *myc* and gene III sequences. The cloning and manipulation of the plasmid is done in TG1 cells (which are supE, and so produce a fusion protein between the sFv and the gene III product). However, for production of the sFv the plasmid is placed into HB2151 cells, which are supE negative, and so stop translation at the amber codon. The sFv-pHEN plasmids are maintained in TG1 cells, and are transferred to HB2151 cells immediately prior to expression. This can be done either by conventional transformation techniques, or more conveniently by infection with phage derived from the TG1 host cells.[68]

Maintenance of Bacteria. Owing to the leakiness of the *lacZ* promoter in these vectors, all the plasmid-carrying vectors are grown and maintained in broth (LB or 2X TY as appropriate) containing 1% glucose. This represses the *lac* promoter, thereby preventing background production of the sFv which, as it is somewhat toxic to the cells, can lead to selection of variants.

Induction of Protein Production. An overnight culture of HB2151 cells containing the appropriate plasmid are added to 500 to 1000 ml of 2X TY medium containing 50 µg/ml ampicillin and 1% glucose. The flasks are grown, with vigorous shaking, at 37°C until the $OD_{600} = 0.8$. At this time the cultures are spun down and washed in either PBS or LB broth. They are then resuspended in the same volume of 2X TY containing 1 mM IPTG and ampicillin (no glucose). The cultures are then shaken overnight at room temperature. An alternative protocol for induction is called for in some situations where the washing of the cells prior to induction with IPTG may be difficult. In such cases it may be possible to grow the bacteria in medium containing just 0.1% glucose. When the bacteria reach $OD_{600} = 0.8$, add IPTG to 1 mM and transfer the flask to room temperature to complete induction.

Time and Temperature Optimization. To maximize the yields of sFv, and to solve expression problems, it can be useful to optimize the time and temperature of the induction. Small cultures (1 to 3 ml) of the cells are grown and induced with 1 mM IPTG at either 37°C or at room temperature for varying lengths of time (1, 2, 4, and 16 hours). The cultures are then spun down and the pellet and supernatants tested for the presence of sFv by either Western blotting and probing with the anti-*myc* antibody, or by dot blotting small samples (1 to 5 µl) onto nitrocellulose and probing with the same antibody.

Isolation of Secreted sFv from Culture Medium. In cases where the active sFv is found in the culture medium, it is possible to isolate it by simply spinning down the cells, passing the supernatant through a 0.45 µm filter, and using a suitable purification protocol for the sFv, such as antigen or metal chelate affinity chromatography.

IV. Refolding Protocols for Periplasmic sFv Inclusion Bodies. These procedures are designed to take advantage of the presence of intact disulfides in at least some of the sFv expressed as periplasmic inclusion bodies. Thus recourse may also be taken to parts of Protocols I through III, with the modification that no reducing agent or redox couple should be used. The following procedure seems to be particularly effective in recovering active sFv from periplasmic inclusion bodies, even if these other procedures prove ineffective.

A. Inclusion bodies isolated from the cell pellet.
1. Resuspend pellet in cold 50 mM Tris, 1 mM EDTA, 100 mM KCl, 0.1 mM PMSF pH 8.0 (200 ml/liter original buffer).
2. Place cells in "Beadbeater" (Biospec Products, Bartlesville, OK) with 50% (v/v) 0.1 mm glass beads. Cool apparatus with ice and lyse using three 1 min pulses separated by 1 min of cooling time between pulses. Alternatively, the cells can be lysed by sonication using a suitable probe sonicator.
3. Spin samples down (25,000 g, 20 min at 4°C).
4. Resuspend the pellet in 100 ml 7.5 M GuHCl (alternatively, one may try 9 M urea instead of GuHCl). Gently mix at room temperature for 1 to 2 hours. For preparation of material by metal chelate affinity chromatography, see Protocol V.
5. Spin 25,000 g, 20 min at 4°C (keep at 20°C if using urea, as it can crystallize in the cold).
6. With several changes of buffer, dialyze the supernatant against 0.1 M Tris, 0.4 M arginine, 2 mM EDTA pH 8.0 at 4°C. (N.B., when using 8 M urea, start dialysis with buffer at room temperature to prevent precipitation of urea.) During this time a white flocculent precipitate will build up in the dialysis tubing; this normally does not contain the sFv material. The arginine may act as a stabilizing co-solvent, encouraging the correct refolding of the protein.[82] The use of arginine has previously been noted to increase the yield of sFv-toxin fusion proteins refolded from bacterial inclusion bodies, apparently by suppressing the formation of large covalent aggregates.[83]
7. Spin at $25,000 \times g$ for 20 min at 4°C.
8. Purify sFv from supernatant using appropriate techniques. (N.B., When working up a new sFv we keep all pellets and supernatants to check, by dot blotting, for the presence of sFv material. To date we have not found significant levels of material in fractions to be discarded.)

B. Inclusion bodies isolated by metal chelate affinity chromatography.
1. Resuspend pellet in cold 50 mM Tris, 1 mM EDTA, 100 mM KCl, 0.1 mM PMSF pH 8.0 (200 ml/liter original buffer).
2. Place cells in "Beadbeater" (Biospec Products, Bartlesville, OK) with 50% (v/v) 0.1 mm glass beads. Cool apparatus with ice and lyse using three 1-min pulses separated by 1 minute of cooling time. Alternatively, the cells can be lysed by sonication using a suitable probe sonicator.
3. Spin samples down (25,000 g 20 min at 4°C).
4. Resuspend pellet in 6 M GuHCl, 0.1 M NaH_2PO_4, pH 8.0.
5. Purify sFv material by metal chelate affinity chromatography (see below).
6. Dialyse purified material against buffer containing 0.4 M arginine, as described above.

V. Protocol for sFv Purification by Metal Chelate Affinity Chromatography. In initial studies we attempted to isolate sFv that had already been refolded, and so was present in the 0.1 M Tris, 0.4 M arginine 2 mM EDTA pH 8.0 buffer. However, this failed as it appeared that the arginine buffer leached the Ni^{2+} off the immobilized chelate on the column. We now isolate the material while it is in denaturing solution, following removal of the insoluble material. This protocol was adapted, with permission, from the manufacturer's booklet (*The QIAexpressionist*, 1992. Qiagen, Chatsworth, CA) and has proven suitable for purifying sFv fusion proteins containing an His_6-peptide between the C-terminal end of the sFv and the *myc* peptide.[19]

1. Wash 5 ml Ni^2+-NTA beads (Qiagen, Chatsworth, CA or Hilden, Germany) with 6 M GuHCl, 0.1 M NaH_2PO_4, pH 8.0.
2. Incubate beads with material containing sFv-H_6 at 4°C for 2 hours (gently shaking or tumbling).
3. Spin samples down (25,000 g, 20 min at 4°C) and wash beads 2× in 6 M GuHCl, 0.1 M NaH_2PO_4, pH 8.0. (This step is optional, but it can speed up preparation of column).
4. Add beads to empty column, wash column 2× column volume 6 M GuHCl, 0.1 M NaH_2PO_4, pH 8.0.
5. Elute with step gradient of imidazole (10, 50, 100, 150, 200, 500 mM) in 6 M GuHCl, 0.1 M NaH_2PO_4, pH 8.0.
6. Check eluted fractions by dot blotting for the presence of sFv and pool fractions containing sFv, which normally elutes between 50 and 150 mM imidazole, and refold according to Protocol IV.
7. Wash beads once with 1 M imidazole, 6 M GuHCl, 0.1 M NaH_2PO_4, pH 8.0.
8. Wash beads three times with 6 M GuHCl, 0.1 M NaH_2PO_4, pH 8.0, store at 4°C in this buffer. The beads can be reused for the purification of the same sFv.

Isolation of Active sFv. Although the above procedures are capable of isolating pure sFv material, not all the material is likely to be active. Indeed, for U7.6 sFv we have shown that the majority of the periplasmically secreted sFv is aggregated after refolding and does not show specific activity. Much of this material is crosslinked by disulfide bonds, which probably result from high levels of sFv production overwhelming the secretory apparatus of the cell. Where antigen is available, the problem can be overcome by affinity purification. However, we have found that simple size-exclusion chromatography separates aggregated material from monomeric sFv, and it can be used to isolate a population of active sFv. In initial studies, using an sFv directed against the hapten DNP, we were able to demonstrate that all the material present in the monomeric peak was active sFv.[19] We have used this technique to isolate active OKT9 sFv with specificity for the human transferrin receptor.[19] For size-exclusion chromatography, we have used a 1.6 × 50 cm Superdex 75 column (Pharmacia LKB Biotechnology) on a Pharmacia FPLC system. The column is run in 0.1 M Tris, 0.4 arginine, 2 mM EDTA pH 8.0. The flow rate is 2 ml/min and 4 ml fractions are collected. The protein peaks can be determined by OD_{280}, while the location of the sFv material can be determined by dot blotting.

CONCLUSION

In this chapter we have outlined relatively simple methods for refolding sFv made as inclusion bodies in the cytoplasm or periplasm of *E. coli*. These methods are often capable of producing relatively high yields of active sFv (5 to 130 mg active sFv/liter fermented cells), and can readily be adapted to allow the isolation of active sFv with any specificity. The yields will depend on the nature of the sFv produced, but may potentially be enhanced by optimizing fermentation conditions. A major problem with these refolding methods is that prior to fractionation, the sFv produced is in active and inactive forms. Some may be aggregated, either by interchain disulfide bonds, nonvalent aggregation, or specific formation of tandem Fv interactions that yield dimers or higher polymers.[58,59] However, we have described methods that frequently overcome formation of these side-products by the application of dilute folding conditions and size-exclusion chromatography. In particular cases, it may prove possible to increase the proportion of correctly folded material by fine adjustment of refolding conditions, and for periplasmic secretion either by coexpressing chaperonin molecules or by using different promoters. Particular folding methods will vary in their utility for specific sFv proteins, but there should be continued use of sFv folding methodology, for example, to complement secretion methods and to produce fusion proteins that may be toxic to their host cell. In addition, a recent investigation has shown that genetically altered forms of *E. coli* may allow cytoplasmic disulfide bond formation of expressed proteins,[84] which may yield significant improvements in the refolding yields of single-chain Fv analogues and fusion proteins.

Acknowledgments

The authors gratefully acknowledge the long-term support of the National Cancer Institute (National Institutes of Health, U.S. Public Health Service), particularly from the Biological Response Modifiers Program of the NCI for research at Creative BioMolecules (grants CA 39870, CA 52323, and U01 CA 51880). Additional funding was provided by our respective institutions and the Bristol-Myers Squibb Pharmaceutical Research Institute. We are also indebted to our colleagues on the anti-digoxin sFv program, Professor Edgar Haber (Harvard School of Public Health, Harvard Medical School, and Massachusetts General Hospital) and Dr. Meredith Mudgett-Hunter (Massachusetts General Hospital and Harvard Medical School).

REFERENCES

1. Huston, J.S., Levinson, D., Mudgett-Hunter, M., Tai, M.-S., Novotny, J., Margolies, M.N., Ridge, R.J., Bruccoleri, R., Haber, E., Crea, R., and Oppermann, H. 1988. Protein engineering of antibody binding sites: Recovery of specific activity in an anti-digoxin single-chain Fv analogue produced in *Escherichia coli. Proc. Natl. Acad. Sci. USA* 85:5879–83.
2. Bird, R.E., Hardman, K.D., Jacobson, J.W., Johnson, S., Kaufman, B.M., Lee, S.-M., Lee, T., Pope, S.H., Riordan, G.S., and Whitlow, M. 1988. Single-chain antigen-binding proteins. *Science* 242:423–26.
3. Huston, J.S., McCartney, J., Tai, M.-S., Mottola-Hartshorn, C., Jin, D., Warren, F., Keck, P., Oppermann, H. 1993. Medical applications of single-chain antibodies. *International Reviews of Immunology* 10:195–217.
4. Huston, J.S., Keck, P., Tai, M.-S., Jin, D., McCartney, J., Stafford III, W.F., Mudgett-Hunter, M., Oppermann, H., Haber, E. 1993. Single-chain immunotechnology of Fv analogues and fusion proteins. In *Immunotechnology*. Gosling, J., and Reen, D., eds. London: Portland Press, pp. 47–60.
5. Huston, J.S., Tai, M.-S., McCartney, J., Keck, P., and Oppermann, H. 1993. Antigen recognition and targeted delivery by the single-chain Fv. *Cell Biophysics* 22:189–224.
6. Tai, M.-S., Mudgett-Hunter, M., Levinson, D., Wu, G.-M., Haber, E., Oppermann, H., and Huston, J.S. 1990. A bifunctional fusion protein containing Fc-binding fragment B of staphylococcal protein A amino terminal to antidigoxin single-chain Fv. *Biochemistry* 29:8024–30.
7. McCartney, J., Lederman, L., Drier, E., Cabral-Denison, N., Wu, G.-M.,

Batorsky, R., Huston, J.S., and Oppermann, H. 1991. Biosynthetic antibody binding sites: Development of a single-chain Fv model based on anti-dinitrophenol IgA myeloma MOPC 315. *Journal of Protein Chemistry* 10:669–83.

8. Chaudhary, V.K., Queen, C., Junghans, R.P., Waldmann, T.A., FitzGerald, D.J., and Pastan, I. 1989. A recombinant immunotoxin consisting of two antibody variable domains fused to *Pseudomonas* exotoxin. *Nature* 339:394–97.

9. McCafferty, J., Griffiths, A.D., Winter, G., and Chiswell, D.J. 1990. Phage antibodies: filamentous phage displaying antibody variable domains. *Nature* 348:552–54.

10. Savage, P., So, A., Spooner, R.A., and Epenetos, A.A. 1993. A recombinant single chain antibody interleukin-2 fusion protein. *Br. J. Cancer* 67:304–10.

11. Huston, J.S., Tai, M.-S., Mudgett-Hunter, M., McCartney, J., Warren, F., Haber, E., and Oppermann, H. 1991. Protein engineering of single-chain Fv analogues and fusion proteins. In *Molecular Design and Modelling: Concepts and Applications, Part B*. Langone, J.J., ed. *Methods in Enzymology* 203:46–88.

12. Marks, J.D., Hoogenboom, H.R., Bonnert, T.P., McCafferty, J., Griffiths, A.D., and Winter, G. 1991. By-passing immunization: Human antibodies from V-gene libraries displayed on phage. *J. Mol. Biol.* 222:581–97.

13. Fuchs, P., Breitling, F., Dübel, S., Seehaus, T., and Little, M. 1991. Targeting recombinant antibodies to the surface of *Escherichia coli*: Fusion to a peptidoglycan associated lipoprotein. *Bio/Technology* 9:1369–72.

14. Breitling, F., Dübel, S., Seehaus, T., Klewinghaus, I., and Little, M. 1991. A surface expression vector for antibody screening. *Gene* 104:147–53.

15. Nedelman, M.A., Shealy, D.J., Boutin, R., Brunt, E., Seasholtz, J.I., Allen, I.E., McCartney, J.E., Warren, F.D., Oppermann, H., Pang, R.H.L., Berger, J.J., and Weisman, H.F. 1993. Rapid infarct imaging with a technetium-99m antimyosin sFv fragment: Evaluation in a canine model of acute myocardial infarction. *J. Nuclear Med.* 34:234–41.

16. Adams, G.P., McCartney, J.E., Tai, M.-S., Oppermann, H., Huston, J.S., Stafford III, W.F., Bookman, M.A., Fand, I., Houston, L.L., and Weiner, L.W. 1993. Highly specific *in vivo* tumor targeting by monovalent and divalent forms of 741F8 anti-c-*erb*B-2 single-chain Fv. *Cancer Research* 53:4026–34.

17. Yokota, Y., Milenic, D.E., Whitlow, M., Wood, J.F., Hubert, S.L., and Schlom, J. 1993. Microautoradiographic analysis of the normal organ distribution of radioiodinated single-chain Fv and other immunoglobulin forms. *Cancer Research* 53:3776–83.

18. Tai, M.-S., Jin, D., Mudgett-Hunter, M., Haber, E., Oppermann, H., and Huston, J.S., 1993. In preparation.

19. George, A.J.T., Titus, J.A., Jost, C.R., Kuruca, I., Perez, P., Andrew, S.M., Nicholls, P.J., Huston, J.S., and Segal, D.M. 1993. Redirection of T cell mediated cytotoxicity by a recombinant single-chain Fv molecule. Submitted for publication.

20. Whitlow, M., and Filpula, D. 1991. Single-chain Fv proteins and their fusion proteins. *Methods: A Companion to Methods in Enzymology* 2:97–105.

21. McCartney, J.E., Tai, M.-S., Jin, D., Hudziak, R.M., Stafford III, W.F., Liu, S., Bookman, M.A., Adams, G.P., Fand, I., Weiner, L.M., Laminet, A., Houston, L.L., Oppermann, H., and Huston, J.S. 1993. Folding of sFv', single-chain

Fv fusions with carboxyl-terminal cysteine peptides: Formation of (sFv')$_2$ homodimers or heterodimers from 26-10 sFv' to digoxin and 741F8 sFv' to c-*erb*B-2 extracellular domain. In preparation.

22. McCartney, J.E., Tai, M.-S., Oppermann, H., Jin, D., Warren, F.D., Weiner, L.M., Bookman, M.A., Stafford III, W.F., Houston, L.L., and Huston, J.S. 1993. Refolding of single-chain Fv with C-terminal cysteine (sFv'): Formation of disulfide-bonded homodimers of anti-c-*erb*B-2 and anti-digoxin sFv'. *Miami Short Reports* 3:91.

23. Kurucz, I., Jost, C.R., George, A.J.T., Andrew, S.M., and Segal, D.M. 1993. A bacterially expressed single-chain Fv construct from the 2B4 T-cell receptor. *Proc. Natl. Acad. Sci. USA* 90:3830–34.

24. Colcher, D., Bird, R., Boselli, M., Hardman, K.D., Johnson, S., Pope, S., Dodd, S.W., Pantoliano, M.W., Milenic, D.E., and Schlom, J. 1990. *In vivo* tumor targeting of a recombinant single-chain antigen-binding protein. *J. Nat. Cancer Inst.* 82:1191–97.

25. Anand, N.N., Mandal, S., MacKenzie, C.R., Sadowska, J., Sigurskjold, B., Young, N.M., Bundle, D.R., and Narang, S.A. 1991. Bacterial expression and secretion of various single-chain Fv genes encoding proteins specific for a *Salmonella* serotype B O-antigen. *J. Biol. Chem.* 266:21874–79.

26. Laroche, Y., Demaeyer, M., Stassen, J.-M., Gansemans, Y., Demarsin, E., Matthyssens, G., Collen, D., and Holvoet, P. 1991. Characterization of a recombinant single-chain molecule comprising the variable domains of a monoclonal antibody specific for human fibrin fragment D-dimer. *J. Biol. Chem.* 266:16343–49.

27. Pantoliano, M.W., Bird, R.D., Johnson, S., Asel, E.D., Dodd, S.W., Wood, J.F., and Hardman, K.D. 1991. Conformational stability, folding and ligand-binding affinity of single-chain Fv immunoglobulin fragments expressed in *Escherichia coli*. *Biochemistry* 30:10117–25.

28. Takkinen, K., Laukkanen, M.-L., Sizmann, D., Alfthan, K., Immonen, T., Vanne, L., Kaartinen, M., Knowles, J.K.C., and Teeri, T.T. 1991. An active single-chain antibody containing a cellulase linker domain is secreted by *Escherichia coli*. *Prot. Eng.* 4:837–41.

29. Argos, P. 1990. An investigation of oligopeptides linking domains in protein tertiary structures and possible candidates for general gene fusion. *J. Mol. Biol.* 211:943–58.

30. Bernstein, F.C., Koetzle, T.F., Williams, G.J.B., Meyer, E.F., Brice, M.D., Rodgers, J.R., Kennard, O., Shimanouchi, T., and Tasumi, M.J. 1977. The Protein Data Bank: A computer-based archival file for macromolecular structures. *J. Mol. Biol.* 112:535–42.

31. Fischmann, T.O., Bentley, G.A., Bhat, T.N., Boulot, G., Mariuzza, R.A., Phillips, S.E.V., Tello, D., and Poljak, R.J. 1991. Crystallographic refinement of the three-dimensional structure of the FabD1.3-lysozyme complex at 2.5 Å resolution. *J. Biol. Chem.* 266:12915–20.

32. Saul, F.A., Amzel, L.M., and Poljak, R.J. 1978. Preliminary refinement and structural analysis of the Fab fragment from human immunoglobulin New at 2.0 Å resolution. *J. Biol. Chem.* 253:585–97.

33. Marquart, M., Deisenhofer, J., Huber, R., and Palm, W. 1980. Crystallographic

refinement and atomic models of the intact immunoglobulin molecule Kol and its antigen-binding fragment at 3.0 Å and 1.9 Å resolution. *J. Mol. Biol.* 141: 369–91.

34. Satow, Y., Cohen, G. Padlan, E.A., and Davies, D.R. 1986. Phosphocholine binding immunoglobulin Fab McPC 603, an x-ray diffraction study at 2.7 Å. *J. Mol. Biol.* 190:593–604.

35. Sheriff, S., Silverton, E.W., Padlan, E.A., Cohen, G.H., Smith-Gill, S.J., Finzel, B.C., and Davies, D.R. 1987. Three-dimensional structure of an antibody-antigen complex. *Proc. Natl. Acad. Sci. USA* 84:8075–79.

36. Padlan, E.A., Silverton, E.W., Sheriff, S., Cohen, G.H., Smith-Gill, S.J., and Davies, D.R. 1989. Structure of an antibody-antigen complex: Crystal structure of the HyHEL-10 Fab-lysozyme complex. *Proc. Natl. Acad. Sci. USA* 86:5938–42.

37. Suh, S.W., Bhat, T.N., Navia, M.A., Cohen, G.H., Rao, D.N., Rudikoff, S., and Davies, D.R. 1986. The galactan-binding immunoglobulin Fab J539: An x-ray diffraction study at 2.6-Å resolution. *Proteins: Structure, Funct., Genetics* 1:74–80.

38. Lascombe, M.-B., Alzari, P.M., Boulot, G., Saludjian, P., Tougard, P., Berek, C., Haba, S., Rosen, E.M., Nisonoff, A., and Poljak, R.J. 1989. Three-dimensional structure of Fab R19.9, a monoclonal murine antibody specific for the p-azobenzenearsonate group. *Proc. Natl. Acad. Sci. USA* 86:607–11.

39. Herron, J.N., He, X., Mason, M.L., Voss, E.W. Jr., and Edmundson, A.B. 1989. Three-dimensional structure of a fluorescein-Fab complex crystallized in 2-methyl-2,4-pentanediol. *Proteins: Structure, Funct., Genetics* 5:271–80.

40. Freund, C., Ross, A., Guth, B., Plückthun, A., and Holak, T.A. 1993. Characterization of the linker peptide of the single-chain Fv fragment of an antibody by NMR spectroscopy. *FEBS Letters* 320:97–100.

41. Tanford, C. 1968. Protein denaturation. *Advances in Protein Chemistry* 23: 121–282.

42. Volk, W.A., Benjamin, D.C., Kadner, R.J., and Parsons, J.T. 1991. *Essentials of Medical Microbiology.* Philadelphia: J.B. Lippincott Company.

43. Rapoport, T.A. 1992. Protein transport across the endoplasmic reticulum membrane: facts, models, mysteries. *FASEB Journal* 5:2792–98.

44. Bardwell, J.C.A., McGovern, K., and Beckwith, J. 1991. Identification of a protein required for disulfide bond formation *in vivo*. *Cell* 67:581–89.

45. Landry, S.J., and Gierasch, L.M. 1991. Recognition of nascent polypeptides for targeting and folding. *Trends in Biochemical Sci.* 16:159–63.

46. Söderlind, E., Lagerkvist, A.C.S., Duenas, M., Malmborg, A.-C., Ayala, M., Danielsson, L., and Borrebaeck, C.A.K. 1993. Chaperonin assisted phage display of antibody fragments on filamentous bacteriophages. *Bio/Technology* 11:503–507.

47. Haber, E. 1964. Recovery of antigenic specificity after denaturation and complete reduction of disulfides in a papain fragment of antibody. *Proc. Natl. Acad. Sci. USA* 52:1099–1106.

48. Whitney, P.L., and Tanford, C. 1965. Recovery of specific activity after complete unfolding and reduction of an antibody fragment. *Proc. Natl. Acad. Sci. USA* 53:524–32.

49. Buckley III, C.E., Whitney, P.L., and Tanford, C. 1963. The unfolding and refolding of a specific univalent antibody fragment. *Proc. Natl. Acad. Sci. USA* 50:827–34.
50. Noelken, M.E., and Tanford, C. 1964. Unfolding and renaturation of a univalent anti-hapten antibody fragment. *J. Biol. Chem.* 239:1828–32.
51. Whitney, P.L., and Tanford, C. 1965. Properties of the soluble product obtained from reoxidation of reduced fragment I from rabbit 7S gamma-immunoglobulin. *J. Biol. Chem.* 240:4271–76.
52. Hochman, J., Gavish, M., Inbar, D., and Givol, D. 1976. Folding and interaction of subunits at the antibody combining site. *Biochemistry* 15:2706–10.
53. Cheadle, C., Hook, L.E., Givol, D., and Ricca, G.A. 1992. Cloning and expression of the variable regions of mouse myeloma protein MOPC315 in *E. coli*: Recovery of active Fv fragments. *Molecular Immunology* 29:21–30.
54. Givol, D. 1991. The minimal antigen-binding fragment of antibodies—Fv fragment. *Molecular Immunology* 28:1379–86.
55. Rowe, E.S., and Tanford, C. 1973. Equilibrium and kinetics of the denaturation of homogeneous human immunoglobulin light chain. *Biochemistry* 12:4822–27.
56. Rowe, E.S. 1976. Dissociation and denaturation equilibria and kinetics of a homogeneous human immunoglobulin Fab fragment. *Biochemistry* 15:905–16.
57. Lesk, A., and Chothia, C. 1982. *J. Mol. Biol.* 160:325–42.
58. Holliger, P. Prospero, T., and Winter, G. 1993. "Diabodies": Small bivalent and bispecific antibody fragments. *Proc. Natl. Acad. Sci. USA* 90:6444–48.
59. Whitlow, M.D., Wood, J.F., Hardman, K., Bird, R.E., Filpula, D., and Rollence, M. 1993. Multivalent antigen-binding proteins. *International Application, World Intellectual Property Organization* WO 93/11161, 1–76.
60. Skerra, A., Pfitzinger, I., and Plückthun, A. 1991. The functional expression of antibody Fv fragments in *Escherichia coli*: Improved vectors and a generally applicable purification technique. *Bio/Technology* 9:273–78.
61. Takacs, B.J., and Girard, M.-F. 1991. Preparation of clinical grade proteins by recombinant DNA technologies. *J. Immunol. Meth.* 143:231–40.
62. Gill, S.C., and von Hippel, P.H. 1989. Calculation of protein extinction coefficients from amino acid sequence data. *Anal. Biochem.* 182:319–26.
63. Babul, J., and Stellwagen, E. 1969. Measurement of protein concentration with interference optics. *Anal. Biochem.* 28:216–21.
64. Kabat, E.A., Wu, T.T., Perry, H.M., Gottesman, K.S., and Foeller, C. 1991. *Sequences of Proteins of Immunological Interest*, 5th edition. U.S. Department of Health and Human Services, Public Health Service, National Institutes of Health, Publication No. 81-3242.
65. Orlandi, R., Güssow, D.H., Jones, P.T., and Winter, G. 1989. Cloning immuno-globulin variable domains for expression by the polymerase chain reaction. *Proc. Natl. Acad. Sci. USA* 86:3833–37.
66. Frohman, M.A., Dush, M.K., and Martin, G.R. 1988. Rapid production of full-length cDNAs from rare transcripts: Amplification using a single gene-specific oligonucleotide primer. *Proc. Natl. Acad. Sci. USA* 85:8998–9002.
67. Kunkel, T.A., Roberts, J.D., and Zakour, R.A. 1987. Rapid and efficient

site-specific mutagenesis without phenotypic selection. *Methods in Enzymology* 154:367–82.

68. Ausubel, F.M., Brent, R., Kingston, R.E., Moore, D.D., Seidman, J.G., Smith, J.A., and Struhl, K., eds. 1989. *Current Protocols in Molecular Biology*. New York: Wiley-Interscience.

69. Studier, F.W., and Moffat, B.A. 1990. Use of T7 RNA polymerase to direct the expression of cloned genes. *Methods in Enzymology* 185:60–89.

70. Buchner, J., and Rudolph, R. 1991. Renaturation, purification and characterization of recombinant Fab-fragments produced in *Escherichia coli*. *Bio/Technology* 9:157–62.

71. Better, M., Chang, C.P., Robinson, R.R., and Horwitz, A.H. 1988. *Escherichia coli* secretion of an active chimeric antibody fragment. *Science* 240: 1041–43.

72. Carter, P., Kelly, R.F., Rodriguez, M.L., Snedecor, B., Covarrubias, M., Velligan, M.D., Wong, W.L.T., Rowland, A.M., Kotts, C.E., Carver, M.E., Yang, M., Bourell, J.H., Shephard, H.M., and Henner, D. 1992. High level *Escherichia coli* expression and production of a bivalent humanized antibody fragment. *Bio/Technology* 10:163–67.

73. Skerra, A., and Plückthun, A. 1988. Assembly of a functional immunoglobulin Fv fragment in *Escherichia coli*. *Science* 240:1038–41.

74. Plückthun, A., and Skerra, A. 1989. Expression of functional antibody Fv and Fab fragments in *Escherichia coli*. *Methods in Enzymology* 178:497–515.

75. Ward, E.S. 1992. Secretion of T cell receptor fragments from recombinant *Escherichia coli* cells. *J. Mol. Biol.* 224:885–90.

76. Ward, E.S., Güssow, D., Griffiths, A.D., Jones, P.T., and Winter, G. 1989. Binding activity of a repertoire of single immunoglobulin variable domains secreted from *Escherichia coli*. *Nature (London)* 341:544–46.

77. Gandecha, A.R., Owen, M.R.L., Cockburn, B., and Whitelam, G.C. 1992. Production and secretion of a bifunctional staphylococcal protein A: antiphytochrome single-chain Fv fusion protein in *Escherichia coli*. *Gene* 122:361–65.

78. Kohl, J., Rüker, F., Himmler, G., Razazzi, E., and Katinger, H. 1991. Cloning and expression of an HIV-1 specific single-chain Fv region fused to *Escherichia coli* alkaline phosphatase. *Ann. N. Y. Acad. Sci. USA* 646:106–14.

79. von Heijne, G. 1986. A new method for predicting signal sequence cleavage sites. *Nucleic Acids Research* 14:4683–90.

80. George, A.J.T., 1993. The production of antibodies using phage display libraries. In *Monoclonal Antibodies*. Ritter, M.A., ed. Cambridge: Cambridge University Press. In press.

81. Hoogenboom, H.R., Griffiths, A.D., Johnson, K.S., Chiswell, D.J., Hudson, P., and Winter, G. 1991. Multi-subunit proteins on the surface of filamentous phage: methodologies for displaying antibody (Fab) heavy and light chains. *Nucleic Acids Research* 19:4133–37.

82. Timasheff, S.N., and Arakawa, T. 1989. Stabilization of protein structure by solvents. In *Protein Structure, a Practical Approach*. Creighton, T.E., ed. Oxford: IRL Press. pp. 331–45.

83. Buchner, J., Pastan, I., and Brinkmann, U. 1992. A method for increasing the yield of properly folded recombinant fusion proteins: Single-chain immuno-

toxins from renaturation of bacterial inclusion bodies. *Anal. Biochem.* 205: 263–70.

84. Derman, A.I., Prinz, W.A., Belin, D., and Beckwith, J. 1993. Mutations that allow disulfide bond formation in the cytoplasm of *Escherichia coli. Science* 262:1744–47.

CHAPTER 8

Expressing Antibodies in *Escherichia coli*

Liming Ge, Achim Knappik, Peter Pack, Christian Freund, and Andreas Plückthun

There are many reasons why researchers may wish to work with recombinant antibodies: for answering basic questions of protein architecture, specific binding and protein folding, for applying the modified antibodies in biology and biotechnology (ranging from biosensors to affinity chromatography), or for use in medicine. At the beginning of each project there is the problem of producing the protein in sufficient quantities. This problem will still be encountered, even if the antibody is first selected from a phage library. This chapter will summarize methodology for cloning, expression, purification, and detection of recombinant antibodies, using *Escherichia coli*, the most familiar host in biochemical laboratories.

There are several attractive features about *E. coli*. Manipulations are quite simple and well established, growth is rapid, and new ideas about changes in the protein can therefore be tested in a very short time. Scale-up of fermentations is reasonably straightforward, and very large amounts of protein are thus accessible. The transformation efficiency of *E. coli* is probably unrivaled by any other microorganism; it is this last feature that makes *E. coli* (together with its phages) particularly suitable for constructing and screening libraries.

Two decisions have to be made at the beginning of a recombinant antibody project. The first is which fragment of the antibody is desired and the second is the method of expression. Both points have been discussed in more detail elsewhere,[1–3] and therefore only a brief synopsis is given here.

CHOICE OF THE ANTIBODY FRAGMENT TO BE EXPRESSED

The antigen binding site is made up of the variable domains V_L and V_H. The smallest fragment containing the complete binding site is therefore the Fv fragment (Fig. 8–1).[4] Several investigations have shown that the Fv fragment has indeed the full intrinsic antigen binding affinity of one binding site of the whole antibody. The relative affinity of V_H and V_L for each other can be fairly low or sufficiently high to be stable, however, depending on the particular sequence of the antibody.[1, 5] This may result in dissociation of the Fv fragment into its components in some antibodies, while others do not have this problem. On the other hand, there does not seem to be a severe kinetic problem of the two domains "finding" each other, as the heterodimers *do* form in *E. coli*, provided they are stable.[6]

The reversible dissociation of the Fv fragment can be counteracted in several ways. One may chemically cross-link the Fv fragment,[5] but this

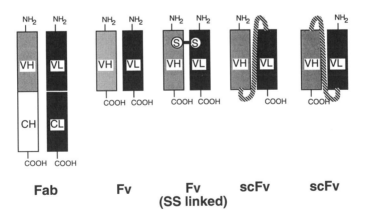

Figure 8–1. Monovalent fragments of antibodies that can be functionally expressed in *E. coli*. The disulfide-linked Fv fragment is obtained by engineering.[5] Note that every antibody domain also contains an internal disulfide bond, which is usually crucial for stability, but is not shown here for simplicity. Depending on the antibody class, the two chains of the Fab fragment may also be (naturally) covalently linked.

requires obtaining the Fv fragment in the first place. A better approach is to engineer disulfide bonds into the fragment.[5] The latter strategy has a dramatic effect on increasing the stability against irreversible denaturation,[5] but more research is necessary to evaluate the generality of useful positions in all antibodies[7] and the stabilization effect caused by the various disulfide bonds.

The most popular method of covalently linked V_H and V_L is probably the single-chain Fv fragment (Fig. 8–1), in which the two domains are connected by a genetically encoded peptide linker.[5,8–10] A variety of linker peptides have been tested,[10] and the most frequently used one has the sequence $(Gly4Ser)_3$.[9] Recent NMR evidence[11] suggests that this linker is a flexible entity, which does not change the structure of the variable domains. It is compatible both with the orientation V_H-linker-V_L and V_L-linker-V_H, and both proteins show about the stability in reversible denaturation experiments.[12,13] The peptide linker does not stabilize the Fv fragment to the same extent as the disulfide bonds against irreversible denaturation,[5] and aggregation phenomena of scFv fragments have occasionally been noticed. Furthermore, some proteolysis within or at the edge of the linker is occasionally seen. Nevertheless, the scFv strategy secures the correct stoichiometry of the two domains, and only one expression module for the foreign gene is needed.

Closest to the naturally occurring antibodies is the Fab fragment. The constant domains make important contributions to the interactions between the light and the heavy chain and thus indirectly increase the stability against irreversible denaturation. On the other hand, several Fab fragments are more poorly expressed in functional form in *E. coli* than Fv-fragments and their derivatives[14] (Knappik and Plückthun, unpublished observations). It appears that a higher tendency to aggregate causes this problem.

Whole antibodies probably cannot be efficiently produced in functional form in *E. coli*, because they would lack the glycosylation in the Fc part, which seems to contribute to stability, and its absence would abolish all biological function.[15] In order to preserve the bivalency, which is a very effective means of increasing the functional affinity (avidity) to a surface or polymeric antigen,[16,17] another strategy can be used: the linking of scFv fragments by a small modular dimerization domain in the form of one or two amphipathic helices.[18,19] These "miniantibodies"[18,19] assemble in dimeric form in *E. coli*, and the binding performance of the best of them is indistinguishable from a whole antibody in avidity (see below).

EXPRESSION IN *E. COLI*: OVERVIEW

Two basic strategies can be employed to obtain antibody fragments from *E. coli*. The first is the functional expression of correctly folded fragments by

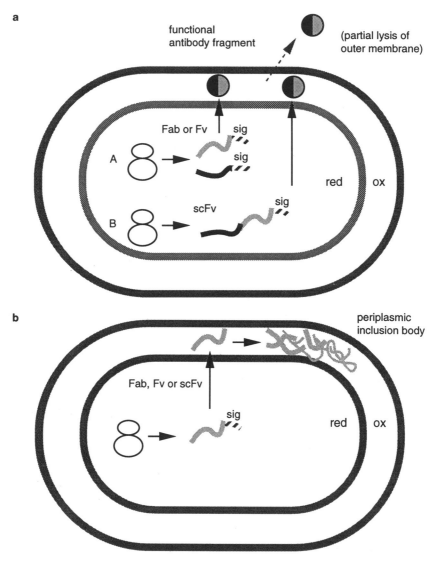

Figure 8–2. Schematic view of the different strategies found to be useful for antibody production in *E. coli*. (a) Functional expression: In this case, the two chains making up the antibody combining site are transported to the oxidizing milieu of the periplasm. In this compartment, there is a disulfide forming enzyme (DSbA) that facilitates the formation of the crucial intramolecular disulfide bonds. In (A), the pathway for two independent chains (as in Fv or Fab fragments is shown. Alternatively (B), both variable domains can be linked to form a continuous polypeptide chain (scFv fragment, see Fig. 8–1), which can also be secreted. (b) Periplasmic inclusion body formation. This is observed for many fragments

(continued)

c

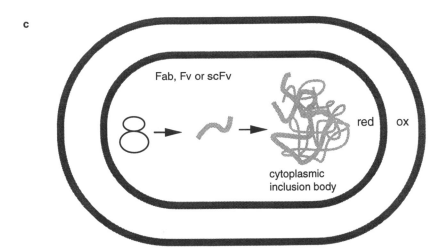

Figure 8–2. *(cont.)* as a side reaction to correct folding, and its extent depends on the sequence and the growth conditions. It can be also exploited preparatively. The protein is apparently transported, processed, and then precipitates. The protein must be refolded *in vitro* as in (c). (c) Cytoplasmic inclusion body formation. In this case, the protein is expressed without a signal sequence under a strong promoter and translation signal. Inclusion body formation appears to be more successful at temperatures of 37°C and higher. The protein must be refolded *in vitro*.

secretion into the periplasmic space,[6, 20] the second the *in vitro* refolding of protein obtained from inclusion bodies, either from the cytoplasm or the periplasm (Fig. 8–2). Before going into details about both approaches, some general remarks about the various strategies are necessary.

The intramolecular disulfide bond of antibody domains appears to be crucial for stability.[21] Most antibody domains do not have a sufficient free energy of folding to tolerate removal of this stabilizing element. Consequently, in order to form the intradomain disulfide bonds correctly, which is the prerequisite to obtain folded antibodies from *E. coli*, a cosecretion of both chains (in the case of Fv or Fab fragments) must be achieved.[6, 20] In the case of the single-chain Fv fragment, secretion also leads to the correct folding and assembly of both chains.[5] Nevertheless, there is not a quantitative yield of folded material, but a competition between folding and aggregation occurs, whose main determinant is the primary sequence (Knappik and Plückthun, unpublished).

To achieve secretion of antibody fragments, both chains must be equipped with signal sequences. A variety of bacterial signal sequences were shown to lead to transport and correct cleavage.[1–3] Notably, the leakage of the antibody protein to the medium appears to be linked to the sequence of the mature antibody tested and also depends on the growth physiology (see below).

In order to cosecrete both chains to the bacterial periplasm, they are advantageously arranged in an artificial operon.[6,20] Thereby, they can be coordinately regulated. This is crucial for maintaining stability of the host-vector system, since the secretion of many (if not most) antibody fragments presents a stress to the cell. For phage display vectors, expression systems comprising two plasmids[22] or two promoters on one plasmid[23] have also been used, but the rigorous plasmid maintenance and the minimization of promoter leakiness is probably more easily controlled in a two-cistron system with only one promoter.

As will be discussed below in more detail, the folding of periplasmic protein does not proceed quantitatively for most antibody fragments.[12,14] The yield of this process depends on external factors such as temperature and growth physiology, but also on the primary sequence and the type of fragment used (Fv, scFv, Fab). It is also possible to isolate the periplasmically precipitated protein and use it as the starting material for refolding experiments.[24–26]

Finally, one may not even attempt to secrete the antibody fragment, but directly aim for cytoplasmic inclusion bodies. They can be refolded *in vitro*, but experiments with a number of fragments have shown that the refolding procedure has to be adapted to the particular fragment under study.[10] Therefore, folding optimization work is almost certainly necessary. Additionally, one must separate correctly folded from soluble, but incorrectly folded, molecules for any critical application.

Most important for deciding which strategy to use is the desired application of the antibodies. If a series of mutants is to be analyzed or a number of binding constants to be established, secretion is almost certainly the faster method. Using properly designed vectors and high-cell-density fermentation, very high-volume yields can now be obtained from secretion,[19,27] and no refolding is necessary. However, if one particular fragment is needed repeatedly in very large quantities, and the work of optimizing the refolding conditions can be justified, the expression from cytoplasmic or periplasmic inclusion bodies may be considered as an alternative. An example is the production of isotope-labeled protein required for NMR investigations: The high cost of the isotopic label may justify the optimization of the refolding protocol.[11]

To evaluate different "expression systems," two points must be kept in mind. First, the only relevant quantity is the amount of *correctly folded, purified* protein obtainable at the end. Second, because of the individuality of the different antibody sequences and their dramatic influence on the yield of *in vivo* and *in vitro* folding, a comparison of two expression systems with two *different* antibody fragments may be close to meaningless. Obviously, the cell density at which a volume yield is reported must be considered as well.

In this chapter, we will go through all steps from cloning an antibody to the actual expression, purification, and detection of the fragment, using the secretion strategy. We will also describe the refolding of an scFv fragment from cytoplasmic inclusion bodies in order to obtain isotopically enriched protein for NMR purposes.

VECTORS FOR ANTIBODY SECRETION

General Considerations

The vectors described here have been in constant "evolution," and there may never be a single solution for every possible expression experiment. Therefore, the latest generation (pIG) vectors (Fig. 8–3) have a modular architecture allowing convenient changes of most elements.

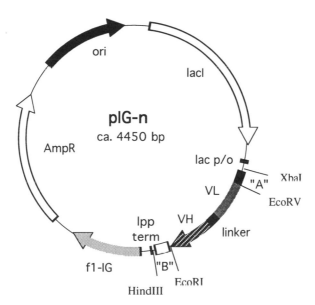

Figure 8–3. (a) Schematic drawing of the pIG vector series. *lac p/o* denotes the *lac* promoter operator, *lacI* the *lac* repressor gene, *lpp term* the terminator of the *E. coli* lipoprotein, *f1-IG* the intergenic region (origin of replication) of the f1 phage, *Amp*[R] the ampicillin resistance and *ori* the origin of replication of the plasmid. For details, see the text and Skerra et al.[34] The "A" and "B" cassettes are explained below. *(continued)*

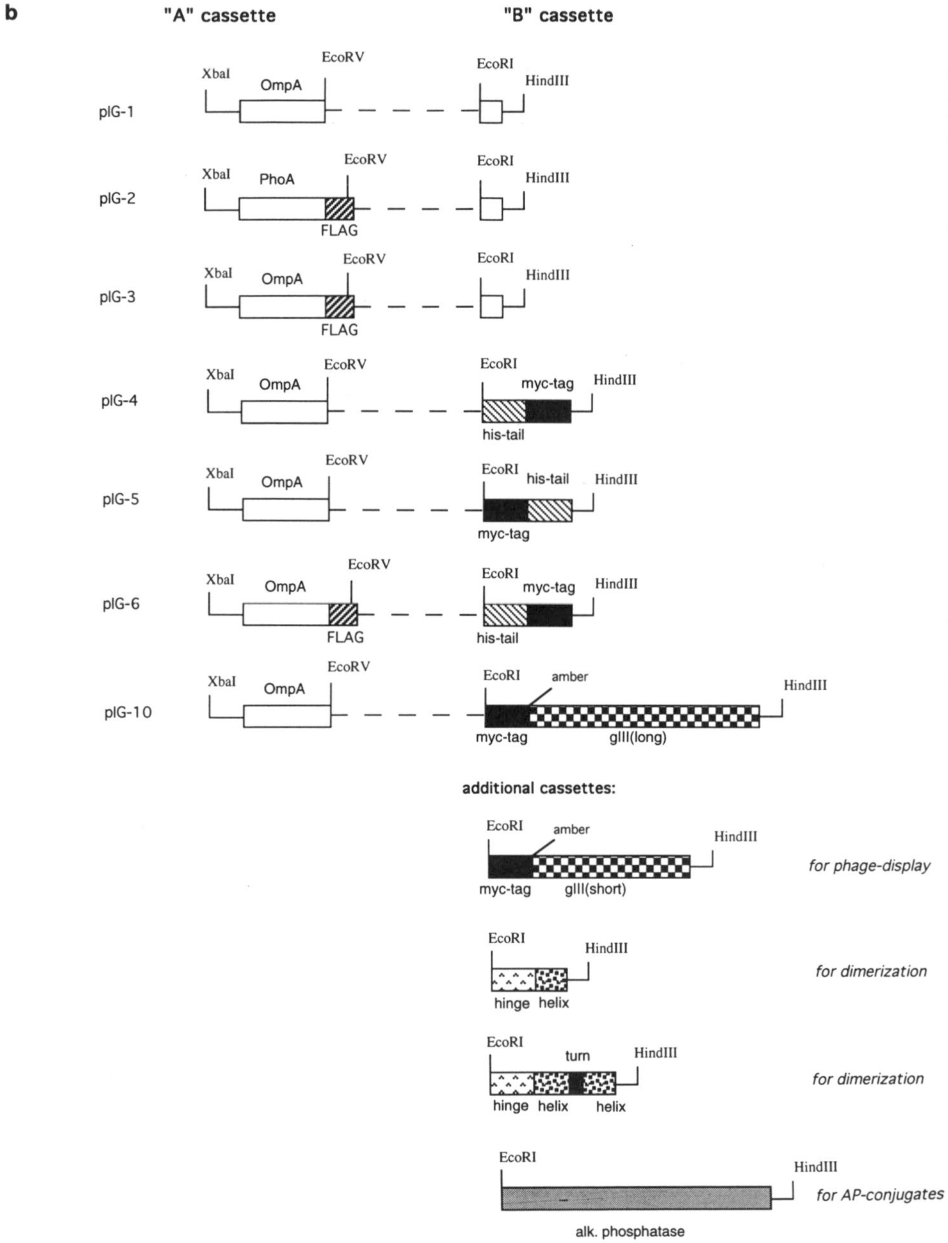

b

"A" cassette   **"B" cassette**

pIG-1

pIG-2

pIG-3

pIG-4

pIG-5

pIG-6

pIG-10

additional cassettes:

for phage-display

for dimerization

for dimerization

for AP-conjugates

Figure 8–3. (*cont.*) (b) Schematic drawing of the N- and C-terminal cloning cassettes ("A" and "B" cassettes) of the pIG vectors, as well as additional cassettes useful in antibody work. The various fragments are not drawn to scale.   (*continued*)

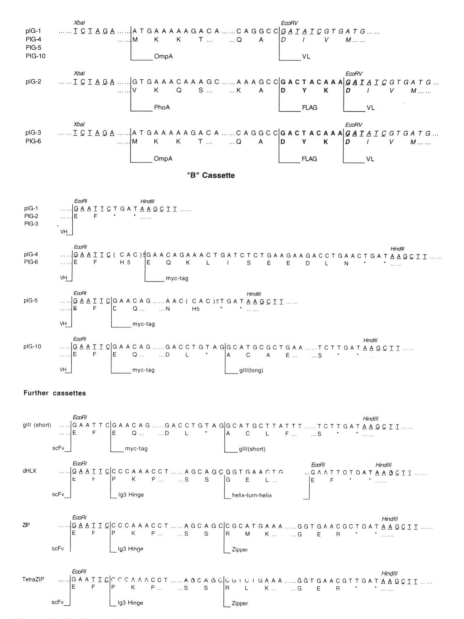

Figure 8-3. (cont.) (c) Detailed drawing of the "A" and "B" cassettes as shown in (b). Only the N- and C-terminal sequence of OmpA, PhoA (pIG-1 to -3) and gene III (gIII) (pIG-10) are shown. The C-terminal Asn (N) residue in the myc-tag is

(continued)

We will first summarize some general features of the vectors. The origin of replication is that of the pUC family of vectors, giving rise to a high copy number,[28] which is, however, much lower at room temperature.[29] The vectors contain the origin of the f1 phage, which allows packaging of the plasmid for site-directed mutagenesis[30] and for use in phage-display of the antibody fragments.[31] Three different antibiotic resistances have been tested. Ampicillin resistance is not advantageous in extended fermentations[19] or at high growth temperatures,[32] since the leakage of periplasmic proteins through the outer membrane of the bacterial cells, resulting from antibody expression, leads to very rapid hydrolysis of the antibiotic by leaked-out β-lactamase. In the shorter shake-flask experiments, and especially at low growth temperature, this resistance can be used, however. The activity of β-lactamase has been very helpful to diagnose the plasmid content and the integrity of the outer membrane of the bacterial cells. Tetracyclin resistance has been used in large-scale experiments.[27] Kanamycin has also been very useful in fermentation experiments,[19] but in selection and cloning experiments, a background problem may appear if streptomycin-resistant cells are used, since they can mutate to become kanamycin resistant.[33]

The promoter favored in this laboratory has been the *lac* promoter,[6,34] since it can be induced independent of any external variables. Particularly, this allows induction experiments to be carried out at any temperature and growth phase or in any strain. Strain independence is secured by the plasmid-encoded *lac* repressor. The *phoA* promoter has also been used in antibody fermentation,[27] but it requires defined media, calculated for phosphate to be depleted at a desired point.

In the dicistronic constructs, we usually prefer two *different* signal sequences.[6,34] While this is not absolutely necessary, it guards against homologous recombination, which has been observed under certain conditions (G. Wall, personal communication), since the expression of antibodies by secretion does constitute a stress for *E. coli*.

The pIG series of vectors (Fig. 8–3) contain an antibody expression cassette flanked by one set of unique restriction sites external to the coding sequence (*Xba*I and *Hind*III) and a different unique set at the edge of the coding sequence, but within it (*Eco*RV and *Eco*RI). This permits convenient shuffling of elements, fused at the N-terminus to the antibody coding

---

Figure 8–3 *(contd.)* replaced by the amber codon (TAG) in pIG-10, such that the expression of the complete fusion protein is only possible in an amber suppressor strain, and the scFv form is obtained in nonsuppressor strain. The underlined bases are the restriction sites of the restriction enzymes indicated above. The bases and amino residues *in italics* are of the scFv. An asterisk (*) denotes a stop codon. The "FLAG" sequence is written in **bold**. Further "B" cloning cassettes are available for mini-antibodies (ZIP, dHLX and tetraZIP), phage-display (gIII-(short)), and alkaline phosphatase fusion (AP) and are explained in the text.

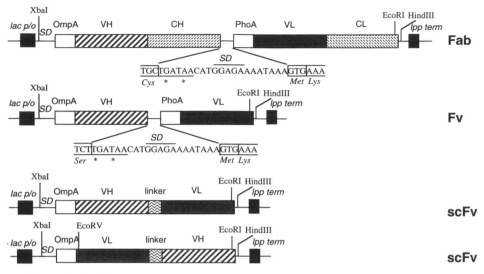

Figure 8-4. Schematic arrangement of the genes for functional expression of other antibody fragments in these vectors. The independent chains of the Fab fragment or the Fv fragment are expressed in a dicistronic operon, with the intergenic region shown. The two different orientations of the scFv fragment are also shown, both of which have been shown to function. *p/o* denotes a promoter/ operator structure, *SD* the Shine-Dalgarno sequence, and *lpp term* the lpp transcription terminator. For details, see text and Skerra et al.[34]

sequence (the "A" cassette containing different signal sequences and N-terminal "tag" sequences), or C-terminal to the antibody sequence (the "B"-cassette containing C-terminal "tag" sequences, fusion domains for enzymatic activity, e.g., alkaline phosphatase or protein III for phage display, or an oligo-his tail for purification). Furthermore, the "B" cassette may contain the dimerization elements for making bivalent or bispecific "miniantibodies."[18, 19] Having the option of using either a C-terminal and/or an N-terminal "tag" sequence for detection increases the flexibility for constructing fusion proteins of any kind and comparing different fragments with the same tag. The internal restriction sites (*Eco*RV and *Eco*RI) also permit the convenient ligation of PCR products. Further internal restriction sites (see below) permit domain exchange in the expression vectors of the pHJ series.

The vectors are equally suited to accommodate Fab, Fv, or single-chain Fv fragments (Fig. 8-4). The unlinked fragments (Fab and Fv) are arranged as a dicistronic operon, controlled under one promoter. The intergenic region is shown, and the upstream region has been described.[34] In Fig. 8-5, an example of the pHJ series is shown. Here the goal was to make domain exchange very convenient, e.g., to exchange variable and constant domains and interconvert Fab, Fv and scFv fragments. Since scFv fragments are

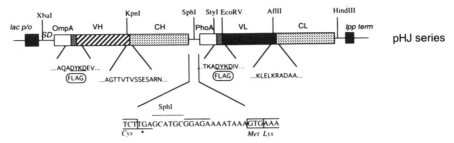

Figure 8–5. Schematic drawing of the pHJ vector series, which was created to facilitate domain switches as well as versatile detection of all chains. They are compatible with the pIG vectors. Both chains carry FLAG epitopes (see text) for easy detection. As an example, the Fab fragment is shown; Fv and scFv vectors carry analogous restriction sites to convert between these antibody fragments conveniently.

currently used in most laboratories, we will take those as an example and describe their cloning, expression, purification, and detection.

pIG Vectors

The pIG vector series is derived from pASK-lisc,[34] where the unique *Eco*RV site in the *lacI* gene in the natural gene was removed by site-directed mutagenesis. As the first vector in the series (Fig. 8–3), we have introduced a single *Eco*RV site behind the OmpA signal sequence, and there is a unique *Eco*RI site downstream from the *Eco*RV site. The genes for scFv fragments (in V_L-linker-V_H arrangement) can be easily inserted between these sites.

For detection of the scFv fragments, it is preferable to clone the fragments into the vectors containing a detection "tag": pIG-2 (with the PhoA signal sequence) and pIG-3 (with the OmpA signal sequence) contain a short version of the FLAG epitope[35–37] sequence (Asp-Tyr-Lys-Asp) (see below) and can thus be used for this purpose. Since the majority of mouse V_κ starts with the sequence Asp-Ile or Glu-Ile, we could conveniently use the last Asp residue in the shortened FLAG sequence as the first amino acid of V_L (hence the V_L-linker-V_H arrangement) and *Eco*RV (GAT ATC, encoding Asp-Ile) was used as the N-terminal cloning site for the scFvs. Therefore, the scFv constructs from pIG-2 and pIG-3 need only an additional three amino acids at the N-terminus for sensitive immunological detection (see below).

For combining detection and purification on large scales, the vectors pIG-4 or PIG-5 are recommended. Instead of the N-terminal FLAG, they contain a C-terminal his_5-tail/myc-tag or myc-tag/his_5 tail. pIG-6 contains both FLAG tag and his_5-tail/myc-tag. The scFv fragments in these constructs containing the oligo-histidine stretch can be purified in one step by immobilized metal affinity chromatography (IMAC) (see below). By using the latter construct, we have routinely purified scFv fragments of several antibodies and their mutants. The use of the myc-tag for immunological detection has been reported,[38–39] and the use of the FLAG tag in very sensitive Western blots is described below.

Another vector in this series is pIG-10, which has been used for phage-display of the ScFv fragments.[31] Production of functional scFv after phage selection is also possible by transforming a non-amber-suppressor strain of *E. coli*. The scFv fragments produced contain a C-terminal myc-tag for detection.

The advantages of choosing *Eco*RV and *Eco*RI for scFv cloning are their robustness as restriction enzymes and the relatively low occurrence of those sites in mouse and also in human variable domains (Table 8–1). By using *Eco*RI as the C-terminal cloning site, we can also insert any C-terminal tail for modification of the scFvs (Fig. 8–3).

Obtaining the Genes of the Antibody

At the beginning of any antibody project, there are several typical starting scenarios.

1. A hybridoma line exists as the source of the antibody genes, but they have not yet been cloned. In this case, it is necessary to first isolate mRNA and then obtain cDNA using primers specific for the constant regions or located at the C-terminus of the variable domains. A PCR can be carried out using either set of primers in conjunction with primers specific for the 5'-end of the variable domain. For the cDNA synthesis, the poly-d$(T)_n$ primer used in most kits can be used, when a PCR with specific primers follows.

2. The antibody has already been cloned, either in a nonexpression vector or a eukaryotic vector or as a fragment other than the one desired. In this case, the sequence would already be known and precise custom primers should be used. In most instances, PCR is the fastest way to reclone a gene if restriction sites are not compatible. It is strongly recommended that any PCR is followed by DNA sequencing, and that a polymerase with low error rate is used.

3. The sequence of the antibody is known, but the gene is not easily available. In this case, a total synthesis of the genes is most practical.[41] If the genes of a related antibody are available, the new gene can be made by site-directed mutagenesis,[42] making use of the f1 origin in the plasmids.

4. Only the antigen is available. While we will not describe the construction of libraries, as this is done elsewhere in this volume, such construction is possible with display vectors such as pIG-10. The

Table 8–1. Frequency of Restriction Enzyme
Recognition Sites in Variable Domains of Human and
Mouse Antibodies

Enzyme	V_H (%)		V_L (%)	
	Human	Mouse	Human	Mouse
Aat II	8.9	0.9	0	0.3
Afl II	0.4	2.4	0	0.4
Apa I	11.7	0.6	12.2	0
Apa LI	1.2	2.5	1.0	1.4
Asc I	0	0	0	0
Ase I	2.0	20.8	1.0	2.9
Bam HI	16.1	8.3	11.2	9.9
Bcl I	0.8	0.4	6.1	1.8
Bgl I	0.4	0.3	3.1	0.3
Bsp HI	2.0	0.6	0	0.9
Bst EII	47.6	8.5	18.4	23.1
Bst XI	5.6	2.5	1.0	1.5
Bsu 36I	1.6	3.2	16.3	6.4
Cla I	0	1.4	0	0.1
Drd I	5.6	25.7	2.0	2.0
Eag I	36.7	0.1	0	0
Eco RI	0	6.1	9.2	0
Eco RV	6.9	8.1	1.0	6.9
Hind III	0.8	4.8	1.0	3.7
Hpa I	0	0.2	0	0.1
Kpn I	0	1.9	64.3	56.4
Mlu I	0.8	0	0	0
Nar I	0	3.2	3.1	0.1
Nco I	2.4	6.2	0	5.6
Nde I	4.4	2.0	0	0.9
Nhe I	0.4	0.1	1.0	0
Not I	0	0	0	0
Nru I	0.4	0	0	0
Pac I	0	0	0	0
Pfl MI	20.2	13.4	6.1	19.5
Pme I	0	0.1	0	0.4
Pst I	32.3	42.6	74.5	53.4
Pvu II	79.8	19.0	7.1	1.9

(continued)

Table 8–1. (*Cont.*)

Enzyme	V_H (%)		V_L (%)	
	Human	Mouse	Human	Mouse
Sac I	10.5	5.3	1.0	4.7
Sac II	11.7	0.2	0	0
Sal I	1.6	0.2	0	0.1
Sfi I	0	0	0	0
Sph I	2.4	0.3	5.1	0.4
Sma I	8.9	0.6	7.1	1.0
Spe I	1.6	0.9	0	6.2
Sty I	19.8	30.7	73.5	6.1
Tth111 I	14.5	5.0	77.6	30.4
Xba I	0	1.2	0	3.8
Xho I	1.2	7.4	0	0.5

Note: 1271 V_H, 785 V_L of mouse, 164 V_H and 98 V_L of human immunoglobulin sequences in the computer readable Kabat data base were scanned with the restriction sites listed using a word processor. An additional 84 human germline V_H sequences (Marks et al., 1993) were scanned with the same set of restriction sites using the UWGCG MAPSORT program. The frequencies given in the table indicate the percentage of sequences having *at least* one occurrence (but possibly more) of the respective restriction site.

expression and purification of a defined fragment is then carried out as described here.

PCR Assembly

N and C-terminal Primers

We are describing the use of a degenerate N-terminal primer (SC-1) to amplify several murine monoclonal antibodies for expression in the pIG vectors as an example. This type of degeneracy can certainly be extended, and the expanded primer set can be used for direct scFv cloning or repertoire cloning. There are several tabulations of useful PCR primers,[43–50] but consultation of on-line databases is recommended to take advantage of the many new sequences that are continuously being reported.

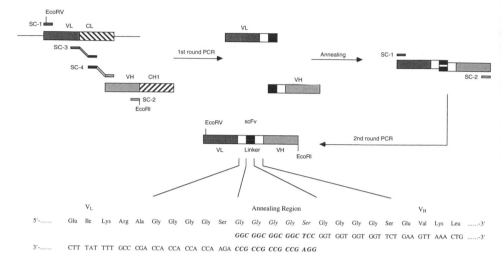

Figure 8–6. Schematic representation of the assembly reaction of the scFv. The bases that are involved in the final assembly of the individual V_L and V_H fragments to yield the scFv (annealing region), as well as the amino acids they code for, are shown in *italics*.

N- and C-Terminal Primers

SC-1: 5'- GCC *GAT ATC* **GT(GT) ATG AC(CA) CAG TCT CCA GCA** -3'

SC-2: 5'- ATG *GAA TTC* **TGA GGA GAC TGT**-3'

The restriction sites (*Eco*RV for SC-1 and *Eco*RI for SC-2) are shown in *italics*. The sequences for scFv are in **bold**. Note that due to the *Eco*RV cloning site, the first two amino acids of the scFv are always Asp-Ile.

Linker Primers

We have used the most common linker sequence (Gly4Ser)$_3$.[9] Since this is a repeated sequence, the primers have to be chosen such that incorrect annealing (leading to, say (Gly4Ser)$_2$) is avoided. We chose to use different codons at the annealing site to fine-tune the annealing temperature of the PCR (Fig. 8–6). Because this annealing is independent of the internal V_H or V_L sequences, the pairing of V_H and V_L is completely random, and therefore this strategy is also suitable for repertoire cloning.

Sequence of the Linker Primers

SC-3: 5'- *GGA GCC GCC GCC GCC* **AGA ACC ACC ACC ACC AGC** CCG TTT TAT TTC-3'

SC-4: 5'- *GGC GGC GGC GGC TCC* **GGT GGT GGT GGT TCT** GAA GTT AAA CTG GT-3'

The sequences shown in italics are involved in the anneling of the amplified V_H and V_L fragments ($T_m = 58°C$). The linker sequences are shown in **bold**.

Cloning and Assembly Procedure

In the following, the assembly of an scFv fragment is given as an example. The assembly of other fragments, using other vectors described in the text, is carried out analogously. The methods generally follow ref. 52.

Protocol 1: PCT Cloning of an scFv Fragment

(A) PCR amplifications of individual domains
1. Mix the following in order:

> 5 μl 10X PCR buffer (0.2 M Tris-HCl, pH 8.8, 0.1 M KCl, 0.1 M $pNH_4)_2SO_4$, 20 mM $MgSO_4$, 1% Triton X-100)
> 0.5 μl 100X BSA (10 mg/ml)
> 1 μl of a solution 10 mM in each dNTP
> 0.2 μg of the DNA encoding the antibody fragment
> 1 μl of primer (50 pmol/μl stock solution) SC-1 and SC-3 for V_L and SC-2, SC-4 for V_H respectively
> 1 μl of vent polymerasc (2 u/μl) (New England Biolabs)
> H_2O to give final volume of 50 μl

2. Overlay the mixture with approximately 50 μl of mineral oil (Sigma, heavy whitc oil).
3. 30 cycles of PCR: 92°C, 1 min.; 60°C, 1 min.; 72°C, 1 min. We use a Perkin Elmer DNA Thermal Cycler TC1.

> As a negative control, the DNA template solution or the PCR primers are omitted. The 0.2 μg of the DNA listed above is given for using a cloned gene as the template. When starting from mRNA, we use the RNA extraction kit (No. 27-9270-01), the mRNA purification kit (No. 17-9258-02) and the first strand cDNA synthesis kit (No. 27-9261-01) from Pharmacia.

(B) PCR assembly of the V_H and V_L genes to an scFV fragment
1. Mix the following in order:

> 5 μl 10X PCR buffer
> 0.5 μl 100X BSA (10 mg/ml)
> 1 μl of 10 mM dNTP
> 1 to 1.5 μl of V_H and V_L mixture (in 1:1 proportion, as estimated from the agarose gel of reaction mixture from step 3 above)
> 1 μl of vent polymerase
> H_2O to give final volume of 50 μl

2. Overlay the mixture with approximately 50 µl of mineral oil.
3. 2 cycles of OCR: 92°C, 1 min.; 60°C, 1 min.; 72°C, 1 min.
4. Add 1 ml of primer SC-1 and SC-2 under the mineral oil, respectively.
5. Continue PCR for 30 cycles under the same conditions.

As a negative control, the V_H and V_L mixture or the PCR primers are omitted. Although it is possible to assemble the scFv fragment using the crude PCR mixture of V_H and V_L as the template, it usually yields better results in our hands when the V_H and V_L PCR-bands are first purified (Fig. 8–6).

Protocol 2: Cloning of scFv into the pIG Vectors

(A) Purification of PCR product
1. The PCR mixture containing the assembled scFv gene is purified on a 1% LMP agarose gel using TAE as the running buffer.
2. The band with the correct molecular weight as detected under long-wavelength UV light is excised (see Fig. 8–6) and mixed with TEN buffer (0.1 M Tris-acetate, pH 7.5, 0.5 M NaCl, 5 mM EDTA).
3. The suspension is heated to 65°C until the agarose is completely molten and dissolved in the buffer (approximately 5 min.).
4. The solution is extracted 2X with Roti-Phenol (Roth), 1X with phenol-chloroform-isoamylalcohol (25:24:1) and 2X with chloroform-isoamylalcohol (24:1). Each organic layer can be back-extracted with a small volume of TE buffer.
5. To the final aqueous layer, 5 M NaCl is added to give a final concentration of 0.3 M. Two volumes of absolute ethanol are added and the suspension is left at −20°C for over 2 hr (it is preferable to keep it overnight at this temperature).
6. The suspension is centrifuged (14,500 × g, 4°C, 45 min), washed with 70% ethanol (kept at −20°C) and dried (in a speed-vac).
7. The pellet is dissolved in 20 µl of TE.

(B) Restriction digest of the PCR product and the cloning vector
1. Mix the following:

   5 µl of buffer "H" (Boehringer-Mannheim)
   approximately 0.2 µg of the vector or 0.5 mg of the PCR product
   2 µl of *Eco*RV (20 u/µl)
   2 µl of *Eco*RI (20 u/µl)
   H_2O to give a final volume of 50 µl

2. The mixtures are incubated at 37°C overnight.
3. The fragments of the desired size are purified via LMP agarose gel and extracted with phenol-chloroform (see steps 4 to 7 above).

(C) Ligation and transformation of competent *E. coli* cells
1. Mix the following on ice:

> 2 μl of 10X ligase buffer (containing 10 mM ATP)
> 50 fmol of the *Eco*RV–*Eco*RI fragment of the vector
> 150 fmol of the *Eco*RV–*Eco*RI digested PCR fragment
> 1 μl of concentrated T4 DNA ligase (2,000 u/μl, New England Biolabs units)
> H$_2$O to give a final volume of 20 μl

> As a control, the PCR fragment is omitted.

2. The mixtures are incubated at 16°C overnight.
3. 5 μl of each mixture is used to transform 200 μl of thawed (or freshly prepared) competent JM83 cells from a frozen stock. The JM83 cells are made competent by the "Simple Efficient Method" (SEM).[40]
4. Ten percent of the transformed cells (approximately 100 μl) is spread onto an LB agar plate (containing streptomycin and ampicillin), and the rest is centrifuged and most of the supernatant is removed. The cell pellet is resuspended in the remaining medium and spread on another LB plate, containing the antibiotics.
5. The agar cultures are incubated at 37°C for approximately 12 hr.

(D) Screening the correct clones
1. Ten to 20 colonies are picked (depending on the number of colonies on the control plate) and are used to inoculate 2 ml of LB media (containing the antibiotics).
2. The cultures are shaken at 37°C for 6 to 12 hrs.
3. The plasmids are isolated following the mini-boiling method.[53] Alternatively, commercial DNA preparation kits (e.g., Quiagen from Diagen GmbH) can be used.
4. The screening of the correct clones is carried out by *Eco*RV, *Hind*III double digest (see Fig. 8–3).

Expression by Secretion: Procedure

As an example, the expression and purification of an scFv fragment is described. Here, immobilized metal affinity chromatography (IMAC) is described as a general purification method, but antigen-affinity chromatography is another powerful purification method.

Protocol 3: Expression and Purification of Recombinant scFv Fragments

1. An overnight culture of *E. coli* JM83 (50 ml) containing the plasmid encoding the scFv fragment is used to inoculate 2 l of LB medium

(25 µg/ml streptomycin and 100 µg/ml ampicillin). The culture is shaken at room temperature (RT) until $OD_{550} = 0.6$ is reached (2 to 3 hr.).

2. 2 ml of 1 M IPTG solution (1 mM final concentration) is added and the shaking is continued for another 4 hrs.

3. The culture is harvested by centrifugation (Sorval GS-3, 5,000 rpm, 4°C, 30 min.) and the supernatant is thoroughly removed.

4. The cell pellets are suspended in 15 ml of TBS buffer (0.1 M Tris-HCl, pH 8.0, 1 M NaCl), and more TBS is added to reach a final volume of 20 ml.

5. 0.5 M EDTA (pH 8.0, 40 µl) is added (final concentration of 1 mM), and the suspension is stirred vigorously on ice for at least half an hour.[67]

6. The suspension is centrifuged (Sorval SS-34, 15,000 rpm, 4°C, 30 min) and the supernatant is carefully collected.

7. The supernatant is loaded onto a Diagen Ni-NTA agarose column (3 ml) preequilibrated with TBS buffer at 4°C. The supernatant is allowed to pass through the column with a flow rate of approximately 0.5 ml/min.

8. After the supernatant has completely entered into the gel bed, the column is washed with TBS buffer until the baseline is reached (about 20 column volumes) or overnight.

9. A washing solution containing 8 mM imidazole in TBS buffer (60 ml) is applied, followed by an 8 to 130 mM imidazole gradient (120 ml), and 5 ml fractions are collected.

10. Each fraction or every other fraction is analyzed by 12.5% silver-stained SDS-PAGE.

11. The fractions containing the pure scFv fragments are collected and concentrated by ammonium sulfate precipitation. The precipitate is collected (centrifuged at $15,000 \times g$, SS-34, 4°C, 30 min).

12. The precipitated scFv fragment is dissolved in 1 ml of TBS buffer and dialyzed against 2×2 l of TBS buffer. The protein solution is stored at 4°C (0.03% sodium azide is added to prevent bacterial growth) or at -20°C.

The *E. coli* strain JM83 (*ara*, Δ[*lac, pro*], *str*A, *thi*-1 [Φ80d*lac*ZΔM15])[51] has been found useful for expression experiments under shake flask conditions, since it shows comparatively little leakage of periplasmic proteins or lysis before and after induction. Many other strains tested do not perform as well, while a few others do. The genetic basis for this relative robustness is unclear; we found no obvious correlation to the genetic markers indicated. In the controlled culture conditions of a high-cell-density fermentation, leakage and lysis are suppressed,[19] perhaps because growth is artificially slowed by the feeding regime, giving the cells the time for more extensive cell wall synthesis. The considerations for choosing a strain in fermentation are therefore different ones: It should be prototrophic (to avoid the feeding of special nutrients) and have no defects preventing the growth to high cell densities.[19] Some workers find the introduction of a phage resistance useful.

Optimizing the Yield in Secretion

The major goal in expression is to optimize the yield. For this purpose, it is necessary to define the bottleneck in the overall process. There are several areas where a problem could conceivably lie: (1) plasmid copy number, (2) amount and stability of mRNA, (3) translation, (4) transport, (5) folding and assembly, and (6) protein degradation. We will now discuss the relevance of each of these points to antibody secretion.

1. Plasmid copy number: Most modern vectors, especially those based on modified ColE1 origins as in the pUC series,[28] have high copy numbers. However, this copy number is a function of temperature.[29] At the lower temperature recommended for secretion, this copy number is about equal to pBR322. One problem not completely understood is the spontaneous plasmid loss from *E. coli*, on which secreted proteins impose a stress. It appears that this is not due to the decrease in the steady state copy number,[53] but rather to a total loss of plasmid in a subpopulation of cells.[19] Stable antibiotics are helpful to combat this phenomenon before induction, and constitutive plasmid-encoded functions (e.g., the antibiotic resistance protein) can be used to measure this quantitatively. However, since growth is slowed or even stops after induction, plasmid loss during the expression phase cannot be prevented with antibiotics.[19]

2. Amount of mRNA: The amount of mRNA is mostly determined by initiation frequency and degradation rate. The initiation rate (i.e., the promoter strength) is crucial, but easily remedied by the choice of any suitable "strong" promoter, as this is now well understood.[54] There is no indication that promoter strength is dependent on the gene itself. In principle, any strong promoter should be suitable, but other criteria such as complete repressibility narrow down the choices.

The mode of induction is also crucial, but for protein folding reasons. Antibody folding proceeds better at low temperature, and thus a heat-inducible promoter is less suitable for functional expression. Apparently, even if the temperature is lowered to 25°C after the heat pulse, some nucleation of protein precipitation can occur, decreasing the amount of soluble protein in some systems (Wülfing and Plückthun, unpublished observations). The degradation of mRNA in *E. coli* is only beginning to be understood.[55]

Therefore, even if an antibody gene were chemically synthesized completely, as has been done,[56] it is not clear what features should be avoided or accumulated within the coding region to increase the half-life of the mRNA. The sequence outside the coding region is also important, but there is probably still no better strategy than to use 5' regions (i.e., the promoter and Shine-Dalgarno sequence) and 3' regions (i.e., the transcription terminator) of well-expressed genes to guard against exonucleolytic attack.

3. Translation: Translation efficiency is mostly determined by translation initiation.[57] Generally, elongation plays less of a role, as the amounts achievable with secretory systems will reach other limits much sooner than that of the translation elongation.

Despite intensive research, translation initiation is still not rationally understood, although several trends have become apparent. Since translation initiation efficiency is a complex function of mRNA secondary structure, Shine-Delgarno sequence, its distance from the AUG start codon, but also the primary sequence of the 5'-untranslated region itself and even that of the beginning of the translated region,[57] cassettes of other well-expressed genes are normally used, and the whole sequence from the promoter to the start codon is left unchanged. In the case of a transported gene, the signal sequence may also come from the same gene as the Shine-Dalgarno sequence. To increase translation efficiency further, a two-cistron strategy has been used in our laboratory, in which a strongly expressed gene (e.g., β-galactosidase) is present only as a very short peptide, immediately followed by a stop codon and another Shine-Dalgarno sequence preceding the signal sequence of the protein of interest.[34, 58] The intention is to gather ribosomes via the first and second ribosome binding site, each one optimized in its natural context. However, other efficient upstream regions are now available, which may be even more potent than the two-cistron approach. In using the extremely potent upstream region of gene 10 of phage T7 with a secreted protein, occasional problems of plasmid rearrangements have been noticed (Freund and Plückthun, unpublished).

4. Transport: The expression limit is probably not restricted by translocation through the membrane. When plasmids containing the w.t. *lac* promoter and the *lacUV5* promoter are compared, the amount of soluble protein remains remarkably constant. Yet, the amount of processed but insoluble protein dramatically increases (Knappik and Plückthun, unpublished experiments).[14] In one example,[21] it has been directly demonstrated that this insoluble protein can be degraded by externally added protease after spheroplasting the cells. Therefore, this insoluble and processed protein must have been transported and then precipitated. Only much smaller amounts of precursor can be seen in these experiments, which are soluble and cannot be degraded by externally added protease, and thus must be cytoplasmic. These observations suggest that it is periplasmic folding, not transport, that is the bottleneck of antibody expression.

5. Folding and assembly: Taking all data together, it appears that periplasmic protein folding most frequently limits the yield of functional expression of antibody fragments. The following observations support this notion: (1) in the same vector, antibodies with different primary sequences differ dramatically in the distribution between periplasmic soluble and periplasmic precipitated protein (Knappik and Plückthun, unpublished experiments), (2) the same is true for different fragments of the same antibody,[12,14] (3) variations in promoter strength usually increase the amount of periplasmic precipitated proteins, but not of folded protein.[12 14]

Empirically, the simplest measure to improve the folding yield is to lower the growth temperature of the cells to about room temperature. Probably the distribution of folding intermediates between pathways of folding and aggregation is more favorable at lower temperature. Depending on the primary sequence of the antibody, these effects may be more or less pronounced.

The most effective way to improve folding may be to change the primary sequence of the antibody. Analyzing the sequences of well-expressed and poorly expressed antibodies as a basis, it was indeed found to be possible (1) to improve the ratio of folded to unfolded protein and (2) to diminish the lysis of the cells, usually caused by the onset of antibody expression, by simple point mutations of the antibody (Knappik and Plückthun, in preparation). However, the mechanism of how these mutations exert their effects is not yet known.

Two slow steps of antibody folding, which is taking place in the periplasm of *E. coli*, could conceivably be rate limiting: disulfide formation and proline *cis-trans* isomerization. Both steps occur during the folding process of antibody fragments and, for both, periplasmic proteins exist in *E. coli* that are known to catalyze these processes: proline *cis-trans* isomerase (rotamase) and disulfide isomerase (DsbA).[59,60] Nevertheless, the overexpression of neither was found to have a dramatic effect, nor was there any synergistic effect noted when both were coordinately overexpressed.[12] It is possible that the aggregation tendency of the folding intermediates just cannot be overcome by catalyzing these steps. This does not exclude that some antibody fragments that suffer less from aggregation of early folding intermediates (because of their primary sequence) are further helped along the folding pathway by these putative folding catalysts. Interestingly, antibody fragments *do* make use of the *E. coli* disulfide formation machinery, as no functional fragments can be obtained in the absence of the DsbA gene.[21] However, the availability of large amounts of DsbA does not help further.

6. Degradation: No comprehensive study on the effect of *E. coli* proteases on series of antibody fragments has yet been reported. It is likely that many recombinant fragments are being somewhat degraded by proteases, but to various extents, and depending on the sequence. Occasionally, there are degradation products seen on SDS-PAGE; however, small peptide fragments may be removed much faster than they are made, and they probably do not accumulate.[61] Protease-deficient strains have been tested in our lab, yet without dramatic effects on the yields of folded antibody fragments (Schroeckh and Plückthun, unpublished experiments). On the other hand, even bivalent miniantibodies with very extended hinge regions between the antibody domain and the dimerization domain can be prepared from *E. coli* with only small amounts of degradation in the hinge region.[18,19]

DETECTION: USING THE FLAG "TAG" FOR IMMUNODETECTION OF ANTIBODY FRAGMENTS

In many experiments involving the expression of antibody fragments in *E. coli* it is desirable to detect and quantify the products using standard immunological methods. Therefore, a system that allows highly sensitive and specific detection regardless of the particular antibody fragment expressed would be necessary. The costly and time-consuming production of antisera against particular fragments can thus be avoided.

The FLAG epitope was originally described as consisting of a highly charged and therefore soluble eight-amino-acid peptide (DYKDDDDK), which is recognized by a commercially available monoclonal antibody in a calcium-dependent manner.[35,36] The fusion of the peptide sequence to any antibody fragment allows the rapid and sensitive detection of the expressed protein by immuno-blotting or ELISA, and even one-step purifications using an anti-FLAG affinity column.[35-37] There is no observable cross-reactivity with *E. coli* proteins present in the crude extract.

The C-terminal four amino acids of the FLAG sequence (DDDK) were originally designed as an enterokinase cleavage site to allow the specific removal of the tag after purification. We found that these four amino acids are not necessary for the sensitive and specific recognition of the anti-FLAG antibody, and they can be omitted for all applications where there is no need for the removal of the FLAG peptide,[37] leading to the four-amino-acid tag DYKD. We then investigated the influence of the amino acid at the fifth position, which is an aspartic acid in the original FLAG epitope used as the immunogen,[35,36] and we found that the sensitivity of recognition is increased about ten-fold if this position contains a glutamic acid instead of an aspartic acid. A change of the positions six to eight gave only minor differences in sensitivity of detection.[37] Therefore, we suggest the five-amino-acid tag

DYKDE as the optimal sequence. So far, this is the shortest high-affinity tag sequence for which an antibody is commercially available. However, even when the fifth amino acid is Ile, as results from fusing Asp-Tyr-Lys to the N-terminus of the V_L domain in the pIG vectors (Fig. 8–3), very good sensitivity is obtained. This design creates an appendage of only three additional amino acids.[37]

The FLAG sequence can be fused either to the N- or C-terminus of a given antibody fragment. The fusion to the N-terminus has several advantages:

1. It was found by competition ELISAs that the anti FLAG antibody binds three to four orders of magnitude better under conditions where the α-amino group of the first amino acid is freely accessible.[35, 36] The expression of antibody fragments in native form by transporting them to the periplasmic space of *E. coli*[6, 20] generates such a free N-terminus, since the signal sequence is cleaved off after transport. We found that the FLAG sequence fused between the signal sequence and the mature part of the antibody fragment maintains correct processing of the signal peptide after transport, leading to the desired free N-terminal FLAG peptide.[37]

2. The FLAG sequence at the N-terminus is stable and is not removed by *E. coli* proteases, which was confirmed by N-terminal sequencing of purified antibody fragments produced in *E. coli*. Furthermore, at least the short FLAG sequence does not interfere with the binding of the antibody, regardless of which particular fragment is used (Fab, Fv, or scFv fragment) and regardless of which chain carries the FLAG (light chain or heavy chain or both).[37]

3. After cloning and successful expression of an antibody fragment in *E. coli*, there is often the need of constructing several variants of this antibody, for example, connecting the two chains to an scFv, constructing a Fab fragment from an Fv fragment by insertion of the constant domains or vice versa, or fusing the antibody gene to other genes (protein III of phage M13, alkaline phosphatase, toxins) or peptides (*his-tag* for purification or helix peptides for dimerization). All of these constructions involve one or both C-termini of the antibody chains. A FLAG peptide at the N-terminus does not interfere with additional cloning steps at the C-terminus and facilitates direct comparisons of these constructs.

The only disadvantage of the N-terminal location of the epitope appears to be that in functional ELISAs (where the plate is coated with antigen, antibody fragment is bound and detected with an anti-*tag* antibody), antigen binding may make the N-terminal epitope inaccessible to the anti-FLAG antibody. We prefer C-terminal *myc-tag* sequences[38,39] for this purpose (Fig. 8–3).

Protocol 4: Western Blot Using the FLAG Peptide as Detection Tag

This procedure is designed to detect as few as 1 ng of an antibody fragment containing the FLAG sequence with the sequence DYKDE at its N terminus. Therefore, 10 μl of an *E. coli* culture with an $OD_{550} = 1$ (about 10^6 cells) expressing the fragment can be loaded directly on the gel, or an equal amount can be used after harvest and fractionation of the *E. coli* culture. Generally, the amount of antibody fragment loaded on the gel should be in the range of 1 to 100 ng, if this protocol is used.

1. Perform the SDS gel electophoresis and blot the gel using standard protocols. Both nitrocellulose or PVDF membranes can be used.
2. All following steps are carried out at room temperature. Block the membrane by soaking it for 30 min in 50 ml blocking solution (1% PVP-40 in TBST) with shaking.
3. Wash the membrane 3 times with 50 ml TBST.
4. Dilute 0.75 μg per lane of the anti-FLAG antibody M1 in blocking solution (15 μg antibody in 50 ml blocking solution for a gel with 20 slots). Soak the membrane for 60 min in this antibody solution. The solution can be stored at 4°C (add 0.05% thimerosal to prevent bacterial growth) and used several times.
5. Wash the membrane 3 times with 50 ml TBST.
6. Perform a second blocking step by soaking the membrane for 30 min in 50 ml 1.5% gelatine in TBST.
7. Wash the membrane 3 times with 50 ml TBST.
8. As the second antibody, use an Fc-specific anti-mouse antiserum conjugated to horseradish peroxidase (POD). Dilute the antiserum in 50 ml blocking solution. To avoid a high background, a dilution up to 1:25,000 is recommended. Soak the memrane for 45 min in this solution.
9. Wash the membrane 3 times with 50 ml TBST, then 3 times with 50 ml TBS. These final washing steps also reduce the background and can be extended up to several hours without losing the intensity of the signal.
10. Detect the POD-conjugate by using enhanced chemoluminescence with luminol as substrate and an X-ray film as detection medium. Follow the protocol given by the supplier. Start with an exposure time of one minute and then reduce or increase the time accordingly.

The TBST listed in step 2 consists of 50 mM Tris Cl, pH 7.4; 15 mM NaCl; 1 mM $CaCl_2$; and 0.05% Tween-20. TBS is TBST without Tween. PVP-40 is polyvinylpyrrolidone, MW 40,000 Da.

The POD in step 8 together with luminol as substrate gives a high sensitivity for the detection. If different detection systems are used, the amount of antibody fragment loaded on the gel should be increased. The optimal dilution depends on the antiserum used and should be checked by performing serial dilutions from 1:2,500 to 1:25,000.

PURIFICATION OF NATIVE ANTIBODY FRAGMENTS

General Considerations

Purification of whole antibodies has usually relied on classical chromatograph, antigen-affinity chromatography, or affinity chromatography using bacterial immunoglobulin-binding proteins such as proteins A, B, G, or L.[62–65] However, the usefulness of this strategy for Fv or scFv fragments is fairly limited, as these bacterial proteins bind mostly to the constant domains, and only few subgroups of V domains are recognized.[64,66]

However, using affinity tails, any fragment can now be purified by rather convenient and reproducible procedures, and—in contrast to immunoaffinity purifications—this technology can be carried out on a very large scale. The most convenient strategy is probably the use of a stretch of histidines at the C-terminus[34,37] (Fig. 8–3). This has been successfully applied to scFv fragments (Protocol 3) either of the form V_H-linker-V_L-his$_5$ or V_L-linker-V_H-his$_5$, to V_L domains (V_L-his$_5$), and also a Fab fragments, where the his$_5$ tail was fused only on C_H1. Since the theavy chain of most Fab fragments is practically insoluble, if not paired with a light chain, this amounts to a purification of assembled Fab fragments (Knappik, Bauer, and Plückthun, unpublished), even when the two chains are not covalently linked. In the mouse kappa V_L domain, the last two amino acids Arg-Ala (number 108 and 109 according to Kabat) are replaced by histidines in our standard constructs, and only three additional His residues had to be added to the end. X-ray crystallography showed that this had no influence at all on the structure of the V_L domain.[67] In the case of V_H, the histidines were added behind residue 113 (Kabat numbering), separated by Glu-Phe for cloning purposes (Fig. 8–3).

If the yield of a particular fragment is very poor or contains some unusually sensitive sites, it is possible that small amounts of a putative copurified protease become noticeable. Having a better ratio of antibody fragment to protease contaminant, this degradation is insignificant. Nevertheless, this copurification can occur both on Ni^{2+}-NTA and Zn^{2+}-IDA, irrespective of whether the procedure is carried out under nondenaturing or denaturing conditions.[67] This suggests that this protease can itself be easily refolded. Its identity is not yet clear, and it cannot be inhibited by the protease inhibitors EDTA or PMSF. Yet, for most fragments studied, this has not been a problem.

Factors Influencing IMAC Purifications

Influence of the Histidine Tail Sequence

We recommend the use of his_5 and his_6 tails. Other tails for IMAC have been investigated with antibody domains,[34] but the retardation was found to be worse than with his_5. A dynamic equilibrium involving all juxtaposed imidazoles of the tail may contribute to the binding and favor oligo-his sequences as the best ligand (Hochuli, personal communication). Longer oligo-histidine tails may bind better, but appear to lead to lower yields of purified protein, perhaps because of proteolytic degradation. We cannot rigorously exclude problems with the transport through the membrane using very long histidine tails, but no toxic effects nor significant precursor bands were observed with a his_9 tail in the case of the scFv fragment.

Influence of the Chromatography Conditions

A number of chromatography conditions were previously tested using V_L domains as a model.[67] It was found that the buffer composition itself is of secondary importance and can thus be adapted to other requirements of the procedure. We prefer elution with an imidazole gradient at constant pH. The choices of the ligand and the metal, however, are very important, as they are interrelated. Iminodiacetic acid (IDA) is used most advantageously with Zn^{2+}, as Ni^{2+} leads to a long tailing of host proteins and thus exceedingly long washes. Nitrilotriacetic acid (NTA), on the other hand, must be used with Ni^{2+}, since the recombinant protein elutes too early with Zn^{2+}. If the expression of a particular fragment is very poor (e.g., because of severe folding problems), it is possible that homogeneity cannot be achieved in a single step. In such cases, a fractionated ammonium sulfate precipitation or an additional ion exchange chromatography step have been found useful.

SPECIAL ANTIBODY FRAGMENTS: MINIANTIBODIES

Bivalency is a very efficient means of increasing the functional affinity of an antibody to a surface or to a polymeric antigen. Its physical basis[16,17] is that, once the first binding site is bound to the surface, the local concentration of the second is dramatically increased. This can be quantified, based on simplifying assumptions such as neglecting the energy to "bend" either antigen or antibody, inhomogeneous surface distributions, or already occupied sites. Importantly, this gain from a second binding site is inversely proportional to the distance of the binding sites, provided this distance is long enough to ever reach a second antigen. Furthermore, it is proportional to the surface concentration of epitopes. Nature has presumably optimized the geometry of antibodies for the antigens typically encountered. Yet, no successful attempts to make functional bivalent whole antibodies in *E. coli* have been reported so far. Part of the problem is that, at least in IgG, and C_H2 domains contact each other only via the oligosaccharide residues, which are of course absent in *E. coli*.

Functional analogs of bivalent whole antibodies have been designed that assemble in *E. coli*.[18,19] They have been termed "miniantibodies" because of their small size. They are based on scFv fragments, linked via a hinge region to a small dimerization domain, which consists of amphipathic helices (Fig. 8–7). The hinge region has been taken from the long upper hinge of mouse IgG3, giving the miniantibody enough flexibility for the dimerized binding sites to adapt to the same surface.

Two different principles of dimerization domains have been tested. One is based on parallel coiled-coil helices, as they occur in eukaryotic transcription factors (leucine zippers) (Fig. 8–7(a)). However, this was not found to be the optimal design. These constructs have lower functional affinities, and there have been problems with nonspecific binding and, consequently, scattering of binding data.

The best construct found so far is based on anti-parallel amphipathic helices, arranged as a helix-turn-helix module (Fig. 8–7(b), dHLX), presumably forming a four-helix bundle.[19] A larger percentage of the hydrophobic surface of each helix is shielded in the 4-helix bundle than in the coiled-coil helices. This construct (the sequence is given in Fig. 8–7(c)) was found to give rise to an identical functional affinity (avidity) as a whole antibody and to be very stable. It could be produced in high-cell density fermentation at about 200 mg/l,[19] in a model construct with the phosphorylcholine binding scFv fragment of the antibody MCPC603.

Most experiments have been carried out with single-chain Fv fragments of the orientation V_H-linker-V_L, and in this context, there was only negligible proteolysis in the hinge. These miniantibodies are most efficiently constructed from synthetic DNA encoding the dimerization domain introduced between the *Eco*RI and *Hin*dIII site of the expression vectors ("B" cassette, Fig. 8–3).

The expression procedures of the miniantibodies follow the expression of all other fragments (Protocol 3). The assembly of the molecules occurs spontaneously in the bacterial periplasm, and functional dimers can be directly isolated. However, purification has to rely on antigen-affinity chromatography or N-terminal tags, because additional C-terminal tags behind the dimerization domains were found to lead to proteolytic instability.

a. coiled coil derivatives

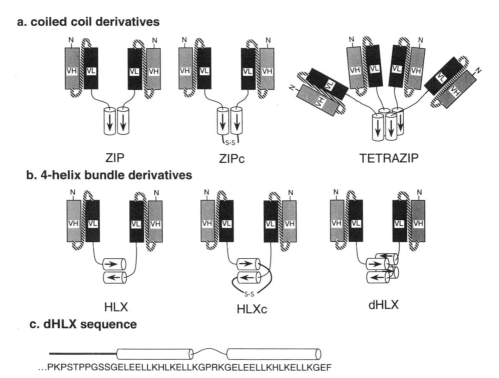

ZIP   ZIPc   TETRAZIP

b. 4-helix bundle derivatives

HLX   HLXc   dHLX

c. dHLX sequence

...PKPSTPPGSSGELEELLKHLKELLKGPRKGELEELLKHLKELLKGEF

Figure 8–7. Bivalent fragments that have been shown to assemble in *E. coli*. In each case, an scFv fragment is connected to a hinge region followed by an amphipathic helix. In (a), a parallel coiled-coil helix from a leucine zipper is used, without (ZIP) or with (ZIPc), a short extension ending in a cysteine. On the right (TETRAZIP), the sequence of the zipper leads to tetramerization by exchanging the amino acids at the interface (Pack and Plückthun, in preparation). In (b) the helix comes from a 4-helix bundle design by deGrado and co-workers. On the top left (HLX) only one helix is fused, but the predominant molecular species are dimers. In the middle (HLXc), they are connected by a peptide, which ends in a cysteine. On the right, two helices are fused in tandem and a 4-helix bundle is probably obtained (dHLX), as very stable dimers are formed *in vivo*. For details, see Pack and Plückthun,[18] Pack et al.,[19] and Pack and Plückthun, submitted. (c) Amino acid sequence of the hinge and the helix-turn-helix module used for dimerization in the dHLX construct. The helical regions are indicated diagrammatically.

EXPRESSION OF ANTIBODY FRAGMENTS AS INCLUSION BODIES

General Considerations

The production of antibody proteins as cytoplasmic inclusion bodies in *E. coli* is also possible as an alternative. All types of antibody fragments (Fab,

Fv, scFv, and even the chains for the whole antibody) have been produced this way,[8–11,68–70] and a variety of strains, plasmids, and promoters have been used. The use of the T7 system[71] as a particularly strong but regulatable system was found useful by several investigators.[10,11]

Usually, the inclusion body approach is carried out using genes not encoding signal sequences. Therefore, the antibody fragments stay in the cytoplasm and largely precipitate. Since precipitation is desired, it is useful to do exactly what needs to be avoided when secreting the antibody, namely to grow the cells at higher temperature, e.g., 37°C.

Alternatively, one may also isolate that portion of the secreted protein which precipitates after transport to the periplasm. This has been described for scFv fragments[24–26] and Fab fragments.[72] At higher temperatures (37°C), the protein can still be transported, but the folding in the periplasm is often severely impaired, although apparently not for all antibody sequences.[27] In the oxidizing milieu of the periplasm, some of the precipitated protein already has disulfide linkages, but it is not known what percentage of molecules have them, and how many of them are correct.

A vector for cytoplasmic inclusion body formation is shown in Fig. 8–8. Cloning of the antibody fragments (without signal sequences) can be carried out analogously to Protocols 1 and 2. For expression experiments, the plasmid must be transformed into *E. coli* BL21 (DE3),[71] which carries the T7-RNA polymerase in the chromosome under the control of the lacUV5 promoter.

Protocol 5: Refolding of scFv Fragments from Inclusion Bodies

1. Grow BL21 (DE3) cells containing the plasmid of interest at 37°C. In isotope labeling studies, the limiting component would lead to early starvation, and growth is carried out only to an OD_{550} of 0.5. In other experiments without limiting isotopic nutrient, much higher ODs can be used. Induce for about 4 h. This recipe is given for 2 liters of cells.
2. Centrifuge the cells at 3,000 × g for 10 min, and resuspend them in 100 ml of 10 mM TrisHCl, 2 mM $MgCl_2$, pH 8.0.
3. Lyse the cells by two-fold passage through a French Pressure Cell at 1.1 kbar.
4. Treat the lysate with DNAse (approximately 5,000 u) and RNAse (approximately 1,000 u) for 30 min at 37°C.
5. Add 0.5 M EDTA, pH 8.0 (20 ml) and Triton X-100 (1 ml) and store the solution in an ice-bath for aboout 30 min.
6. Centrifuge the sample at 16,000 × g for 10 min and wash the pellet with 0.5 M GdmCl, 0.1 M TrisHCl, 20 mM EDTA, pH 6.8 once and then two more times with 0.1 M TrisHCl, 5 mM EDTA, pH 6.8.
7. Solubilize the pellet in 50 ml of 5.5 M GdmCl, 0.2 M Tris, 2 mM EDTA, 0.14 M DTE, pH 9.5 and stir under argon at room temperature for 2 h.

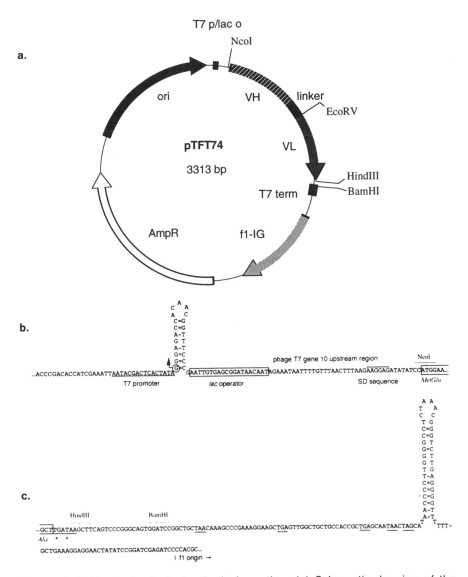

Figure 8–8. Vector for inclusion body formation. (a) Schematic drawing of the pTFT74 vector. *T7p/lac o* denotes the *T7* promoter with the *lac* operator, *T7 term* the phage T7 terminator of gene 10, *f1-IG* the intergenic region (origin of replication) of the f1 phage, *Amp^R* the ampicillin resistance and *ori* the origin of replication of the plasmid. The vector does not contain a *lacI* gene, and no other *E. coli* terminator is present beside the T7 terminator behind the antibody gene. (b) Details of the promoter/operator region. The promoter, operator, and Shine-Dalgarno sequence of phage T7 gene 10 are indicated, the transcriptional start is circled, and a putative hairpin structure is shown.[71] (c) Details of the terminator region. The end of the antibody gene is shown, stop codons are underlined, and restriction sites are overlined. A putative hairpin structure is shown.[71].

8. Centrifuge at 30,000 × g for 30 min to remove insoluble material, and dialyze the supernatant three times against 1 l of the same buffer not containing DTE.

9. Renaturation of the protein now takes place after diluting the solubilisate into 5 l of renaturation buffer (0.8 M arginine, 0.2 M Tris, 2 mM EDTA, 0.2 mM reduced glutathione, 1 mM oxidized glutathione, 0.2 mM benzamidine hydrochloride, 0.2 mM ε-caproic acid, pH 9.5). We have also included hapten at 100-fold the K_D value in renaturing hapten-binding antibodies. **It is crucial to let this reaction continue for more than 50 h at 10°C, as it is very slow.** It is useful to add the denatured protein in several small aliquots to the refolding buffer. At higher temperatures, more precipitation occurs.

10. Concentrate the protein with an Amicon RA2000 cell to a final volume of 300 ml.

11. Dialyze the solution against buffer for affinity chromatography (e.g., 0.2 M sodium borate, 0.4 M NaCl).

12. Purify functional scFv by antigen affinity chromatography, if possible.

Factors Influencing *in vitro* Refolding

The refolding of antibodies is not principally different from that of other disulfide containing proteins.[73–75] The disulfide formation must be kinetically catalyzed and thermodynamically allowed, and at the same time, aggregate formation must be suppressed. The ratio of reduced to oxidized glutathione has to be optimized for the particular antibody to be refolded, covering the range of 1:5 to 5:1 molar ratios. Because of the importance of S–S formation, it is useful to carry out refolding at high pH, in order to speed up the disulfide exchange reactions, since the reactive species is the thiolate anion.

The aggregation of folding intermediates is a severe problem and an important side reaction lowering the yield *in vitro* and *in vivo*. Thus, rather low protein concentrations have to be used. It is useful to add the unfolded protein to the refolding mix in aliquots, since the folded protein has a much higher solubility. Additionally, additives such as arginine are useful, as they appear to increase the solubility of intermediates.[73–75] Too low a final protein concentration, on the other hand, may lead to gigantic volumes, and prevents chain association when refolding heterodimeric Fab fragments.

CONCLUSIONS

The expression of various antibody fragments in *E. coli* has been a useful technology in the study and continued engineering of antibodies. It greatly lowers the activation barrier to work with recombinant and engineered antibodies and should therefore be autocatalytic in the further development of the field. Many of the initial difficulties of scale-up, sufficient yields, and simple work-up procedures are being solved now and should add to the advantages of the technology. While there are still many fundamental questions of protein folding, structure, and stability unsolved, which have a direct bearing on the expression technology itself, there is justified hope that the rational understanding will catch up with the technological advances, and they will cross-fertilize.

REFERENCES

1. Plückthun, A. 1991. Antibody engineering: Advances from the use of *Escherichia coli* expression systems. *Biotechnology* 9:545.
2. Plückthun, A. 1992. Mono- and bivalent antibody fragments produced in *Escherichia coli*: engineering, folding and antigen binding. *Immunolog. Reviews* 130:151.
3. Plückthun, A. Antibodies from *Escherichia coli*. In *Handbook of Pharmacology*. Springer-Verlag. In press.
4. Givol, D. 1991. The minimal antigen-binding fragment of antibodies—Fv fragment. *Mol. Immunol.* 28:1379.
5. Glockshuber, R., Malia, M., Pfitzinger, I., and Plückthun, A. 1990. A comparison of strategies to stabilize immunoglobulin Fv-fragments. *Biochemistry* 29:1362.
6. Skerra, A., and Plückthun, A. 1988. Assembly of a functional immunoglobulin Fv fragment in *Escherichia coli*. *Science* 240:1038.
7. Plückthun, A. 1993. Stability of engineered antibody fragments. In *Stability and Stabilization of Enzymes*. van den Tweel, W.J.J., Harder, A., and Buitelaar, R.M., eds. Elsevier. 81–90.
8. Bird, R.E., Hardman, K.D., Jacobson, J.W., Johnson, S., Kaufman, B.M., Lee, S., Lee, T., Pope, S.H., Riordan, G.S., and Whitlow, M. 1988. Single-chain antigen-binding proteins. *Science* 242:423.
9. Huston, J.S., Levinson, D., Mudgett-Hunter, M., Tai, M., Novotny, J., Margolies, M.N., Ridge, R.J., Bruccoleri, R.E., Haber, E., Crea, R., and Oppermann, H. 1988. Protein engineering of antibody binding sites: recovery of specific activity in an anti-digoxin single-chain Fv analogue produced in *Escherichia coli*. *Proc. Natl. Acad. Sci. USA* 85:5879.
10. Huston, J.S., Mudgett-Hunter, M., Tai, M.S., McCartney, J., Warren, F., Haber, E., and Oppermann, H. 1991. Protein engineering of single-chain Fv analogs and fusion proteins. *Meth. Enzymol.* 203:46.
11. Freund, C., Ross, A., Guth, B., Plückthun, A., and Holak, T. 1993. Characterization of the linker peptide of the single-chain Fv fragment of an antibody by NMR spectroscopy. *FEBS Lett.* 320:97.
12. Knappik, A., Krebber, C., and Plückthun, A. 1993. The effect of folding catalysts on the *in vivo* folding process of different antibody fragments expressed in *Escherichia coli*. *Biotechnology* 11:77.

13. Pantoliano, M.W., Bird, R.E., Johnson, S., Asel, E.D., Dodd, S.W., Wood, J.F., and Hardman, K.D. 1991. Conformational stability, folding, and ligand-binding affinity of single-chain Rv immunoglobulin fragments expressed in *Escherichia coli. Biochemistry* 30:10117.
14. Skerra, A., and Plückthun, A. 1991. Secretion and *in vivo* folding of the Fab fragment of the antibody McPC603 in *Escherichia coli*: influence of disulphides and cis-prolines. *Protein Eng.* 4:971.
15. Shin, S.U., Wright, A., Bonagura, V., and Morrison, S.L. 1992. Genetically engineered antibodies. Tools for the study of diverse properties of the antibody molecule. *Immunol. Rev.* 130:87.
16. Karush, F. 1978. The affinity of antibody: range, variability and the role of multivalence. In *Immunoglobulins.* Litman, G.W., and Good, R.A., eds. Plenum. 85–116.
17. Crothers, D.M., and Metzger, H. 1972. The influence of polyvalency on the binding properties of antibodies. *Immunochemistry* 9:341.
18. Pack, P., and Plückthun, A. 1992. Miniantibodies: use of amphipathic helices to produce functional flexibly linked dimeric Fv fragments with high avidity in *Escherichia coli. Biochemistry* 31:1579.
19. Pack, P., Kujau, M., Schroeckh, V., Knüpfer, U., Wenderoth, R., Riesenberg, D., and Plückthun, A. 1993. Improved bivalent miniantibodies with identical avidity as whole antibodies, produced by high cell density fermentation of *Escherichia coli. Biotechnology.* 11:1271.
20. Better, M., Chang, C.P., Robinson, R.R., and Horwitz, A.H. 1988. *Escherichia coli* secretion of an active chimeric antibody fragment. *Science* 240:1041.
21. Glockshuber, R., Schmidt, T., and Plückthun, A. 1992. The disulfide bonds in antibody variable domains: Effects on stability, folding *in vitro*, and functional expression in *Escherichia coli. Biochemistry* 31:1270.
22. Collet, T.A., Roben, P., O'Kennedy, R., Barbas III, C.F., Burton, D.R., and Lerner, R.A. 1992. A binary plasmid system for shuffling combinatorial antibody libraries. *Proc. Natl. Acad. Sci. USA* 89:10026–30.
23. Barbas III, C.F., Kang, A.S., Lerner, R.A., and Benkovic, S.J. 1991. Assembly of combinatorial antibody libraries on phase surfaces: the gene III site. *Proc. Natl. Acad. Sci. USA* 88:7978–82.
24. Colcher, D., Bird, R., Roselli, M., Hardman, K.D., Johnson, S., Pope, S., Dodd, S.W., Pantoliano, M.W., Milenic, D.E., and Schlom, J. 1990. *In vivo* tumor targeting of a recombinant single-chain antigen-binding protein. *J. Natl. Cancer Inst.* 82:1191.
25. Gibbs, R.A., Posner, B.A., Filpula, D.R., Dodd, S.W., Finkelman, M.A.J., Lee, T.K., Wroble, M., Whitlow, M., and Benkovic, S.J. 1991. Construction and characterization of a single-chain catalytic antibody. *Proc. Natl. Acad. Sci. USA* 88:4001.
26. Whitlow, M., and Filpula, D. 1991. Single-chain Fv proteins and their fusion proteins. *Methods: A Companion to Meth. Enzymol.* 2:97.
27. Carter, P., Kelley, R.F., Rodrigues, M.L., Snedecor, B., Covarrubias, M., Velligan, M.D., Wong, W.L.T., Rowland, A.M., Kotts, C.E., Carner, M.E., Yang, M., Bourell, J.H., Shepard, H.M., and Henner, D. 1992. High level *Escherichia*

coli expression and production of bivalent humanized antibody fragment. *Biotechnology* 10:163.

28. Yanish-Perron, C., Vieira, J., and Messing, J. 1985. Improved M13 phage cloning vectors and host strains: nucleotide sequences of the M13mp18 and pUC19 vectors. *Gene* 33:103.

29. Lin-Chao, S., Chen, W.T., and Wong, T.T. 1992. High copy number of the pUC plasmids results from a Rom/Rop-suppressible point mutation in RNAII. *Mol. Microbiol.* 6:3385.

30. Yuckenberg, P.D., Witney, F., Geisselsoder, J., and McClary, J. 1991. Site-directed mutagenesis using uracil-containing DNA and phagemid vectors. In *Directed Mutagenesis, A Practical Approach.* McPherson, M.J., ed. Oxford: IRL Press. 27–48.

31. Hoogenboom, H.R., Marks, J.D., Griffiths, A.D., and Winter, G. 1992. Building antibodies from their genes. *Immunol. Rev.* 130:41.

32. Plückthun, A., and Skerra, A. 1989. Expression of functional antibody Fv and Fab fragments in *Escherichia coli. Meth. Enzymol.* 178:497.

33. Kleckner, N., Bender, J., and Gottesman, S. 1991. Uses of transposons with emphasis on Tn10. *Meth. Enzymol.* 204:139.

34. Skerra, A., Pfitzinger, I., and Plückthun, A. 1991. The functional expression of antibody Fv fragments in *Escherichia coli*: Improved vectors and a generally applicable purification technique. *Biotechnology* 9:273.

35. Prickett, K.S., Amberg, D.C., and Hopp, T.P. 1989. A calcium-dependent antibody for identification and purification of recombinant proteins. *Bio-Techniques* 7:580.

36. Hopp, T.P., Prickett, K.S., Price, V.L., Libby, R.T., March, C.J., Ceretti, D.P., Urdal, D.L., and Conlon, P.J. 1988. A short polypeptide marker sequence useful for recombinant protein identification and purification. *Biotechnology* 6:1204.

37. Knappik, A., and Plückthun, A. 1994. An improved affinity tag based on the FLAG peptide for the detection and purification of recombinant antibody fragments. *Biotechniques* 17:754–761.

38. Munro, S., and Pelham, H.R.B. 1986. An Hsp70-like protein in the ER: Identity with the 78 kd glucose-regulated protein and immunoglobulin heavy chain binding protein. *Cell* 46:291.

39. Ward, E.S., Güssow, D., Griffiths, A.D., Jones, P.T., and Winter, G. 1989. Binding activities of a repertoire of single immunoglobulin variable domains secreted from *Escherichia coli. Nature* 341:544.

40. Inoue, H., Nojima, H., and Okayama, H. 1990. High efficiency transformation of *Escherichia coli* with plasmids. *Gene* 96:23.

41. Prodromou, C., and Pearl, L.H. 1992. Recursive PCR: A novel technique for total gene synthesis. *Protein Eng.* 5:827.

42. Glockshuber, R., Stadlmüller, J., and Plückthun, A. 1991. Mapping and modification of an antibody hapten binding site: A site-directed mutagenesis study of McPC603. *Biochemistry* 30:3049.

43. Orlandi, R., Güssow, D.H., Jones, P.T., and Winter, G. 1989. Cloning immuno-globulin variable domains for expression by the polymerase chain reaction. *Proc. Natl. Acad. Sci. USA* 86:3833.

44. Sastry, L., Alting-Mees, M., Huse, W.D., Short, J.M., Sorge, J.A., Hay, B.N., Janda, K.D., Benkovic, S.J., and Lerner, R.A. 1989. Cloning of the immunological repertoire in *Escherichia coli* for generation of monoclonal catalytic antibodies: construction of a heavy chain variable region-specific cDNA library. *Proc. Natl. Acad. Sci. USA* 86:5728.

45. Larrick, J.W., Danielsson, L., Brenner, C.A., Abrahamson, M., Fry, K.E., and Borrebaeck, C.A.K. 1989. Rapid cloning of rearranged immunoglobulin genes from human hybridoma cells using mixed primers and the polymerase chain reaction. *Biochem. Biophys. Res. Commun.* 160:1250.

46. Jones, S.T., and Bending, M.M. 1991. Rapid PCR-cloning of full-length mouse immunoglobulin variable regions. *Bio/Technology* 9:88.

47. Persson, M.A.A., Caothien, R.H., and Burton, D.R. 1991. Generation of diverse high-affinity human monoclonal antibodies by repertoire cloning. *Proc. Natl. Acad. Sci. USA* 88:2432.

48. Marks, J.D., Tristem, M., Karpas, A., and Winter, G. 1991. Oligonucleotide primers for polymerase chain reaction amplification of human immunoglobulin variable genes and design of family-specific oligonucleotide probes. *Eur. J. Immunol.* 21:985.

49. Campbell, M.J., Zelenetz, A.D., Levy, S., and Levy, R. 1992. Use of family specific leader region primers for PCR amplification of the human heavy chain variable region gene repertoire. *Mol. Immunol.* 29:193.

50. Kettleborough, C.A., Saldanha, J., Ansell, K.H., and Bendig, M.M. 1993. Optimization of primers for cloning libraries of mouse immunoglobulin genes using the polymerase chain reaction. *Eur. J. Immunol.* 23:206.

51. Vieira, J., and Messing, J. 1982. The pUC plasmid, an M13mp7-derived system for insertion mutagenesis and sequencing with synthetic universal primers. *Gene* 19:259.

52. Sambrook, J., Fritsch, E.F., and Maniatis, T. 1989. *Molecular Cloning.* Cold Spring Harbor Laboratory Press.

53. Lopilato, J., Bortner, S., and Beckwith, J. 1986. Mutations in a chromosomal gene of *Escherichia coli* K-12., pcnB, reduce plasmid copy number of pBR322 and its derivatives. *Mol. Gen. Genet.* 204:285.

54. Knaus, R., and Bujard, H. 1990. Principles governing the activity of *E. coli* promoters. In *Nucleic Acids and Molecular Biology*, Vol. 4, Eckstein, F., and Lilley, D.M.J., eds. Springer-Verlag Berlin.

55. Bouvet, P., and Belasco, J.G. 1992. Control of RNAseE mediated RNA degradation by 5′ terminal base pairing in *E. coli. Nature* 360:488.

56. Plückthun, A., Glockshuber, R., Pfitzinger, I., Skerra, A., and Stadlmüller, J. 1987. Engineering of antibodies with a known three-dimensional structure. *Cold Spring Harbor Symp. Quant Biol.* 52:105.

57. McCarthy, J.E.G., and Gualerzi, C. 1990. Translational control of prokaryotic gene expression. *Trends Genet.* 6:78.

58. Schoner, B.E., Belagaje, R.M., and Schoner, R.G. 1990. Enhanced translational efficiency with two cistron expression system. *Meth. Enzymol.* 185:94.

59. Liu, J., and Walsh, C.T. 1990. Peptidyl-prolyl *cis-trans*-isomerase from *Escherichia coli*: A periplasmic homolog of cyclophilin that is not inhibited by cyclosporin A. *Proc. Natl. Acad. Sci. USA* 87:4028.

60. Bardwell, J.C.A., McGovern, K., and Beckwith, J. 1991. Identification of a protein required for disulfide bond formation *in vivo*. *Cell* 67:581.
61. Maurizi, M.R. 1992. Proteases and protein degradation in *Escherichia coli*. *Experientia* 48:178.
62. Boyle, M.D.P., ed. 1990. *Immunoglobulin Binding Proteins*, Vol. I. San Diego, Ca: Academic Press.
63. Boyle, M.D.P., and Reis, K.J. 1987. Bacterial Fc receptors. *Biotechnology* 5:697.
64. Nilson, B.H.K., Solomon, A., Björck, L., and Akerström, B. 1992. Protein L from *Peptostreptococcus magnus* binds to the κ light chain variable domain. *J. Biol. Chem.* 267:2234.
65. Faulmann, E.L., Duvall, J.L., and Boyle, M.D.P. 1991. Protein B, a versatile bacterial Fc-binding protein selective for human IgA. *Biotechniques* 10:748.
66. Inganäs, M., Johansson, S.G.O., and Bennich, H.H. 1980. Interaction of human polyclonal IgE and IgG from different species with protein A from *Staphylococcus aureus*: Demonstration of protein-A-reactive sites located in the Fab_2 fragment of human IgG. *Scand, J. Immunol.* 12:23.
67. Lindner, P., Guth, B., Wülfing, C., Krebber, C., Steipe, B., Müller, F., and Plückthun, A. 1992. Purification of native proteins from the cytoplasm and periplasm of *Escherichia coli* using IMAC and histidine tails: A comparison of protocols. *Methods: A companion of Meth. Enzymol.* 4:41.
68. Boss, M.A., Kenten, J.H., Wood, C.R., and Emtage, J.S. 1984. Assembly of functional antibodies from immunoglobulin heavy and light chains synthesised in *E. coli. Nucl. Acids Res.* 12:3791.
69. Cabilly, S., Riggs, A.D., Pande, H., Shively, J.E., Holmes, W.E., Rey, M., Perry, L.J., Wetzel, R., and Heyneker, H.L. 1984. Generation of antibody activity from immunoglobulin polypeptide chains produced in *Escherichia coli. Proc. Natl. Acad. Sci. USA* 81:3273.
70. Wood, C.R., Boss, M.A., Patel, T.P., and Emtage, S.J. 1984. The influence of mRNA secondary structure on expression of an immunoglobulin heavy chain in *Escherichia coli. Nucl. Acids Res.* 12:3937.
71. Studier, F.W., Rosenberg, A.H., Dunn, J.J., and Dubendorff, J.W. 1990. Use of T7 RNA polymerase to direct expression of cloned genes. *Meth. Enzymol.* 185:60.
72. Shibui, T., Munakata, K., Matsumoto, R., Ohta, K., Matsushima, R., Morimoto, Y., and Negahari, K. 1993. High-level production and secretion of a mouse-human chimeric Fab fragment with specificity to human carcino embryonic antigen in *Escherichia coli. Appl. Microbiol. Biotechnol.* 38:770.
73. Rudolph, R. 1990. Renaturation of recombinant, disulfide-bonded proteins from "inclusion bodies." In *Modern Methods in Protein and Nucleic Acid Research*. Tschesche, H., ed. Berlin: Walter de Gruyter. 149–71.
74. Buchner, J., and Rudolph, R. 1991. Renaturation, purification and characterization of recombinant Fab-fragments produced in *E. coli. Biotechnology* 9:157.
75. Buchner, J., Pastan, I., and Brinkmann, U. 1992. A method for increasing the yield of properly folded recombinant fusion proteins: single-chain immunotoxins from renaturation of bacterial inclusion bodies. *Anal. Biochem.* 205:263.

CHAPTER 9

Vectors and Approaches for the Eukaryotic Expression of Antibodies and Antibody Fusion Proteins

*Sherie L. Morrison, M. Josefina Coloma,
Daniel Espinoza, Alice Hastings, Seung-Uon Shin,
Letitia A. Wims, and Ann Wright*

Immunoglobulin (Ig) molecules are unique because of their wide range of binding specificities coupled with many different effector functions. In part because of their exquisite specificity, antibodies have great potential as therapeutic agents. Hybridomas are a major source of antibodies; however, hybridomas are largely of rodent origin and the immune response to rodent antibodies precludes their long-term use in man. DNA-mediated transfection and Ig gene expression provides an important new tool to address the shortcomings of the hybridoma technology and can be used to produce both chimeric rodent/human and totally human antibodies with improved functional properties.[1]

EXPRESSION OF GENOMIC CLONES OF Ig VARIABLE REGIONS

Functional antibodies are encoded by V_H and V_L genes that have been rearranged and juxtaposed proximal to the heavy- and light-chain constant region genes expressed by the antibody-producing cell lines. However, within the functional antibody genes a large intron separates the V regions from the constant region. This intronic region provides a convenient site for genetic engineering and for the joining of variable regions to diverse constant regions. Since the intron will not be a part of the final antibody molecule, there is great latitude in the manipulations that can be performed without having an impact on the functional properties of the resulting antibody.

The strategy used to identify and clone expressed V-region genes from the genome of antibody-producing cell lines relies on the identification of the gene rearrangement that is required for antibody gene expression. After constructing genomic libraries in either bacteriophage lambda or plasmids, specific DNA probes, usually containing the J regions, are used to screen the library. Only rearranged variable regions are positioned adjacent to J regions. Normally, myeloma and hybridoma antibody-producing cell lines contain more than one rearranged V_H and V_L, but since usually only one encodes the functional antibody, care must be taken to identify the productive rearrangement. Sequence analysis will frequently distinguish the productive from the aberrantly rearranged V regions; alternatively, an expression vector can be constructed containing the cloned V-genes, transfected into a myeloma cell, and the resulting protein product, if any, analyzed.

Vectors for the Expression of Genomic Clones

To create a transfectoma cell line synthesizing a novel antibody, both immunoglobulin heavy and light chains must be transfected into the same recipient cell line. Although both chains can be contained in a single vector, it is usually more convenient to construct separate light- and heavy-chain transfection vectors, which can be simultaneously transferred and expressed. The vectors routinely used for the production of chimeric antibodies with human kappa and gamma constant regions were initially designed for use in protoplast fusion. The unique characteristic of these vectors is that they compatibly replicate and amplify in *E. coli* host cells. Therefore, each plasmid

Figure 9–1. Vectors for the expression of variable regions cloned from genomic libraries. The prokaryotic selectable markers (Cm^R and Amp^R) are shown by cross-hatching. The eukaryotic selectable markers (neo and gpt) are shown as lightly stippled. The direction of their transcription is indicated. The remaining sequences derived from pSV2 are shown by the double, unfilled line. Murine noncoding sequences are indicated by a narrow line; murine exons are solid filled boxes. Human noncoding sequences are medium-width, cross-hatched lines. Human exons are indicated by broad diagonal lines. The direction of transcription of the immunoglobulin coding regions is indicated by the arrows within the circle of the vector. Enhancer regions are indicated by narrow diagonal lines. Restriction sites discussed within the text are indicated.

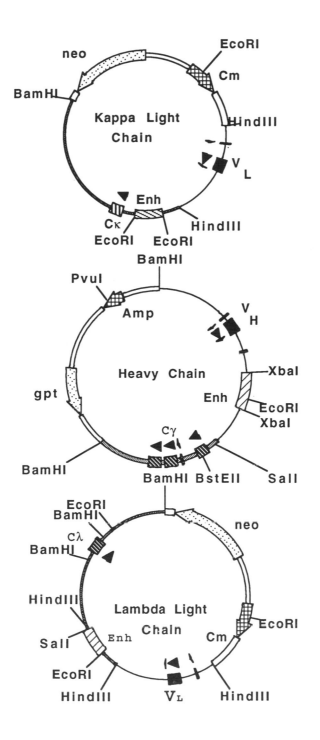

can be manipulated separately, but still maintained together in a single *E. coli* bacterium. Although these vectors were designed to facilitate simultaneous gene transfer by protoplast fusion, they can also be used to produce DNA for use in electroporation or lipofection.

The kappa light-chain transfection vector, pSV184ΔHneo (Fig. 9–1), is derived from the pACYC184 plasmid and contains a chloramphenicol resistance gene, which can be used to maintain the plasmid in *E. coli*.[2] A *neo* gene under the control of the SV40 early promoter makes it possible to select mammalian cells for the ability to grow in medium containing G418. The human kappa constant region with the human intronic enhancer is cloned as a Hind III-BamH I fragment. The genomic rearranged V_L is usually cloned into this vector as a Hind III fragment and provides the promoter used for expression.

The immunoglobulin heavy chain transfection vector, pSV2ΔHgpt (Fig. 9–1), contains an ampicillin resistance gene that is used to maintain the plasmid in *E. coli*. The mycophenolic acid resistance gene (*gpt*) under the control of the SV40 early promoter is used to select for the presence of this vector in mammalian cells by growing cells in medium containing hypoxanthine, mycophenolic acid, and xanthine (HXM). The heavy-chain vector is constructed so that any V_H gene can be inserted into the EcoR I restriction enzyme site to create an antibody with a human constant region. The genomic variable region provides a functional promoter and the murine intronic enhancer is present in the intron. Within the vector the human IgG constant region genes are flanked by Sal I and BamH I restriction enzyme sites; the constant regions are therefore present as a cloning cassette, making it straightforward to change the isotope of the expressed heavy chain. Intervening sequences also separate the several domain exons in the heavy-chain constant region gene. To further facilitate manipulation of the antibody genes, additional unique restriction endonuclease cleavage sites have been introduced into the introns separating the exons encoding the different domains.[3] Using these vectors, one can exchange domains between isotypes and produce antibody molecules with either duplications or deletions of domains.

The light- and heavy-chain vectors contain dominant biochemically selectable markers that select using different biochemical pathways. While dual selection for both markers can be used to guarantee the presence of both vectors, it is usually sufficient to select only for one.

Because the vast majority of murine antibodies are kappa, initial vector construction focused on the production of antibodies with the human kappa constant region. More recently vectors have also been constructed for the expression of the human lambda constant region (Fig. 9–1). In these vectors the 2.8 kb Hind III-EcoR I fragment containing the Cλ1 gene[4] was used to replace the human kappa constant region, leaving the kappa intronic enhancer intact. Cloning was done such that sequence 3′ of the human IgG3 polyadenylation site are also present. As a result of manipulation, the human lambda constant region is on a Sal I-BamH I fragment compatible with many of the other expression vectors.[4]

Things to Consider

While the vectors described above were originally designed for the production of murine/human chimerics, they can also be used for the expression of V regions of any species. Using human variable regions, they can be used to make completely human antibodies. The constant region can also be changed and the vectors used to express Ig from other species such as the rabbit.[5]

VECTORS FOR CLONING AND EXPRESSING VARIABLE REGIONS CLONED USING PCR

Because of the technical difficulties posed by cloning into bacteriophage lambda, attention has increasingly focused on the use of polymerase chain reaction (PCR) based approaches to obtain variable regions for expression. Clearly, the easier it is to acquire variable regions, the more will be available for diverse applications. PCR-based cloning takes advantage of the conservation of sequence seen for both the framework and leader segments of the variable regions. The framework regions, probably because of their role in maintaining the basic architecture of the variable region, have limited polymorphism. While priming within the frameworks using conserved sequences has been used for variable cloning,[6,7,8] the framework residues are more than just a scaffold upon which to hang the variable regions, and amino acid changes in the framework regions have been shown to have an impact on antigen binding;[9,10] therefore, priming within this region is not the optimum approach for the production and expression of variable regions with unchanged binding properties.

Like the framework regions, the hydrophobic leader sequences that direct the heavy and light chains into the secretory pathway are also conserved among the immunoglobulins. Examination of the murine leader sequences contained within the data base compiled by Kabat et al.,[11] enabled us to design a family of degenerate primers predicted to prime the majority of different variable regions;[13] similar families of redundant primers have been designed by others for both human and murine variable regions.[6,12] Unlike the framework regions, the leader sequence is not a part of the functional antibody molecule, so amino acid substitutions introduced into the leader sequences by imprecise priming do not have any impact on antigen recognition by the antibody molecule; many different substitutions should be tolerated as long as the secretion of the heavy and light chains is not impaired. By using degenerate oligomers as primers, PCR-based approaches can be used to clone variable regions from antibody-producing cell lines without any prior information about the amino acid or nucleic acid sequence of the antibody.[6,7,8,13]

Construction of Vectors

A new family of vectors has been produced that facilitates the cloning and expression of immunoglobulin variable regions cloned by PCR. The vectors are designed to express the cloned variable regions joined to human constant regions and, as discussed above, the variable regions are cloned by priming in the leader sequence so that no amino acid changes will be introduced into the mature antibody molecule.

Both the heavy-chain and light-chain vectors utilize a murine V_H promoter. To provide an appropriate cloning site, the V_H promoter from the anti-dansyl hybridoma 27.44 was provided with an EcoR V (GATATC) site between nucleotides -10 to -5.: **aAcATtCACC *ATG*** → **gAtATcCACC *ATG***. These changes leave the upstream regulatory elements and transcription start sequences intact. To generate a cloning site in the human constant regions, an Nhe I site (GCTAGC) was created at the first two amino acids of CH1: GCc tcC → GCt agC. Using this approach, a restriction site can be introduced into all four human gamma isotypes and human IgA, leaving the amino acid sequence unchanged (Ala Ser). In the basic cloning vector a polylinker sequence is present between EcoR V and Nhe I so that the cloning backbone cannot synthesize an H chain, and that heavy chain can only be produced after a variable region is successfully cloned. Since this vector does not contain an intron separating V_H from C_H, the 1 kb EcoR I fragment with the heavy-chain intronic enhancer was inserted into the EcoR I site in pSV2 5' of the heavy-chain gene.

To express a functional heavy chain, a V_H with an EcoR V site at its 5' end and an Nhe I site at its 3' end is generated from the cell line of interest using PCR. The V_H can either be cloned into a T-A vector[14] or, after digestion with the appropriate restriction enzyme, into Bluescript KS($+$) into which an Nhe I site has been introduced at the Sma I site or directly into the expression vector. The vectors are constructed so that the isotype of the heavy chain can be changed by cleaving with Nhe I and BamH I (BamH I must be a partial digest) and exchanging the constant region with the other human constant regions with an Nhe I site at a comparable location. The vectors use pSV2 as backbone, so it is straightforward to exchange the different selectable markers of pSV2 or similar vectors. In Eco*gpt* an EcoR V site is located between nucleotides 381 and 386 within the coding region.[15] We changed the sequence GAtATC to GAcATC thereby eliminating the

EcoR V site but leaving the amino acid sequence unchanged. Therefore, in all of the vectors, regardless of the selectable marker, the EcoR V site used for cloning is unique. The generic heavy-chain vector is shown diagramatically in Fig. 9–2.

Unlike the heavy chain in which the V_H is directly fused with C_H1, the light-chain vector is constructed so that V_L must splice to C_L. It utilizes the same V_H promoter with an EcoR V cloning site as does the heavy-chain vector. To provide a 3' cloning site, *in vitro* mutagenesis was used to introduce a Sal I site into the intron 3' of $J_\kappa 5$. In the final expression

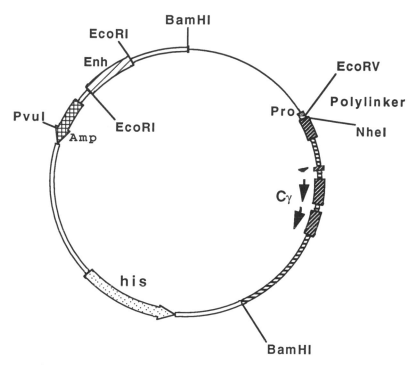

Figure 9–2. Vector for the expression of heavy-chain variable regions cloned by PCR. The prokaryotic selectable marker, Amp^R, is shown by cross-hatching. The eukaryotic selectable marker, his, is shown as lightly stippled. The direction of their transcription is indicated. The remaining sequences derived from pSV2 are shown by the double, unfilled line. Murine noncoding sequences are indicated by a narrow line with the position of the promoter (Pro) indicated. No murine exons are present in the cloning vector; instead a polylinker sequence lies between the EcoR V and Nhe I cloning sites. Human noncoding sequences are medium-width, diagonal lines. Human exons are indicated by broad diagonal lines. The direction of transcription of the immunoglobulin coding regions is indicated by the arrows within the circle of the vector. Enhancer regions are indicated by narrow diagonal lines. Restriction sites discussed within the text are indicated.

plasmid, this Sal I site is located in the intron 5' of human C_κ. Variable regions are cloned as EcoR V-Sal I fragments into the expression vector; the final expression vector is shown in Fig. 9–3. Several additional variations of the light-chain vector can also be produced: most notably the eukaryotic selectable marker can be changed. The markers that we have used are *gpt*,[16] *neo*,[17] and *his*.[18].

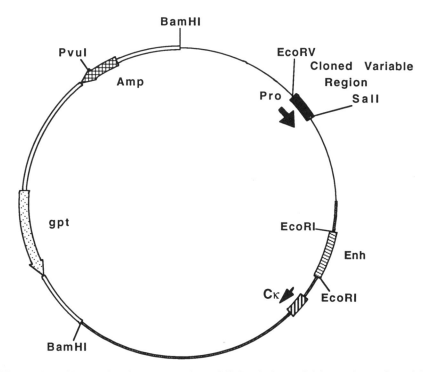

Figure 9–3. Vector for the expression of light-chain variable regions cloned by PCR. The prokaryotic selectable marker, Amp[R], is shown by cross-hatching. The eukaryotic selectable marker, gpt, is shown as lightly stippled; site-directed mutagenesis has been used to remove the EcoR V site from within gpt. The direction of transcription is indicated. The remaining sequences derived from pSV2 are shown by the double, unfilled line. Murine noncoding sequences are indicated by a narrow line with the position of the promoter (Pro) indicated; the murine V_L with the fused leader sequence is indicated by the solid filled box. The specificity of the vector is changed by replacing the variable regions following cleavage with EcoR V and Sal I. Human noncoding sequences are medium-width, cross-hatched lines. The human light-chain exon is indicated by broad diagonal lines. The direction of transcription of the immunoglobulin coding regions is indicated by the arrows within the circle of the vector. Enhancer regions are indicated by narrow diagonal lines. Restriction sites discussed within the text are indicated.

Primers

Two strategies exist for variable-region cloning. Redundant primers specific for both the leader sequence and the J regions can be used; however, while this approach allows the direct cloning of variable regions without any sequence information, it potentially can introduce sequence changes into the resulting antibody molecule because there are several possible amino acid sequences for the J regions of both the H and L chains. Murine J-region primers are listed in Table 9–1. In an alternative approach, variable-region cloning proceeds in two steps. For the first step, primers hybridizing to the leader sequence (5′ primer) and to the constant region immediately downstream of the V-J region (3′ primer) are used to clone the variable regions for sequence analysis of the J regions. Using the sequence information generated, primers appropriate for the 3′ end of the variable regions are used so that they can be cloned into the expression vector without any change in amino acid sequence. Redundant leader primers are always used for cloning because they will not lead to the incorporation of any sequence changes into the final antibody molecule. All primers should be provided with a restriction site flanked by three additional bases to protect the site and facilitate enzyme digestion.

Leader region primers contain a ribosome recognition site[19] and an

Table 9–1. J Region Primers for V Region Amplification and Cloning into PCR Expression Vectors

Murine Heavy-Chain J Region Primers

J_H1	GGG<u>GCTAGC</u> TGA GGA GAC GGT GAC CGT GGT
J_H2	T G A
J_H3	C A A AG A
J_H4	T A

Murine Light-Chain J Region Primers

Antisense of VLJ 1, 2, 4
5′ AGC<u>GTCGAC</u>TTACG TTT (TG)AT TTC CA(GA) CTT (GT)GT CCC

Antisense of VLJ5
5′ AGC<u>GTCGAC</u>TTACG TTT CAG CTC CAG CTT GGT CCC

Note. The sequence of primers for all murine JH is shown. The complete primer for J_H1 is shown; for the others only positions of difference are indicated. For the heavy-chain primers, the Nhe I site (bold and underlined) is used for cloning into the expression vectors. One light-chain J region primer will prime $J_\kappa1$, 2, and 4 without introducing any amino acid changes; the other primer primes $J_\kappa5$. $J_\kappa3$ is a pseudogene. They are provided with a splicing signal and a Sal I site (underlined) for cloning.

EcoR V site 5' of the start codon to allow for cloning adjacent to the promoter. Degeneracies in the primers increase the number of different leader region sequences that can be amplified. A set of 5' primers provides for the amplification of the vast majority of the murine V_L reported by Kabat et al.,[11] similarly, a set of 5' primers will amplify the majority of V_H (Tables 9–2 and 9–3).

Table 9–2. Leader Region Primers for Murine Light-Chain Variable Region Amplification (5' Sense)

MLALT1.RV
5' GGG GATATC **CACC** ATG GAG ACA GAC ACA CTC CTG CTA T

MLALT2.RV
5' GGG GATATC **CACC** ATG GAT TTT CAG GTG CAG ATT TTC AG

MLALT3.RV
5' GGG GATATC **CACC** ATG (GA)AG TCA CA(GT) AC(TC) CAG GTC TT(TC) (GA)TA

MLALT4.RV
5' GGG GATATC **CACC** ATG AGG (GT)CC CC(AT) GCT CAG (CT)T(CT) CT(TG) GG(GA)

MLALT5.RV
5' GGG GATATC **CACC** ATG AAG TTG CCT GTT AGG CTG TTG

Note: Set of 5 primers designed to hybridize to the N terminus of murine light-chain leader sequences. Three 5' Gs protect the EcoR V restriction site, which is underlined; the ribosome binding sequence identified by Kozak in bold type. Degeneracies at a single position are shown in parentheses. For unknown sequences all primers are used simultaneously.

Table 9–3. Leader Region Primers for Murine Heavy-Chain Variable Region Amplification (5' Sense)

MHALT1.RV
5' GGG GATATC **CACC** ATG G(AG)A TG(CG) AGC TG(TG) GT(CA) AT(CG) CTC TT

MHALT2.RV
5' GGG GATATC **CACC** ATG (AG)AC TTC GGG (TC)TG AGC T(TG)G GTT TT

MHALT3.RV
5' GGG GATATC **CACC** ATG GCT GTC TTG GGG CTG CTC TTC T

Note: Set of 3' primers designed to hybridize to the N terminus of heavy-chain leader sequences. Three 5' Gs protect the EcoR V restriction site, which is underlined; ribosome binding sequence is bold. Degeneracies at a single position are shown in parentheses. For the amplification of unknown sequences all primers are used simultaneously.

Table 9–4. 3' Primers for cDNA Synthesis and V Region Amplification

---

Primer for Synthesis of Light-Chain V Region cDNA

OLIGO dT.R1.XBA.H3
5' GCCGGAATTCTAGAAGC(T)$_{17}$

L Chain Constant Region[a]

MC$_k$ AS.XBA
5' GCG TCTAGA ACT GGA TGG TGG GAA GAT GGA

Constant Region Primer for Synthesis of Heavy-Chain V Region cDNA[b]

MgC.C$_H$1 AS
5' AGG TCTAGA A(CT)C TCC ACA CAC AGG (AG)(AG)C CAG TGG ATA GAC

---

[a] The primer is designed to hybridize to amino acids 122 to 116 of the murine kappa constant region; the Xba I site is underlined.
[b] C$_H$1 antisense primer designed to hybridize at amino acids 130 to 120 in C$_H$1 of all murine Ig except IgC3. The Xba I site used for cloning is underlined.

To obtain a product for cloning in a sequencing vector, the 3' primers hybridize to the constant region 20 bases downstream of the V-J region and contain an Xba I site for subcloning purposes. A single primer can be used for all murine gamma isotypes except IgG3. Since over 95% of murine antibodies are κ, only a primer for κ chain is shown (Table 9–4). Because of the limited sequence heterogeneity of the J regions, successful priming can also be achieved with any of the alternative primers shown for the H and L chains (Table 9–4). The murine J$_κ$ region is particularly well conserved with only one amino acid change (residue 106 is Ile in J$_κ$ 1, 2, 4, Leu in J$_κ$5; J$_κ$3 is a pseudogene) occurring within the priming region (amino acids 101 through 107). Murine J$_H$ is more diverse than J$_κ$ and several of the last seven amino acids differ. There are three different amino acids at position 108: Thr in J$_H$1 and 2, Leu in J$_H$3, and Ser in J$_H$4. At position 109, J$_H$1,3 and 4 have Val, J$_H$2 is Leu. At position 113, Ser is present in J$_H$1, 2, and 4; Ala is present in J$_H$3. To avoid the possible introduction of sequence changes, it is preferable to first clone then sequence the J regions and then design 3' primers for the second round of PCR reactions with the actual sequence of the J regions. The J regions contribute to CDR3 of both the heavy and light chain, and CDR3 frequently forms large numbers of contacts with antigen.[20] For the V$_H$ primer, an Nhe I site allows the direct ligation of the V$_H$ to the first two amino acids of the C$_H$1 of the gamma constant regions. The V$_L$ primer has a Sal I site following the splicing signal for cloning into the expression vectors.

RNA Preparation and cDNA Synthesis

To obtain RNA for PCR of the variable regions, approximately 5×10^5 cells from an antibody-producing cell line are pelleted, washed twice with PBS, and either quick frozen in liquid nitrogen and kept at $-70°C$ for later use or extracted immediately. The single step guanidinium/phenol method is used for RNA extraction.[21] All instruments and containers must be autoclaved and rinsed with diethyl pyrocarbonate-treated water to avoid degradation due to RNases. To extract the RNA, 0.5 ml of solution D (Solution D: 0.1 M 2-mercaptoethanol, 4 M guanidinium thiocyanate, 25 mM Na citrate pH 7, 0.5% sarcosyl), 50 μl of 2 M Na acetate pH 4, 0.5 ml phenol (dH_2O equilibrated) and 100 μl of chloroform:isoamylalcohol (49:1) are added to the cell pellet and mixed by inversion after each addition. The extraction is left on ice for 15 min and centrifuged at $13,000 \times g$ for 20 min at 4°C. The upper aqueous phase containing the RNA is removed to a new tube and precipitated with 2 volumes of cold absolute ethanol for 2 hr at $-70°C$. After two 70% ethanol washes, the RNA pellet is dried briefly and resuspended in 20 μl of 0.5% SDS.

For cDNA synthesis, 9 μl of the total RNA from 5×10^5 cells are annealed with 2 μl of 3' primer (1 mg/ml) at 60°C for 10 min. For light-chain V region amplification, an oligo dT primer is used, whereas for the amplification of heavy-chain V regions a C_H1 antisense primer is used (Table 9–4). After annealing, the samples are cooled on ice, 4 μl of first strand cDNA buffer (50 mM Tris pH 8.3, 50 mM KCl, 10 mM $MgCl_2$, 1 mM DTT, 1 mM EDTA, 0.5 mM spermidine), 1 μl of RNAse inhibitor (Promega, Madison, WI), 2 μl of 10 mM dNTPs, and 2 μl of prediluted 1:10 AMV Reverse Transcriptase (Promega, Madison, WI) added and the reaction incubated for 1 h at 42°C. The cDNA can be kept at $-20°C$ until used for PCR.

When using a mixture of primers, an equimolar amount of each primer is used in the PCR reaction. Alternatively, when first attempting the amplification of unknown sequences, a series of independent reactions can be set up, each using a different leader primer. This approach allows for the independent amplification of the different V region families and provides information as to which family is being used by the cell line being amplified. It is especially helpful for V_L amplification and provides one approach to minimizing the problem of the aberrant L-chain transcript present in most hybridoma cell lines (see below). PCR reactions are performed in a volume of 100 μl using 2 μl of cDNA, 0.5 μl Taq polymerase (Cetus Corporation, Emeryville, CA), 200 μM each, dNTP, 1 μM of each primer, 100 μg BSA in 10 mM Tris pH 8, 1.5 mM $MgCl_2$, 50 mM KCl. The reaction solution is overlaid with 50 μl of mineral oil. Typically, PCR is carried out for 30 cycles with 1 min denaturing (94°C), 1 min annealing (55°C), 1.5 min extension (72°C), and a final extension of 10 min. The size of the PCR products is verified by agarose gel electrophoresis in a 2% TAE (0.04 M Tris-acetate, 0.002 MEDTA, pH 8.0) agarose gel stained with ethidium bromide. When using the leader and constant region primers, the correct products are approximately 380 base pairs for the light-chain and 420 base pairs for the heavy-chain variable region.

Subcloning and Sequencing

After the PCR reaction the oil is removed by chloroform extraction; samples can be stored at 4°C. Products are either directly cloned into a T-A vector (a vector digested with a restriction enzyme yielding a blunt end and tailed with dideoxythymidine triphosphate using terminal transferase prepared following the procedure of Holton and Graham,[14] vectors are also commercially available from InVitrogen, San Diego, CA) or gel purified, cut with the appropriate restriction enzymes, and cloned into a plasmid such as Bluescript or pUC previously cut with the same enzymes. To facilitate cloning of V_H, an Nhe I cloning site was introduced into the Sma I site of Bluescript using linkers.

For T-A cloning 3 μl of the PCR product is ligated with 50 ng of T-A vector in a 15 μl reaction for 4 to 12 hours at 16°C. For sticky end ligations 200 ng of plasmid digested with the appropriate enzymes is ligated with 200 to 400 ng of digested PCR product in a 20 μl reaction. Five μl of the ligation is used for transformation of competent *E. coli* cells. Colonies are picked, miniprep DNA analyzed, and the apparently correct clones sequenced. Dideoxynucleotide chain termination sequencing is carried out using T7 DNA polymerase (Pharmacia, Uppsala, Sweden, or Sequenase, US Biochemical Corp., Cleveland, OH) according to the manufacturer's protocol. Several independent clones from different PCR reactions must always be sequenced in both directions to obtain the consensus sequence and to identify potential errors introduced by PCR.

Things to Consider

Two problems are frequently encountered when using this approach for expression. One problem is a result of the error-prone nature of the Taq polymerase; it is not uncommon to find a nucleotide substitution in the variable region, and these substitutions can lead to amino acid changes. It is therefore necessary to isolate and sequence several independent clones of each variable region to ensure that the correct sequence has been placed in the expression vector. A second problem arises from the fact that the nonproducing myelomas used to generate the majority of all murine monoclonal cell lines contain an aberrant kappa transcript.[22] This transcript contains a single base deletion leading to a frameshift and premature termination, so it does not encode a light-chain protein. However, the V_L region cloned using leader and J primers is identical in size to that of a functional V_L. Since this V_L is primed using Oligomer LALT1.RV, it is possible by performing independent reactions with each leader primer to avoid its priming if the expressed V_L is primed with a different oligomer.

The vectors described here represent a significant advance over others described previously because of their versatility and ease of use. cDNAs have been cloned into genomic expression vectors using restriction endonuclease cleavage sites fortuitously located within variable-region exons.[8,23] This approach has limited applicability because of the infrequent occurrence of the appropriate sites. The approach of providing the variable regions cloned as cDNAs with splice junctions is more versatile[24] but required linkers specific for each V and was designed for use with already cloned cDNAs. The current vectors aim to integrate the cloning and expression steps and minimize the need to synthesize variable-region-specific primers and linkers.

VECTORS FOR THE EXPRESSION OF Ig-FUSION PROTEINS

Ig fusion proteins have the advantage of joining the antibody combining specificity and/or antibody effector functions with molecules contributing unique properties. The ability to produce this family of proteins was first demonstrated when c-myc was substituted for the Fc of the antibody molecule,[25] but many examples now exist. Ab fusion proteins can be achieved in several different ways. In one approach non-Ig sequences are substituted for the variable region; the molecule replacing the V region provides specificity of targeting with the antibody contributing properties such as effector functions and improved pharmacokinetics. Examples include IL-2 and CD4. Alternatively, non-Ig sequences can be substituted for or attached to the constant region. The resulting molecules retain the binding specificity of the original antibody but gain characteristics from the attached protein. Depending on the position of the substitution, different antibody-related effector functions and biologic properties will be retained.

A series of vectors has now been produced that permits the fusion of proteins at different positions within the Ab molecule, thereby facilitating the construction of fusion proteins with different properties. Using these vectors it is possible to produce a family of fusion proteins with molecules of differing molecular weight, valence, and having different subsets of the functional properties of the antibody molecule.

Vectors for the Construction of IgG3 Fusion Proteins

To facilitate the construction of fused genes, site-directed mutagenesis was used to generate unique restriction enzyme sites in the human IgG3 heavy-chain gene at the 3′ end of the C_H1 exon, immediately after the hinge at the 5′ end of the C_H2 exon, and at the 3′ end of the C_H3 exon. The restriction sites thus produced were SnaB I at the end of C_H1 by replacing TtgGTg with TacGTa, Pvu II at the beginning of C_H2 by replacing CAcCTG with CAgCTG, and Ssp I at the end of C_H3 replacing AATgag with AATatt. These manipulations provided a unique blunt-end cloning site at these positions. In all cases the restriction site was positioned so that after cleavage the Ig would contribute the first base of the codon. Human IgG3 with an extended hinge region of 62 amino acids was chosen for use as the immunoglobulin; when present this hinge should provide spacing and flexibility, thereby facilitating simultaneous antigen and receptor binding. An EcoR I site was also introduced at 3′ of the IgG3 gene to provide a 3′ cloning site and polyA addition signal (Fig. 9–4). Although initially designed for use with growth factors, these restrictions sites can be used to position any novel sequence at defined positions in the antibody. Also, using these cloning cassettes the variable region can easily be changed.

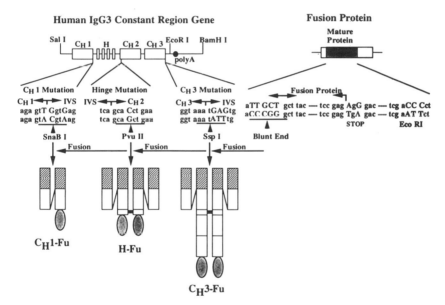

Figure 9–4. Vectors for the expression of IgG3-Fusion proteins. The location and nature of the site-directed mutations required for the introduction of cloning sites at the end of C_H1, after the hinge, and after the end of C_H3 are indicated, and the expected fusion proteins are shown diagrammatically. An EcoR I site was positioned 3′ of the gene so that a polyA addition sequence would be provided. The protein to be fused to the antibody molecule is provided with a blunt end cloning site at its 5′ end; care must be taken to preserve the correct reading frame in the fusion proteins. A termination codon and EcoR I cloning site are provided at the 3′ end of the gene.

Production of Gene to be Fused to the Antibody

As a first step in the production of a fusion protein, a blunt-end restriction site must be introduced at the desired position into the 5′ end of the gene to be fused. In order to maintain the correct reading frame, the site must be positioned so that after cleavage it will contribute two bases to the codon. If the objective is to make a fusion protein with the complete molecule, the restriction site is usually introduced at the position of any post-translational processing, such as after the leader sequence. Alternatively, if the objective is to use only a portion of the protein, the blunt-end site can be introduced at any position within the gene, but attention must always be paid to maintaining the correct reading frame. Additionally, if there is carboxyl-terminal post-translational processing of the fused protein, it is frequently desirable to introduce a stop-codon at this processing site. For example, when rat IGF1 was fused to Ig, a rat IGF1 cDNA was provided with a Pvu II site immediately after its leader sequence, and a stop codon was introduced at the 3′ end position where IGF1 is normally processed by proteolytic cleavage to generate the mature form. When the carboxy-terminus was left intact, incomplete processing was observed following transfection into murine myeloma cells.[26] An EcoR I site must also be introduced 3′ of the gene for cloning purposes.

Variable Region Exchange

The vectors described above were initially designed for the expression of variable regions cloned from genomic libraries and therefore have variable-region cloning sites compatible with the genomic expression vectors. However, the presence of a conserved BstE II restriction site within the C_H1 domain makes it straightforward to join the fused constant region to variable regions cloned in the PCR vectors. Therefore, variable regions cloned both from genomic libraries and by PCR can be used to provide specificity to the fusion proteins.

Things to Consider

A major concern when producing fusion proteins is maintaining the biologic activities of all of the components. The production of fusion proteins with antibodies is facilitated by the domain structure of the antibody, and all of the cloning sites have been positioned immediately following an intact domain. In this configuration the correct folding of the immunoglobulin should be assured. The folding of the attached protein depends on its structure and where it is fused. Whenever structural information is available, it is desirable to produce the fusion at a position that will maintain the structural integrity of the attached protein.

To produce quantities of protein sufficient for functional analysis, it is desirable to have the protein secreted into the medium. While in the examples reported to date, assembled fusion proteins have been assembled and secreted, this remains a concern when designing additional fusion proteins.

VECTORS FOR EXPRESSION IN NON-LYMPHOID EUKARYOTIC CELLS

A large variety of different vectors are available for eukaryotic expression in non-lymphoid cells. The majority of these utilize viral controlling elements. Recently we had occasion to produce antibody molecules in Chinese hamster ovary cells (CHO) and the Lec glycosylation mutants derived from them. For that purpose we constructed a new family of vectors containing chimeric Ig genes under the control of the SV40 early promoter. We have found these vectors to function efficiently in CHO cells. The light-chain vector has also been used in lymphoid cells.

Vector Design for Transfection of CHO Cells

The heavy- and light-chain constructs used to produce chimeric antibodies were modified to allow the expression of immunoglobulin genes in non-lymphoid cells by replacing the immunoglobulin-specific promoter from the murine anti-dansyl hybridoma 27-44 and enhancer with the SV40 promoter and enhancer. To construct the heavy-chain vector, the 2.1 kb BamH I fragment containing the leader and Ig promoter was subcloned into the BamH I site of pBR322. Complete digestion with EcoR I and partial digestion with Hind III removed the Ig promoter and the EcoR I, BamH I fragment of pBR322 but retained the leader sequence. The SV40 promoter/enhancer was obtained from the pSV2*gpt* expression vector, where it is flanked by Pvu II and Hind III restriction sites. The Pvu II site at the 5' end of the SV40 promoter was converted to an EcoR1 restriction site by linker tailing. The EcoR I/Hind III fragment containing the SV40 promoter/enhancer was then cloned into the homologous restriction sites 5' of the antidansyl leader sequence. This intermediate was then digested with BamH I and Sal I, into which sites was cloned the BamH I–Sal I fragment from the heavy-chain construct containing $V_H DNS$ and the J-C intron. The EcoR I–Sal I fragment was then replaced in the expression vector pSV2ΔH*gpt*. The heavy-chain enhancer, which is flanked by two Xba I restriction sites in the J-C intron, was deleted by Xba I digestion and religation. The final vector is shown in Fig. 9–5. The heavy-chain transcription unit was also subsequently cloned into pSV2*his*.

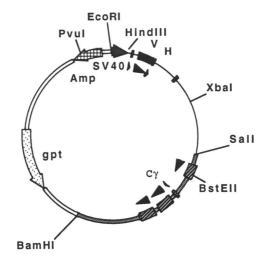

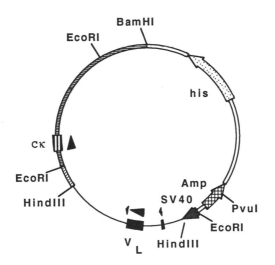

Figure 9–5. Vectors for the expression of immunoglobulin genes using SV40 controlling regions. The prokaryotic selectable marker, Amp^R, is shown by cross-hatching. The eukaryotic selectable markers (his and gpt) are shown as lightly stippled. The direction of their transcription is indicated. The remaining sequences derived from pSV2 are shown by the double, unfilled line. Murine noncoding sequences are indicated by a narrow line; murine exons are solid filled boxes. Human noncoding sequences are medium-width, cross-hatched lines. Human exons are indicated by broad diagonal lines. The direction of transcription of the immunoglobulin coding regions is indicated by the arrows within the circle of the vector. The SV40 early promoter and enhancer is indicated with a darkly cross-hatched arrowhead. Restriction sites discussed within the text are indicated.

To construct the light-chain vector, the Hind III–Sac I fragment containing the V_κ leader and promoter was cloned into pUC19. A Pvu II site 5' of the leader sequence but 3' of the light-chain promoter was converted to a Hind III site by linker tailing. The EcoR I–Sac I fragment encoding the J-C intron but lacking the intronic enhancer was then joined to the Hind III–Sac I fragment in pUC19. The light-chain vector was regenerated by a three-way ligation combining the BamH I–Hind III fragment of pSV2*his* with the SV40 promoter (obtained from the pSV2*his* heavy-chain vectors), the BamH I–EcoR I fragment from the light-chain vector containing C_κ, and the Hind III–EcoR I fragment obtained from the pUC19 cloning intermediate (Fig. 9–5).

Things to Consider

We have had the most success with histidinol selection for CHO or Lec transfectants; the cells are largely resistant to mycophenolic acid and can tolerate large doses of G418 (up to 1 mg/ml). In contrast, the CHO cells are effectively killed by 1 mM or less histidinol depending on the cell line. Moreover, histidinol is considerably less expensive than G418. Whatever is used, it is necessary to determine a killing curve for each cell line.

In our vectors immunoglobulin gene expression is driven by the SV40 promoter and enhancer, and stable transfectants secreting up to 10 μg/ml of antibody have been obtained. Viral promoter/enhancer from cytomegalovirus and Rous sarcoma virus and inducible promoters have also been used successfully by other groups to achieve expression of immunoglobulins and other heterologous proteins in CHO cells.

METHODS OF GENE TRANSFECTION

There are several methods for stably introducing DNA into eukaryotic recipient cells. The initial method of DNA transfer used calcium phosphate precipitates of the DNA; however, the frequency of gene transfection of lymphoid cells using calcium phosphate-precipitated DNA is very low. Instead, we have found protoplast fusion, electroporation, and lipofection to be efficient methods for transfecting lymphoid cells. Electroporation is extremely easy, so it is the method most frequently used in the laboratory. However, occasionally cell lines cannot be transfected using electroporation but can be transfected using either protoplast fusion or lipofection. Stable transfectomas are produced with frequencies of 10^{-3} to 10^{-7} depending on the recipient cell line and the transfection vectors.

Protoplast Fusion

The method used for protoplast fusion of lymphoid cells was modified from the initial description.[27] Using the available vectors, a single immunoglobulin chain or both heavy and light chain can be simultaneously transferred (see first section of this chapter). For protoplast fusion, *E. coli* HB101 containing the appropriate plasmids are grown at 37°C in Luria broth containing 50 µg/ml ampicillin and/or 30 µg/ml chloramphenicol, depending on the vector(s) being used. After overnight growth, 100 ml of Luria broth/1% glucose medium is inoculated with 1 ml of the overnight culture and incubated at 37°C. If the bacteria contain only the pSV2 vectors, the culture is grown to an OD_{600} of 0.5 and chloramphenicol is added to a concentration of 125 µg/ml. If the bacteria contain the pACYC vectors or both vectors, the culture is grown to an OD_{600} of 0.25 and then spectinomycin is added to a final concentration of 150 µg/ml. Spectinomycin and chloramphenicol are kept as 10X sterile-filtered stocks at 4°C. After further overnight incubation, the OD_{600} of the culture is determined, the bacteria are pelleted by centrifugation, placed on ice, and the supernatant removed by aspiration; 1.5 ml of culture are reserved for analysis to verify that the bacteria contain the appropriate vectors.

To prepare the protoplasts, 1.0 ml of cold 20% sucrose in 50 mM Tris, pH 8.0 is added for every 0.1 OD unit of the culture; or 5 mls if the OS is less than 0.5. Gently pipette up and down to mix, and discard any excess volume over 5.0 ml. Prepare a fresh solution of lysozyme at 5 mg/ml in 250 mM Tris, pH 8.0, and add 1 ml while swirling gently. Incubate on ice for 5 min and then slowly add 2.0 ml of cold 250 mM EDTA, pH 8.0 while swirling gently. After a 5 min incubation on ice add 2.0 ml of cold 50 mM Tris, pH 8.0, swirl gently, and incubate at 37°C for exactly 8.5 min. Slowly add 40 ml of Dulbecco's modified Eagle's medium (DMEM) containing 20% sucrose and 10 mM $MgCl_2$ to each bottle while gently swirling. Protoplasts must be allowed to stand at room temperature for at least 30 minutes before use. All solutions used must be sterile.

For fusion, two 5×10^6 myeloma cells are pelleted in a 50 ml polypropylene centrifuge tube by centrifugation at 1000 rpm in an IEC centrifuge, the medium removed, and the cells washed once with DMEM. 5.0 ml of protoblast are added to the pelleted cells, and gently mixed by pipetting up and down. The tubes are then spun at 1600 rpm for 5 min at room temperature, the medium aspirated, and the pellet resuspended by tapping the tube sharply with a finger. Then 0.5 ml of a polyethylene glycol (PEG)-dimethyl sulfoxide (DMSO) solution (41.7% PEG 1500, 12.5% DMSO, 100 mM Tris, pH 8.0 in DMEM) prewarmed to 37°C is added, and the tube is shaken gently for 1.0 min. An additional 0.5 ml of a 50% PEG solution (50% PEG 1500, 100 mM Tris, pH 8.0 in DMEM) prewarmed to

37°C is then added, and after shaking gently for 2 minutes the cells are diluted in 10 ml of prewarmed DMEM and centrifuged at 1600 rpm for 5 min. The supernatant is removed and the cells resuspended in 10 ml of prewarmed growth medium containing 100 µg/ml gentamicin, 100 U/ml Nystatin, and 10% calf serum or other appropriate serum supplement. 2.5×10^6 cells are plated per 96-well microtitre plate. After two days, selective medium is added. After two to four days of selection, half of the medium is removed and an equal volume of fresh selective medium added to each well. All cells are cultured at 37°C, 5% CO_2 in a water-saturated atmosphere.

Transfection by Electroporation

Electroporation provides an effective means to introduce DNA into lymphoid cells, and DNA linearized at unique restriction sites outside of the immuno-globulin genes routinely yields a large number of clones producing antibody. In most cases maxiprep DNA is linearized using Pvu I or its isoschizomer BspC I which cuts within *Amp*; however, linearization is not required for transfection. Prior to use, the DNA is phenol extracted and resuspended in sterile PBS at the appropriate concentration.

The nonproducing myeloma cell lines SP2/0 or P3X63 Ag8.653 are routinely used for transfection. Both heavy- and light-chain vectors can be simultaneously cotransfected into a nonproducing cell line. Alternatively, a cell line producing only the desired light chain can be produced by trans-fection or by isolating a chain loss variant of the appropriate myeloma. The light-chain producing cell line is then used as a recipient for transfection and expression of various heavy chains. Prior to transfection 1×10^7 cells are washed with cold PBS (PBS: 20 mM sodium phosphate, pH 6.8, 0.15 M NaCl), then resuspended in cold PBS at a concentration of 2 to 6×10^6/ml, and 0.9 ml of the cell suspension is placed in a 0.4 cm electrode gap electro-poration cuvette (Bio-Rad, Richmond, CA) with the DNA and incubated on ice for 10 minutes. Routinely, 5 to 10 µg of each vector are added in a total volume of less than 100 µl. For the electrical pulse, the Gene Pulser from Bio-Rad (Bio-Rad, Richmond, CA) is set at a capacitance of 960 µF and 200 V. After the pulse the cells are incubated on ice for 10 minutes, diluted into 15 ml of cold IMDM with 10% calf serum, pelleted by centri-fugation, and resuspended at a concentration of 10^5 cells/ml in IMDM (GIBCO, Grand Island, NY) with 10% calf serum (Hyclone, Logan, UT), gentamicin (100 µg/ml, Sigma, St. Louis, MO), and Nystatin (GIBCO, Grand Island, NY, 100 U/ml). Cells are plated in 96-well microtitre dishes at approximately 125 µl/well. We find a 12-channel Titertek (Flow, Research Triangle Park, NC) convenient for this purpose. Alternatively, 2 drops from a 10-ml pipette delivers the appropriate volume. After two days, an equal volume of selective medium is added. Approximately 3 days later, one-half of the selective medium is removed and replaced with fresh selective medium. Clones are usually visible by 10 to 14 days.

Transfecton by Lipofection

Lipofection has been used to effectively transfer both lymphoid cells and Chinese hamster ovary cells (CHO). The lipofectin reagent, a liposome formulation of a cationic lipid (DOTMA), is available from BRL (Bethesda, MD). For each transfection, 5 to 10 μg of DNA is diluted in 50 μl of sterile water and combined with 50 μl (50 μg) of lipofectin in a polystyrene tube, gently mixed, and incubated for 15 minutes at room temperature.

For the transfection of lymphoid cells, 1 to 5×10^6 healthy recipient cells are washed twice with 3 ml of serum-free IMDM and resuspended in 3 ml of the same medium in a 60 mm petri dish. The Lipofectin-DNA mixture is then added dropwise as uniformly as possible over the plate while gently swirling. After a 24 hr incubation in a water-saturated incubator at 37°C, 3 ml of IMDM with 20% serum and 100 U/ml of Nystatin and 100 μg/ml of gentamicin are added. After 6 to 24 hrs the cells are harvested by centrifugation, resuspended at 10^5 cells/ml in the same medium supplemented with the appropriate selective drug, and plated in 96-well microtiter plates at 100 μl/well. After 2 days, half of the medium is removed by aspiration and replaced with fresh medium.

Chinese hamster ovary cells to be transfected are first grown to 50 to 75% confluence on tissue-culture-treated 60×15 mm Petri dishes. Prior to transfection, the dishes are rinsed three times with serum-free IMDM. A Lipofectin/DNA solution prepared as described above and diluted to a final volume of 3 ml with serum-free IMDM is added to the dish of cells dropwise, with the drops distributed over the surface of the plate. The plate is incubated overnight at 37°C in a humidified 5% CO_2 incubator. The next day an equal volume of IMDM with 10% fetal calf serum is added. On the following day each plate is trypsinized, the cells diluted in 24 ml of selection medium [IMDM supplemented with 10% fetal calf serum, 1% Nystatin (GIBCO, Grand Island, NY), 1% Gentamicin (Sigma, St. Louis, MO), and the appropriate drug] and distributed into 96-well flat-bottomed microtiter plates (Corning, Corning, NY). Fresh selection medium is added three days later. Colonies are generally seen within 10 days.

Things to Consider

For the three procedures detailed above, transfected cells are plated into 96-well plates at a concentration of 10^4 cells/well and selective medium is usually added after 48 hours. The concentration of the drugs required for selection depends on the cell line being used. For P3X63Ag8.653, histidinol was used at 10 mM, mycophenolic acid at 3 μg/ml, and G418 at 1 mg/ml. For SP2/0, histidinol was used at 5 mM, mycophenolic acid at 1 μg/ml, and G418 at 1 mg/ml of activity. However, there is clonal variation in the requirement for selective medium, and cell lines with the same names may require different concentrations. Therefore, it is recommended that when a cell line is used for the first time a killing curve be done to determine the concentration of drug at which there are no surviving clones. Although the vectors are designed so that heavy and light chains are on vectors with different selectable markers, we have found that transfectants singly selected using synthesize both of the transfected genes. Single selection yields a larger number of surviving transfectants than does double section.

IDENTIFICATION AND ANALYSIS OF TRANSFECTANTS

The vectors and procedures described above are designed for the production and identification of stable transfectomas secreting antibodies of the desired specificity and constant region structure. In most instances it is advantageous to have cell lines synthesizing large quantities of antibodies. There is wide clonal variation in production levels, so one approach is to screen large numbers of transfectants to identify those producing the most; ELISA provides a convenient approach. It is also necessary to determine the appropriate structure. SDS-PAGE analysis of the radiolabeled and immuno-precipitated gene products is appropriate.

Screening Procedures

Approximately 12 days after adding selection medium, the supernatants from the growing clones are screened by ELISA to test for the secretion of heavy and light chains. If the appropriate antigen is available, it is used to coat Immulon II 96-well plates; alternatively, if the antigen is not available in sufficient quantities, plates are coated with anti-Ig of the appropriate specificity. To screen transfectomas producing human chimeric antibodies, 50 μl of a 5 μg/ml solution of goat anti-human-Ig in carbonate buffer at pH 9.6 is added to each well, and the plates are covered and incubated either 2 hr at 37° or overnight at 4°C. Plates are flicked in a sink to remove unbound coating solution and washed 3 times with PBS. To block nonspecific binding, 100 μl of 3% BSA in PBS is added to each well, and the plates are covered and incubated either at room temperature for 1 hr or overnight at 4°C. Blocking solution is removed and the plates are washed 3 times in PBS. Plates may be used immediately or the wells filled with 100 μl of 1% BSA in PBS and stored at 4°C. Stored plates are prepared for use by washing 3 times with PBS. To assay for Ig production, 50 µl of supernatants from the transfectants are added and the plates are incubated 1 to 2 hrs at 37° or overnight at 4°C. After washing 3 times with PBS, plates are incubated with goat anti-human κ conjugated with alkaline phosphatase for 1 to 2 hrs at 37°C or 3 hr at room temperature after washing with PBS to remove bound antibody. Wells secreting H and L chains are identified by adding 50 µl of substrate in diethanolamine buffer (0.6 mg/ml p-nitrophenyl phosphate in 1 M diethanolamine, 0.5 mM $MgCl_2$ pH 9.8) and reading the OD at 405 nm in a plate reader. Different combinations of coating and detecting antisera can be used, depending on the isotypes of the transfectants being generated. High producers are expanded for further analysis and selected transfectants subcloned.

Antibody Structural Analysis

Analysis should be designed to determine both the molecular weight of the expressed H and L chains and the structure of the fully assembled and secreted antibody molecule. To determine the nature of the protein being produced, transfectants are biosynthetically labeled using ^{35}S methionine. Routinely, 1×10^6 cells are pelleted at $220 \times g$ for 5 minutes at 4°C, washed twice with labeling medium (high-glucose DME deficient in methionine: Irvine Scientific, Santa Ana, CA), resuspended in 1 ml labeling medium containing 25 µCi ^{35}S-methionine (ICN, Irvine, CA), and allowed to incorporate label for 3 hours at 37°C under tissue culture conditions. Cells are then pelleted, and cytoplasmic extracts are prepared by lysing the cells with 0.5 ml NDET solution (1% Nonidet P-40, 0.4% sodium deoxycholate, 66 mM EDTA, 10 mM Tris pH 7.4). The supernatant is analyzed for the presence of secreted antibodies. For immunoprecipitation, 2.5 µl of polyclonal antiserum (usually rabbit) is added to both the cytoplasmic lysate and the secretions. After 1 hour incubation on ice, 50 µl of IgG Sorb (The Enzyme Center, Inc., Boston, MA) is added. After 15 minutes incubation, the IgG Sorb is either pelleted directly or spun through a sucrose pad of 30% sucrose NDET and 0.3% SDS. The pellet is washed with NDET with 0.3% SDS and then distilled water. 50 µl of sample buffer (25 mM Tris, pH 6.7, 2% SDS, 10% glycerol, 0.008% bromphenol blue) is added, and the pellet is resuspended and placed in a boiling water bath for 2 min. After centrifugation, the immunoprecipitates are analyzed by SDS-PAGE. To determine the polymerization state of the IgG, immunoprecipitates are fractioned on 5% gels made with 0.1 M phosphate [1 M is 59.4 g NaH_2PO_4 and 81.0 g Na_2HPO_4 per liter and acrylamide (prepared using 30% acrylamide, 0.8% bisacrylamide) and 0.1% SDS without treatment with a reducing agent]. To analyze the size of the heavy and light chains synthesized by the transfectants, immunoprecipitates are analyzed on 12% Tris-glycine gels following treatment with 0.15 M 2-mercaptoethanol.

After running the gels until the bromophenol blue is approximately 1 cm from the bottom of the gel (100 to 150 mA/gel for phosphate gels, 150 V for tris-glycine gels), the gels are treated for 10 minutes with staining solution (50% methanol, 7% acetic acid, 0.02% Coomassie blue), and then destained using 7% acetic acid and 5% methanol. Gels may be left in destain overnight. After the gels are dried onto filter paper (Whatman 3MM) they are exposed to X-ray film. The exposure time required depends on the amount of Ig being produced. If enhancement of radioactivity is required, rinse gels once in distilled water, and then add 1 M salicylate and incubated for 30 min prior to drying the gel. Exposure must now be at $-70°C$.

Things to Consider

The analysis described is designed to determine if the heavy and light chains produced by the transfectoma are of the expected sizes and are correctly assembled and secreted. Appropriate molecular weight markers must be included in the gel for comparison; frequently, well-characterized myeloma or transfectoma proteins provide convenient standards. The assembly patterns and polymerization state of the transfectoma antibodies are determined by SDS-PAGE without prior treatment with reducing agents. Examination of the cytoplasmic proteins shows the intermediates in Ig assembly. Most antibodies are assembled using either the pathway $H + L \to HL \to H_2L_2$ or $H + H \to H_2 \to H_2 \to H_2L \to H_2L_2$; the pathway utilized is usually determined by the isotype of the heavy chain.

Generally IgGs are secreted into the serum as H_2L_2, while IgA and IgM form higher polymers. If a large amount of polymers are present, use of 4% Acrylamide-phosphate gels can be useful for the resolution of the polymeric forms. Light chain not assembled with heavy chain is sometimes secreted, but as a general rule heavy chains will not be secreted without an accompanying light chain. Human IgG4 secretes H-L half-molecules; a mutation in its hinge has been observed to lead to the production of only fully assembled IgG4.[28]

Acknowledgments

Work in the laboratory is supported by grants CA 16858 and AI 29470 from the National Institutes of Health, and by grants IM-550 and IM-603 from the American Cancer Society. S.-U. Shin is supported in part by NINDS training grant 1Tew-NSO7356. M.J. Coloma is a Centocor Fellow.

REFERENCES

1. Morrison, S.L. 1985. Transfectomas provide novel chimeric antibodies. *Science* 229:1202.
2. Oi, V.T., and Morrison, S.L. 1986. Chimeric antibodies. *Bio-Techniques* 4:214.
3. Morrison, S.L., Canfield, S., Porter, S., Tan, L.K., Tao, M.-H., and Wims, L.A. 1988. Production and characterization of genetically engineered antibody molecules. *Clin. Chem.* 33:1668.
4. Udey, J.A., and Blomberg, B. 1987. Human λ light chain locus: organization and DNA sequences of three genomic *J* regions. *Immunogenetics* 25:63.
5. Schneiderman, R.D., Hanly, W.C., and Knight, K.L. 1989. Expression of 12 rabbit IgA C alpha genes as chimeric rabbit-mouse IgA antibodies. *Proc. Natl. Acad. Sci. USA* 86:7561.
6. Larrick, J.W., Danielsson, L., Brenner, C.A., Abrahamson, M., Fry, K.E., and Borrebaeck, C.A.K. 1989. Rapid cloning of rearranged immunoglobulin genes from human hybridoma cells using mixed primers and the polymerase chain reaction. *Biochem. Biophys. Res. Comm.* 160:1250.
7. LeBoeuf, R.D., Galin, F.S., Hollinger, S.K., Peiper, S.C., and Blalock, J.E. 1989. Cloning and sequencing of immunoglobulin variable-region genes using degenerate oligodeoxy-ribonucleotides and polymerase chain reaction. *Gene* 82:371.
8. Orlandi, R., Güssow, D.H., Jones, P.T., and Winters, G. 1989. Cloning immunoglobulin variable domains for expression by the polymerase chain reaction. *Proc. Natl. Acad. Sci. USA* 86:3833.
9. Panka, D.J., Mudgett-Hunter, M., Parks, D.R., Peterson, L.L., Herzenberg, L.A., Haber, E., and Margolies, M.N. 1988. Variable region framework differences result in decreased or increased affinity of variant anti-digoxin antibodies. *Proc. Natl. Acad. Sci. USA* 85:3080.
10. Chien, N.C., Roberts, V.A., Grusti, A.M., Scharff, M.D., and Getzogg, E.D. 1989. Significant structural and functional change of an antigen-binding site by a distant amino acid substitution: Proposal of a structural mechanism. *Proc. Natl. Acad. Sci. USA* 86:5532.
11. Kabat, E.A., Wu, T.T., Perry, H.M., Gottesman, K.S., and Foeller, C. 1991. *Sequences of Proteins of Immunological Interest*. U.S. Department of Health and Human Services, 5th edition.
12. Coloma, M.J., Larrick, J.W., Ayala, M., and Gavilondo-Cowley, J.V., 1991. Primer design for the cloning of immunoglobulin heavy-chain leader-variable regions from mouse hybridoma cells using the PCR. *BioTechniques* 11:152.
13. Coloma, M.J., Hastings, A., Wims, L.A., and Morrison, S.L. 1992. Novel vectors for the expression of antibody molecules using variable regions generated by PCR. *Immunol. Methods* 152:89.
14. Holton, T.A., and Graham, M.W. 1990. A simple and efficient method for direct cloning of PCR products using ddT-tailed vectors. *Nucl. Acids Res.* 19:1156.
15. Pratt, D., and Subramani, S. 1983. Nucleotide sequence of the *Escherichia coli* xanthine-quanine phosphoribosyl transferase gene. *Nucl. Acids Res.* 11:8817.
16. Mulligan, R.C., and Berg, P. 1981. Selection for animal cells that express the *Escherichia coli* gene coding for xanthine-guanine phosphoribosyltransferase. *Proc. Natl. Acad. Sci. USA* 78:2072.

17. Southern, P.J., and Berg, P. 1982. Transformation of mammalian cells to antibiotic resistance with a bacterial gene under the control of the SV40 early region promoter. *J. Molec. Appl. Genetics* 1:327.
18. Hartman, S.C., and Mulligan, R.C. 1988. Two dominant-acting selectable markers for gene transfer studies in mammalian cells. *Proc. Natl. Acad. Sci. USA* 85:8047.
19. Kozak, M. 1981. Possible role of flanking nucleotides in recognition of the AUG initiator codon by eukaryotic ribosomes. *Nucl. Acids Res.* 9:5233.
20. Amit, A.G., Mariuzza, R.A., Phillips, S.E.V., and Poljak, R.J. 1986. Three-dimensional structure of an antigen-antibody complex at 2.8 Å resolution. *Science* 233:747.
21. Chomczynski, P., and Sacchi, N. 1987. Single-step method of RNA isolation by acid guanidinium thiocyanate-phenol-chloroform extraction. *Anal. Biochem.* 162:156.
22. Carroll, W.L., Mendel, E., and Levy, S. 1988. Hybridoma fusion cell lines contain an aberrant kappa transcript. *Molec. Immunol.* 25:991.
23. Wallick, S.C., Kabat, E.A., and Morrison, S.L. 1988. Glycosylation of a V_H residue of a monoclonal antibody against a(1 → 6) dextran increases its affinity for antigen. *J. Exp. Med.* 168:1099.
24. Gillies, S.D., Lo, K-M., and Wesolowski, J. 1989. High-level expression of chimeric antibodies using adapted cDNA variable region cassettes. *J. Immunol. Methods* 125:91.
25. Neuberger, M.S., Williams, G.T., and Fox, R.O. 1984. Recombinant antibodies possessing novel effector functions. *Nature* 312:604.
26. Shin, S.-U., and Morrison, S.L. 1990. Expression and characterization of an antibody binding specificity joined to insulin-like growth factor 1: Potential applications for cellular targeting. *Proc. Natl. Acad. Sci. USA* 87:5322.
27. Sandri-Goldin, R.M., Goldin, A.L., Levine, M., and Gloriosos, J.C. 1981. High frequency transfer of cloned *herpes simplex* virus type 1 sequences to mammalian cells by protoplast fusion. *Molec. Cell. Biol.* 1:743.
28. Angal, S., King, D.J., Bodmer, M.W., Turner, A., Lawson, A.D., Roberts, G., Pedley, B., and Adair, J.R. 1993. A single amino acid substitution abolishes the heterogeneity of chimeric mouse/human (IgG4) antibody. *Molec. Immunol.* 30:105.

CHAPTER 10

The Cloning of Hybridoma V Regions for Their Ectopic Expression in Intracellular and Intercellular Immunization

Andrew Bradbury, Francesca Ruberti, Thomas Werge, Viviana Amati, Anna Di Luzio, Stefania Gonfloni, Hennie Hoogenboom, Patrizia Piccioli, Silvia Biocca, and Antonino Cattaneo

We describe in this chapter an experimental strategy to specifically interfere with the function of selected intra- or extracellular gene products in animal cells, where a systematic genetic approach is not feasible. This strategy is based on the ectopic expression of recombinant forms of monoclonal antibodies. As a prelude to this use of antibodies, we have recently shown that immunoglobulins can be efficiently expressed as secreted[15] or intracellular[3] proteins by mammalian nonlymphoid cells. Intracellular expression of functional antibodies has also been demonstrated in yeast.[14,80] In particular, the efficiency of antibody secretion by cells related to the nervous system was shown to be comparable to that of lymphoid cells. As a result we suggested[15,67] that the local secretion of specific monoclonal antibodies by cells of the nervous system could be used in functional and developmental studies on the otherwise intact nervous system of transgenic mice. This approach can be taken one step further by engineering the expression of antibodies in the cell cytoplasm or in other compartments. Thus, we demonstrated that not only can antibody heavy and light chains be synthesized and assembled in the cytoplasm, but that the assembled functional antibody molecules can also be targeted to the nucleus.[3,85]

Following these feasibility studies, performed with test antibodies, the technique was applied, by us and others, to cases of interest in a variety of systems (see below), including transgenic mice and plants (described in Chapter 11).

Among the different methods used to inhibit the function of selected genes in mammalian cells and whole organisms that are presently being pursued, those of gene disruption ("knock out") by homologous recombination,[13] competition with mRNA function by antisense RNA[42] or ribozyme,[32] and expression of dominant negative mutants of a gene[34] appear to be the most promising. Each of these methods, however, suffers from some drawbacks related to intrinsic technical problems or to experimental design, and there may be a strong case for the use of the technique of ectopic antibody expression as an alternative/complementary strategy in defined situations.

Gene disruption by homologous recombination is still difficult to achieve in cultured cell lines[78] and only works efficiently in embryonic stem (ES) cells. This allows mutations to be introduced into the germ line of mice. The obvious advantage is that the knock out is absolute (in homozygous individuals). Disadvantages include possible lack of spatio-temporal control over the knock-out with the present technology, with the consequence that the mutation is expressed in the organism from very early developmental stages. This poses problems of lethality and of interpretation of results observed in the adult (after development). This problem may be overcome in the future by the use of the Cre/loxP site-specific recombination system, as Cre transgenic mice become available.[30]

The stable expression of antisense RNA has not fulfilled initial expectations. It should be noted that the great majority of the papers utilizing this technique describe a phenotype of growth inhibition. This leads one to suspect that some of these effects may be ascribed to double-strand RNA-induced interferon, a point seldom controlled for. The use of ribozymes, albeit promising, is still in its infancy.

Dominant negative mutations have been used with success in some cases. However, the method is not general, as it needs detailed structural and functional information on the target protein and requires that the selected domain of the target protein folds in a similar way to the wild type protein.

The use of ectopic antibody expression, which we describe in this chapter, is in some way similar to the expression of dominant negative mutants of a gene but is more general, in that it exploits the virtually unlimited repertoire of the antibodies. By the judicious use of suitable promoters and intracellular targeting tags, this approach should allow the expression of inhibitory antibody molecules in the appropriate cellular compartment or extracellular space of any tissue of interest.

Expression of antibodies can be controlled at two different levels: a transcriptional one, by the promoter used, allowing expression in different cell types and/or tissues; and a targeting one, by the exploitation of short protein signals that can be inserted at different positions in the recombinant antibody to direct the protein to different cellular compartments of a given cell.

In principle one can envisage different strategies with the antibody/antibody domain being used as a dominant competitor, as an intra- or extracellular anchor, as a vector for degradation, or even as an enzyme (with catalytic antibodies). Post-translational modifications of a given protein could be selectively targeted by the use of suitable antibodies, which would leave the unmodified protein unrecognized. Although conceived of as a way to inhibit biological functions within cells, the technique of intracellular immunization could also be adapted for *in vivo* study of protein interaction. One would expect that *in vivo* footprinting patterns[33] would be considerably altered by the binding of an intracellular antibody (or antibody fragment) to its transcription factor target, whether that binding is inhibitory or not.

Under normal conditions, antibodies are secreted proteins and harbor a hydrophobic leader sequence for secretion. For expression as secreted proteins in nonlymphoid cells, no change to the leader is required. By this means, monoclonal antibodies against proteins of the secretory apparatus have been transiently expressed in *Xenopus* oocytes or in mammalian cells via microinjection of the mRNA isolated from the corresponding hybridoma cell line.[12,17,74,81] Although cytosolic proteins are not accessible to the antibody expressed with this leader, microinjection of antibody protein into the cytoplasm has been succesfully performed. These transient versions of the ectopic antibody expression approach, based on microinjection, will not be further considered as an experimental tool in this chapter, as their use is limited to studies of short-term, fast biological responses which can be assayed in a very small number of cells; however, some transient assays will be described, as a necessary and useful intermediate step of the experimental approach. The cloning of immunoglobulin coding sequences from corresponding hybridoma cell lines, described in this chapter, will not only allow their stable expression, but also their modification into forms suitable for their targeting to intracellular compartments where they will be able to interact with the corresponding antigen.

Probably (and obviously) the most crucial step for the success of this approach is represented by the initial choice of the monoclonal antibody. The properties of isotype, binding affinity, specificity with respect to similar antigens, epitope recognized, and the neutralization of biological functions by the selected monoclonal antibody should be very well characterized at the outset. Also, the relevant activity assays for the selected antibody should be working in the lab. In the future, however, it is conceivable that it will be possible to exploit the intracellular expression of a (maybe restricted or enriched) library of antibody specificities to isolate a previously unknown antibody specificity on the basis of a selectable phenotype.

This chapter is organized in such a way as to describe, in a logical and temporal order, all the steps required, starting from a selected hybridoma cell line, to clone the V region cDNA in a form suitable for its subsequent use in inter- or intracellular immunization. When required, the description and discussion of the experimental protocols and decisions to be taken will be done separately for the expression of intracellular or secreted recombinant antibodies.

CLONING OF HYBRIDOMA V REGIONS

Almost all monoclonal antibodies that can be used for inter- or intracellular immunization have been derived using classical hybridoma technology.[47] This is likely to be the case for some time to come, although we expect that the new technique of phage display (see Chapter 2) will soon provide antibodies with useful affinities from large antibody libraries. The advantage of using antibodies cloned by this method is that the cloned antibody genes are isolated simultaneously with the isolation of the binding specificity, thus avoiding all the steps of cloning, sequencing, and specificity control described in methods 1 through 3 below. Phage display vectors can also be used to clone hybridoma V regions (when those V regions can be amplified by the primers used), although, as discussed below, we feel that they should be used to produce soluble ScFv in bacteria rather than to display the antibody on the surface of phage.

The steps required to use hybridoma V regions in inter- or intracellular immunization (Figures 10–1 and 10–2) involve:

1. Cloning (and sequencing) of the V region cDNA.
2. Checking that the V region cloned produces functional immuno-globulin.
3. Checking that the specificity of the immunoglobulin is correct.
4. Cloning the appropriate targeting sequence in frame with the immuno-globulin V regions in a vector that will provide the necessary transcriptional control elements.

The quickest method, when it works, is to amplify the hybridoma V regions using V region primers, assemble them into single-chain Fv (ScFv) fragments, express them in bacteria, and test the supernatants of bacterial cultures grown in 96-well plates for antigen-binding activity (methods 3.1 through 3.5). We recommend this as a first approach, bearing in mind the potential pitfalls (correct V regions not amplified, ScFv has no activity). Phage display

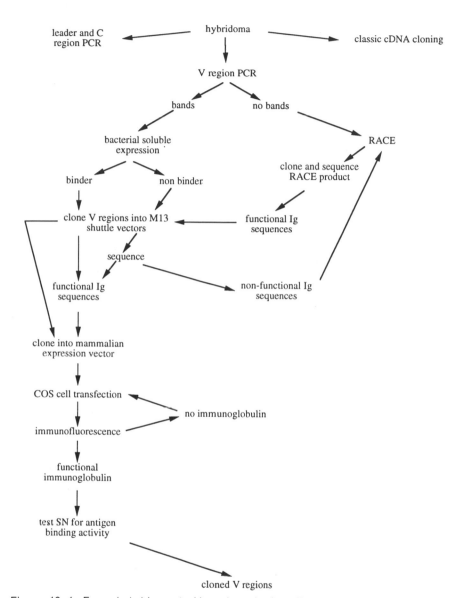

Figure 10–1. From hybridoma to V region cloning. The flow chart depicts the recommended path for cloning hybridoma V regions (described in the text).

is an alternative method, although we feel that if one cannot isolate antigen-binding activity from at least one of 96 different clones, it is likely to be due to a problem intrinsic to the V regions which will not be overcome by using phage display. If cloning by bacterial expression works, the V regions should be recloned into the M13 based vectors (Fig. 10–5 and

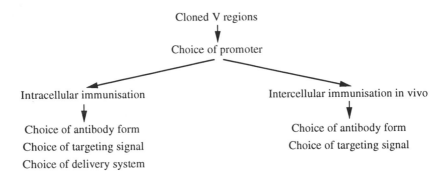

Figure 10–2. From cloned V regions to experimental use (intracellular or intercellular). The flow chart depicts the decisons that need to be made for intra- and intercellular immunization.

Appendix) and then cloned into the appropriate eukaryotic expression vector (Fig. 10–6).[63] Sequencing is optional if clones have been isolated by antigen-binding activity. The multiple cloning steps presently required when cloning by bacterial expression are one disadvantage of this method. We are developing eukaryotic expression vectors that will allow simple cloning from the phage display vectors in a single step.

If V regions are amplified, but no binding activity is obtained (irrelevant V region amplified, ScFv has no activity), we suggest that the V regions arc cloned into the M13 shuttle vectors and sequenced. V regions which should be functional on the basis of amino acid sequence should then be transferred to the mammalian expression vector for testing. If V regions are not amplified, or only nonfunctional V region sequences are obtained (frame-shifts, myeloma, or contaminant V regions), we suggest trying the RACE technique, which will amplify all V regions with the particular isotype, independently of variable sequences, and then transferring the V regions to the shuttle vector by V region PCR. This will be much easier on cloned material, but may require low (37°C) annealing temperatures. These different choices are outlined in Fig. 10–1.

We strongly recommend checking the specificity before embarking upon the cloning of the variable regions of a hybridoma. In the absence of any selective pressure, it is not uncommon for chromosomes that encode immunoglobulin chains to be lost.

Introduction to Method 1: Cloning Hybridoma V Regions by V Region PCR

The 5′ and 3′ ends of immunoglobulin V regions tend to be relatively conserved; as a result, mixtures of degenerate V region primers (see Table 10–1) can be used to amplify V regions from hybridoma RNA or DNA.[63]

Table 10–1. Oligonucleotides for V regions PCR and for RACE

V region PCR primers

1. Primers for cDNA synthesis of mouse immunoglobulin genes

1.1 Heavy-chain primers (all prime at 3′ end of CH1)

MOCG12FOR   CTC AAF TTT CTT GTC CAC CTT GGT GC
(mouse IgG1, IgG2a; rat IgG1, IgG2a, IgG2b)

MOCG2bFOR   CTC AAG TTT TTT GTC CAC CGT GGT GC
(mouse IgG2b)

RACG2cFOR   CTC AAT TCT CTT GAT CAA GTT GCT TT
(rat IgG2c)

MOCG3FOR   CTC GAT TCT CTT GAT CAA CTC AGT CT
(mouse IgG3)

MOCHFOR   TGG AAT GGG CAC ATG CAG ATC TCT
(mouse IgM)

These may be used individually or as an equimolar mixture to prime all heavy chains.

1.2 Light-chain primers (all prime at 3′ end of CL)

CKFOR   CTC ATT CCT GTT GAA GCT CTT GAC
(mouse and rat K)

MOCKFOR   CTC ATT CCT GTT GAA GCT CTT GAC AAT
(mouse K)

RACKFOR   CTC ATT CCT GTT GAA GCT CTT GAC GAC
(rat K)

MOCKFOR and RACKFOR are identical to CKFOR except for the last three bases. If a V region from a rat/mouse hybrid is to be cloned, and the mouse myeloma partner expresses a light chain V region mRNA, it can be excluded by the use of RACKFOR

CL1FOR   ACA CTC AGC ACG GGA CAA ACT CTT CTC
(mouse λ1 λ4; rat λ1)

CL2FOR   ACA CTC TGC AGG AGA CAG ACT CTT TTC
(mouse λ2, λ3; rat λ3)

These may be used individually if the lambda isotype is known or as an equimolar mixture to prime all lambda chains.

2. Primers for PCR amplification of mouse immunoglobulin V-regions

2.1 Heavy-chain primers (mammalian and bacterial vectors)

                           PstI
VH1BACK   AG   GTC CAG <u>CTG CAG</u> GAG TCT GG
                    G A A   C   A

                           BstEII
VH1FOR(-2)   TGA GGA GAC <u>GGT GAC C</u>GT GGT CCC TTG GCC CC

2.2 Light-chain primers 1 (mammaliam vectors)

                           PvuII
VK1BACK   GAC ATT <u>CAG CTG</u> ACC CAG TCT CCA

                       BglII
VK1FOR   G   TTA <u>GAT CTC</u> CAG CTT GGT CCC

2.3 Light-chain primers 2 (mammalian and bacterial vectors)

                           SacI
VK2BACK   GAC ATT <u>GAG CTC</u> ACC CAG TCT CCA

(continued)

Table 10–1. (*cont.*)

XhoI
VK2FOR   G   TTT TAT <u>CTC GAG</u> CTT GGT CCC

2.4 Light-chain primers 3 (phage system, no restriction sites)
VK4FOR (equimolar mix of)
MJK1FONX   CCG TTT GAT TTC CAG CTT GGT GCC
MJK2FONX   CCG TTT TAT TTC CAG CTT GGT CCC
MJK4FONX   CCG TTT TAT TTC CAA CTT TGT CCC
MJK5FONX   CCG TTT CAG CTC CAG CTT GGT CCC

SacI
VK2BACK   GAC ATT <u>GAG CTC</u> ACC CAG TCT CCA
(note that this is identical to that in light chain primers 2.3)

3. ScFv linker assembly primers
MOLINKBACK   GGG ACC ACG GTC ACC GTC TCC TCA
(reverse and complementary to part of VHFOR(-2))

MOLINKFOR   TGG AGA CTG GGT GAG CTC AAT GTC
(reverse and complementary to VK2BACK)

4. Pull through primers to add SfiI/NotI sites
VH1BACKSFI   CAT GCC ATG ACT CGC GGC CCA GCC GGC CAT GGC
                CSA GGT SMA RCT GCA GSA GTC WGG
(The 3′ end of this primer is identical to VH1BACK)

VK4FORNOT (an equimolar mix of)
JK1NOT10   GAG TCA TTC TGC GGC CGC CCG TTT GAT TTC CAG CTT
           GGT GCC
JK2NOT10   GAG TCA TTC TGC GGC CGC CCG TTT TAT TTC CAG CTT
           GGT CCC
JK4NOT10   GAG TCA TTC TGC GGC CGC CCG TTT TAT TTC CAA CTT
           TGT CCC
JK5NOT10   GAG TCA TTC TGC GGC CGC CCG TTT CAT CTC CAG CTT
           GGT CCC
(The 3′ ends of these primers are identical to the MJKFONX series)

RACE primers

5. Primers specific for heavy-chain isotypes (all priming in hinge)
RACEMOG1   TAT GCA AGG CTT ACA ACC ACA
(mouse IgG1)
RACEMOG2a   AGG ACA GGG CTT GAT TGT GGG
(mouse IgG2a)
RACEMOG2b   AGG ACA GGG GTT GAT TGT TGA
(mouse IgG2b)
RACEMOG3   GGG GGT ACT GGG CTT GGG TAT
(mouse IgG3)
RACERAG1   AGG CTT GCA ATC ACC TCC ACA
(rat IgG1)
RACERAG2a   ACA AGG ATT GCA TTC CCT TGG
(rat IgG2a)
RACERAG2b   GCA TTT GTG TCC AAT GCC GCC
(rat IgG2b)

(*continued*)

Table 10–3. (*cont.*)

RACERAG2c   TCT GGG CTT GGG TCT TCT GGG
(rat IgG2c)

Light chain primers can be chosen from the cDNA primers listed above in 1.2).

6. RACE PCR primers

XSCTnTag   GAC TCG AGT CGA CAT CGA TTT TTT TTT TTT TTT TT
Anneals to the poly A tail added by terminal transferase, and provides XhoI, SalI, ClaI sites at the 5′ end.

Heavy chain primers can be chosen from those in (1). In this case cloning will be done using the PstI site in CH1. Alternatively sites can be added to the 5′ end of those primers if PstI is not appropriate for some reason.

6.1 CK primers

CKRABam CAA GTC CGG ATC CCT AAC ACT CAT TCC TGT TGA AGC TCT
              TGA CGA CGG G
(this is identical to CKFOR, except that at the 3′ end it has 6 extra bases to increase its specificity for rat and at the 5′ end it hs a BamHI site)

CKMOBam CAA GTC CGG ATC CCT AAC ACT CAT TCC TGT TGA AGC TCT
              TTA TGG G
(this is identical to CKFOR, except that at the 3′ end it has 6 extra bases to increase its specificity for mouse and at the 5′ end it has a BamHI site)

6.3 Cλ primers

CL1FORBam CAA GTC CGG ATC CAC ACT CAG CAC GGG ACA AAC TCT
              TCT CCA CAG T
(mouse and rat λ1)
CL2FORBam CAA GTC CGG ATC CAC ACT CTG CAG GAG ACA GAC TCT
              TTT CCA CAG F
(mouse λ2, λ3; rat λ2)

CL4FORBam CAA GTC CGG ATC CAC ACT CAG CAC GGG ACA AAC TCT
              TCT CCA CAT G
(mouse λ4)

These may be used individually or in a pooled equimolar mixture. They are identical to the corresponding CLFOR primers, except that at the 3′ end there are 6 extra bases to increase the specificity for each λ and at the 5′ end there is a BamHI site.

These V regions can be amplified directly from hybridoma genomic DNA or from immunoglobulin specific cDNA. The advantage of using cDNA is that the concentration of V region sequences is much higher than that in genomic DNA and that one can discriminate between chains of different origin by the use of appropriate primers; the disadvantage is that extra steps are required. The synthesis of immunoglobulin-specific cDNA may be carried out using the relevant downstream V region PCR primer. We, however, prefer to prime the cDNA synthesis with oligonucleotides corresponding to conserved sequences in the constant regions of the light- and heavy-chain genes (see Table 10–1), in order to introduce a higher level of specificity.

cDNA can also be made by standard methods (oligo dT, hexamer priming). Prior to cloning the PCR product, it is worth fingerprinting the V regions with restriction sites with four base recognition sequences to find out whether more than one chain has been isolated. If this is the case, the sum of the amplified bands will be greater than the expected 300 to 330 bp. We have found that mouse-specific V region primers can also be used to clone rat hybridoma cDNA.[67,85]

Method 1.1: Genomic DNA Preparation

1. Wash 10^6 to 10^7 cells once in PBS, and subsequently in 1 ml buffer A (10 mM Tris-HCl (pH 7.4), 100 mM NaCl, 25 mM EDTA). Dying cells seem to give better amplification, perhaps because the genomic DNA is degraded into smaller pieces.
2. Add 20 µl 10% SDS and 40 µl proteinase K, incubate 3 hr at 50°C.
3. Extract 3 times with phenol/chloroform and precipitate DNA with 3 volumes ethanol.
4. Collect DNA with a capillary and transfer to 1 ml TE.
5. Boil DNA sample 5 min, and quench on ice.
6. Use 50 to 200 ng DNA per PCR reaction.

Method 1.2: RNA Preparation

A large number of commercial kits that yield high-quality material for cDNA synthesis are available for RNA purification. Any of these may be used. We prefer to make total RNA and then purify polyA+ RNA from this. When V region PCR works well (because the oligonucleotides match the V region very well), total RNA may be used for the PCR. In other cases polyA+ RNA should be purified. We use the following protocols, although any of those in standard texts[76] can also be used:

Method 1.2.1: Total RNA Preparation

1. Wash 10^7 cells in PBS.
2. Resuspend the cells in 1 ml buffer GTC (4M Guanidine-Isothio-cyanate, 25 mM Sodium Citrate, 0.5% Sarcoyl, 0.1 M β-mercapto-ethanol).
3. Add 0.1 ml sodium acetate (3 M, pH 5.2), shake.
4. Add 0.5 ml phenol, shake vigorously.
5. Add 0.2 ml chloroform, shake vigorously; leave on ice 15 min.
6. Spin 15 min 10,000 × g.
7. Precipitate aqueous phase with 1 volume of isopropanol; spin 15 min 10,000 × g.

8. Wash pellet in 70% EtOH and allow to air dry.
9. Resuspend pellet in 100 µl DEPC-treated H_2O.
10. Use 5 µl per standard cDNA reaction or proceed to method 1.2.2.

Method 1.2.2: Poly(A)+ RNA Preparation

1. Suspend 1 g of oligo(dT)-cellulose in 0.1 M NaOH.
2. Wash oligo(dT)-cellulose 3X with DEPC H_2O and 3X with binding buffer (20 mM Tris-HCl (pH 7,6), 0.5 M NaCl, 1 mM EDTA, 0.1% Sodium Lauryl Sarcosinate).
3. Heat the RNA sample (100 µg) to 65°C for 5 min. and add 1 volume of 2X binding buffer.
4. Add 10 µl of oligo(dT)-cellulose (10 mg total RNA can be loaded onto 1 ml of packed volume).
5. Incubate 1 hr at room temperature.
6. Wash oligo(dT)-cellulose 4X with binding buffer.
7. Elute poly(A) + RNA with 100 µl elution buffer (10 mM Tris-HCl (pH 7.6), 1 mM EDTA, 0.05% SDS).
8. Precipitate mRNA with 1/10 vol sodium acetate (3 M; pH 5.2) and 2.5 vol ethanol.
9. Recover mRNA by centrifugation, and resuspend in DEPC-treated distilled water.
10. Use 1 to 10 µg mRNA per standard cDNA reaction.

Method 1.3: cDNA Preparation

cDNA can be made simultaneously for both light and heavy chains by incorporating the appropriate primers and using the same stock for amplification of both chains. Either AMV or MuLV reverse transcriptase can be used. Recently some manufacturers have introduced reverse transcriptases which have been genetically modified and which claim to be more effective than the natural version. We have not tried these modified versions.

1. Mix in a sterile tube on ice:

RNA (up to 10 µg mRNA)	10 µl
10 × RT buffer	5 µl
DTT (100 mM)	5 µl
dNTP (5 mM each)	5 µl
Oligo(s) (10 pmoles/µl each)	1 µl
RNAsin (10u/µl)	5 µl
RTase (100 u/µl)	5 µl
H_2O	14 µl

2. Incubate 1 hr at 37°C.

3. Boil 3 min, and then quench on ice.
4. Use up to 10 µl per standard PCR reaction.

The 10 × RT buffer above consists of 0.5 M Tris-HCl (pH 8.2), 0.1 M MgCl$_2$, 1 M KCl. Alternatively, use the 10 × RT buffer provided by the supplier. It may be desirable to analyse the cDNA product by gel electrophoresis and, in some cases, to gel purify the cDNA species of interest. In this case include a tracer (e.g., 10 µCi α[32P] dATP) in the cDNA reaction, to allow identification of individual cDNA species.

Method 1.4: V-Region PCR Amplification

Precautions against PCR contamination (dedicated pipettes and aerosol proof tips, aliquoted reagents, a separate area of the lab for PCR, UV irradiation of samples prior to adding the template and enzyme, etc.) should be observed.

1. Mix the following:

H$_2$O	to 50 µl
cDNA or genomic DNA	5–10 µl
10 × PCR buffer	5 µl
dNTP 5 mM each	2.5 µl
BSA (10 mg/ml)	0.5 µl

    The reaction can be irradiated on a short-wave transilluminator for five minutes at this stage.

For primer (10 pmol/µl)	2.5 µl
Back primer (10 pmol/µl)	2.5 µl

2. Overlay with paraffin oil.
3. Heat to 94°C for 5 min in PCR block, and add 0.5 to 1 µl Taq DNA polymerase (5 u/µl) (Cetus).
4. Perform 30 temperature cycles as follows: 94°C 1 min, 60°C 1 min, 72°C 2 min.
5. Analyze 5 to 10 µl of PCR reaction on 2% agarose gel.

The 10 × PCR buffer above consists of 100 mM Tris-HCl (pH 8.3), 0.5 M, KCl, 15 mM MgCl$_2$. Alternatively, use the 10 × PCR buffer provided by the supplier. The concentration of MgCl$_2$ in the buffer should be checked. BSA is acetylated BSA (New England Biolabs). Other heat stable polymerases which have proofreading activity (e.g. *T. litoralis* [Vent, New England Biolabs] *P. furiosus* (Stakagene) may also be used.

Method 15: Restriction Enzyme Analysis

The PCR product does not need to be purified prior to restriction enzyme digestion unless multiple bands are seen on the PCR, in which case it should be gel purified. Use 2 to 5 µl of a standard PCR reaction and digest with restriction enzymes having four base recognition sequences. BstNI is especially useful for the heavy chain, and HaeIII for the light chain, as these two enzymes are family specific. If it becomes necessary to discriminate between V regions of the same family, we have found the following enzymes useful: AluI, MboI, Sau3AI, RsaI, and TaqI.

Points to Consider, Method 1

There are three things that can go wrong with V region amplification. The first is that nothing is amplified, the second that multiple bands are seen, and the third that a V region band is seen but that it does not encode the specificity of the monoclonal antibody. One encounters the first and second problems immediately, but the third only after sequencing, bacterial expression, or eukaryotic expression. When no bands are seen, the following points should be considered:

1. *The primers do not bind to the ends of the V region due to mutation.* The somatic mutations that the V region genes undergo in order to increase affinity are mainly targeted to the CDRs. Mutations do, however, also occur at low frequency outside the CDRs, including the otherwise conserved sequences used for PCR amplification. We have found this to be a problem when the hybridoma is derived after long-term immunization. We have successfully amplified rat hybridomas with mouse-specific primers,[67, 85] but since few rat V region sequences hve been published, it is also possible that some rat V regions are incompatible with the mouse primers. If V regions fail to amplify, then it is worth trying to decrease the temperature of annealing (range 50 to 65°C), increase the concentration of magnesium (1.5 to 4 mM), or change the DNA polymerase (*T. litoralis*–Vent, or *P. furiosus*–Pfu, or their recombinant variants and those of Taq polymerase).

2. *The light chain is lambda.* Five to 10% of mouse and 1% of rat light chains are not kappa, but lambda. The light-chain V region primers (as well as the cDNA primers) are specific for kappa, and will not amplify lambda. If changing the PCR conditions is not successful, it is worth subtyping the immunoglobulin light chains produced by the hybridoma to confirm that they are kappa.

3. The cDNA is of poor quality. This may arise due to poor-quality mRNA and can be checked by including an analytical cDNA reaction in order to evaluate the quality of the cDNA obtained; however, the power of PCR is such that bands can usually be amplified from most cDNA samples. Should a sample be shown to be of low quality, it would be worth resynthesizing it.

The problem of multiple bands can usually be resolved by increasing the temperature of the annealing step of the PCR or lowering the magnesium concentration. Alternative strategies include adding specificity enhancers (5 to 15% DMSO,[8] 1.25 to 5% formamide,[77] or 10^{-4} to 10^{-5} M TMAC),[38] or trying a different thermostable polymerase.

The problem of irrelevant bands (by irrelevant we mean V regions that do not encode the hybridoma specificity, not other PCR bands that are not V regions), if it is encountered, will be encountered later in the cloning strategy; however, since many of these problems are related to the V region amplification, we will discuss them here. There are a couple of ways in which one can reduce the amplification of irrelevant bands to allow the amplification of the specific band. The first involves the use of immunoglobulin isotype specific primers (priming in the hinge—see Table 10–1), which will make cDNA only from the specific immunoglobulin chain, since (unless one is extremely unlucky) it is unlikely that the contaminant chain has the same hinge. An alternative strategy is to purify the cDNA band from a polyacrylamide gel after labeling (see method 1.3). When it is discovered that a nonfunctional chain has been cloned, the following points should be considered:

1. Does the parental fusion partner express an immunoglobulin chain? Many of the myelomas used as fusion partners express an endogenous antibody chain that continues to be expressed in the hybridoma. In some cases these endogenous chains have a different molecular weight to the hybridoma chain and so can be distinguished by SDS-PAGE electrophoresis. This is especially true of rat-mouse hybrids. In most cases the sequence of the endogenous chain is known and so can be compared with any sequences obtained (it can be found in any of the DNA databases, or alternatively in Kabat et al.[44]). In the case of rat-mouse hybrids, it may be found that the mouse endogenous chain is more easily amplified than the hybridoma. If this is the case, and the sequence of the endogenous chain is known, the use of restriction enzymes that recognise six base sites may allow one to destroy one but not the other, or alternatively, primers that will prime rat but not mouse cDNA synthesis could be used.

2. Is more than one functional heavy and light chain expressed? Some hybridomas are the result of the fusion of multiple spleen cells, giving rise to the possibility that more than one heavy and light immunoglobulin chain is expressed. In this case which of the chains is binding (and which nonbinding) is unknown, so function or binding assays must be carried out on each of the expressed proteins in combination with all the corresponding partner chains in order to identify the correct immunoglobulin V regions.

3. Are nonproductive immunoglobulin messenger RNAs expressed? It is of great importance to remember that a lymphoid cell or a hybridoma may synthesize more than one rearranged heavy/light immunoglobulin *messenger RNA*. Termination of rearrangement in the immunoglobulin loci occurs only after a mature membrane-bound immunoglobulin has been synthesized. Thus, a given cell may produce nonproductive (due to termination codons or frameshifts) immunoglobulin messenger RNAs, in addition to the mRNA that is translated into a mature immunoglobulin chain. These "ghost" messengers can present a real problem when cloning immunoglobulin genes from hybridomas, since they are often very good substrates for V region PCR, even though they do not encode functional polypeptides. This may be because they undergo less somatic mutation. We have, for example, in one case repeatedly amplified a heavy-chain V region that had an out-of-frame mutation in the third CDR.

4. Are important sequences "corrected" by the PCR primers? Antigen affinity and specificity are generally defined by the CDRs, although contributions from non-CDR regions have been identified. If the cloned recombinant antibody does not exhibit the antigen specificity and affinity of the parental hybridoma, it is possible that sequences important for this binding have been altered by the PCR primers. The consensus-based PCR primers may amplify sequences with low homology, thus altering the amino acid sequence in these regions This is a theortical consideration which we have not yet encountered, but which should perhaps be borne in mind if there is no other explanation.

5. Has a contaminant V region been amplified? PCR amplification is so sensitive that, despite precautions, V regions from other hybridomas in the laboratory can easily be amplified. One advantage of sequencing all cloned V regions is that this problem can be eliminated at a relatively early stage. The sequence of newly cloned V regions should be routinely compared against all other known V regions present in the laboratory. An alternative to sequencing is fingerprinting with a number of different four-base-recognizing restriction enzymes.

Introducing to Method 2: Cloning Hybridoma cDNA by RACE

When the V region primers do not work, or amplify chains that are not correct, three alternative approaches can be tried. The first, based on the RACE technique, requires few primers, does not depend on any variability with the V regions, and is the method we use. Alternative methods that have been used with success include standard cDNA cloning[76] and the use of PCR with leader and C region primers.[43,50,51] This latter technique is successful, but requires a relatively large number of primers and cannot be used for hybridomas (such as rat) where few leader sequences are known. We have not used this method.

RACE cloning,[25] as we have modified it for immunoglobulin V region cloning,[91,92] relies on a knowledge of the isotype of the hybridoma immunoglobulin to produce a primer specific for that isotype which is then used to produce cDNA from the mRNA (see Table 10–1). To this relatively specific cDNA, a tail of poly A is then added using terminal transferase. This allows one to perform a PCR reaction using a second primer specific for the Ig constant region (upstream of the cDNA synthesis primer) and another that consists of a string of Ts preceded by useful restriction sites. Once amplified, the V regions are then cloned into standard cloning vectors, such as Bluescript (Stratagene), sequenced, and then transferred back to the M13-based vectors by V region PCR, as described above. To clone into Bluescript, either sites can be incorporated into the primers, or alternatively, conserved sites (such as PstI in rodent CH1) can be used. Table 10–1 gives the sequences of hinge region primers of different heavy-chain isotypes that can be used for cDNA synthesis. If one wishes to incorporate restriction sites, they should be added to the 5' end with a tail of at least 4 nucleotides to allow efficient digestion. The use of hinge-specific primers is not essential, but increases the specificity of the cDNA synthesis and hence the subsequent RACE reaction. The nested primers used for RACE bind at the 3' end of CH1 and are also given in Table 10–1. There is no equivalent to hinge-region primers for the light chain. As a result, irrelevant light chains will also be amplified, although one can distinguish between mouse and rat light chains. The light-chain cDNA synthesis primer is found at the 3' end of the constant region, and the RACE primer overlaps it with a 6 bp extension to preserve specificity and includes a BamHI site at the 5' end.

We have found that the subsequent PCR to clone the V regions into the M13VPCR vectors is best done using relatively high amounts of the cloned V region (1 µl), low annealing temperatures (as low as 37°C in some cases), and a DNA polymerase with proofreading activity (such as *T. litoralis* [Vent, New England Biolabs] or *P. furiosus* [Stratagene]) to reduce errors introduced by PCR. Although this protocol presently requires multiple cloning steps, we are working on ways to reduce these by designing new plasmids that will allow cloning, sequencing, and expression together.

Method 2: Cloning Hybridoma V Regions by RACE

1. Prepare cytoplasmic mRNA from 5×10^6 hybridoma cells using method 1.2.
2. Synthesize cDNA with the following protocol: denature 1 µg of poly(A) mRNA at 65°C for 5 min in DEPC treated water, put on ice, and then add to a mixture containing 5 µl 5 × RT buffer, 10 µl RNasin (Promega), 10 pmol specific primer (see Table 10–1), 250 µM of each of the four deoxynucleotide triphosphates (dNTPs) and 10 U of Moloney murine leukemia virus reverse transcriptase in a total volume of 25 µl. (Although we use this protocol, which is based upon the original published method,[25] we have no reason to believe that the cDNA synthesis protocol in method 1.3 would not work as well.) The reaction mixture is incubated at 42°C for 60 min and then at 52°C for 30 min. After inactivation at 95°C for 5 min the reverse transcription mixture is diluted with 2 ml of 0.1 TE (0.1 TE is 1 mM Tris pH 7.0, 0.1 mM EDTA).
3. Remove excess primer using a Centricon 100 spin filter (20 min at 1000 g, twice). The first retained liquid is collected and diluted to 2 ml before repeating the Centricon concentration. The second is concentrated to 10 µl and used in the following steps.
4. Synthesise a polyA tail at the 5′ end of the cDNA by adding 4 µl 5 × Tailing buffer (supplied by Promega with the enzyme), 4 µl dATP 1 mM and 10 U of Terminal deoxynucleotidyl transferase (Promega). The mix is incubated for 5 min at 37°C and then 5 min at 65°C. The volume of the cDNA/tailing reaction is adjusted to 500 µl.
5. Amplify 10 µl of reaction with Vent polymerase as follows: 1 precycle: 5 min 95°C, 5 min 60°C, 40 min 72°C; 40 cycles: 1 min 95°C, 1 min 60°C, 3 min 72°C.

PCR is performed usng the oligonucleotide XSCTnTag (Table 10–1), which hybridizes to the poly(A) tail added to the 5′ end of cDNA and one oligonucleotide specific for the light or heavy chain (Table 10–1).

Points to Consider, Method 2

We have found this method extremely successful in amplifying V regions that cannot be amplified by V region primers. As in any PCR reaction, varying the annealing temperature, Mg concentration, or the polymerase may improve the quality of the product (see Method 1 for a discussion).

Introduction to Method 3: Expression of Soluble Immunoglobulin Fragments in Bacteria

Once V regions have been amplified, they need to be assembled into single-chain Fv fragments (ScFvs) for expression in bacteria. ScFvs are made up of the heavy- and light-chain V regions connected by a flexible linker. This section discusses the construction of ScFv[1, 5, 6, 41] fragments by a process called PCR assembly (summarized in Fig. 10–3), the cloning into an appropriate *E. coli* expression vector, and the screening to detect binding ScFv fragments by ELISA. This whole procedure takes approximately one week; recently a kit has been produced for the cloning of murine antibody genes (Pharmacia's Recombinant Phage Antibody System). The cloning procedure and expression vector presented here can also be used to make, select, and screen antibody repertoires (see Chapter 2).

Method 3.1: Assembly of ScFv Fragments

The separately amplified VH and VL genes are linked in an assembly step (Method 3.2) to incorporate the linker DNA generated in Method 3.1. The linker DNA has regions of homology with the 3′ end of the amplified VH gene and 5′ end of the amplified VL gene, as well as sequence encoding the $(Gly_4Ser)_3$ linker peptide that joins the two variable chains. In the same PCR, the two outer flanking PCR primers provide restriction sites for cloning. We describe here the assembly of an ScFv fragment. Using the same principle, FAb fragments can be made, for example by linking VHCH1 to VLCL fragments via a linker fragment providing a ribosome binding site and a pelB signal sequence. V genes can also be cloned sequentially; however, the assembly procedure allows cloning using only two restriction sites (four are needed for sequential cloning), which reduces the risk of V genes being cut by the cloning enzymes.

Method 3.1.1: Preparation of the ScFv Linker Fragment

1. For batch preparation of the murine ScFv linker fragment, set up a number of PCR reactions as follows. Mix the following:

H_2O	35.5 μl
10 × PCR buffer	5 μl
5 mM dNTP	2.5 μl
BSA (10 mg/ml)	0.5 μl
MO-LINK-FOR (10 pmol/μl)	2.5 μl
MO-LINK-BACK (10 pmol/μl)	2.5 μl
pSW1-ScFvD1.3 (∼10 ng/μl)	1 μl

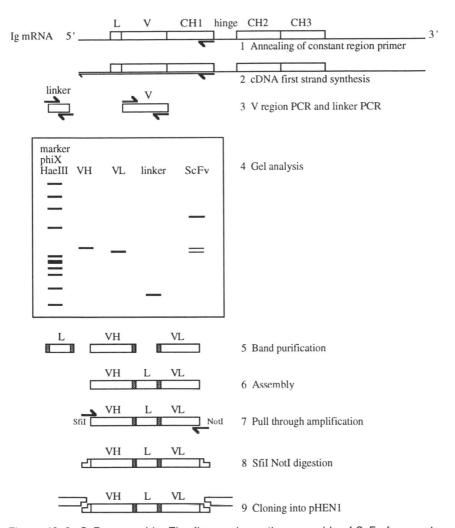

Figure 10–3. ScFv assembly. The figure shows the assembly of ScFv fragments starting with immunoglobulin RNA. Small dark half-arrows represent oligonucleotides annealing to their target sequence. Steps 1, 2 and 3 have been depicted for heavy-chain genes, although similar mechanisms will apply for the light-chain genes (using light-chain constant region primers and light-chain V region primers). Step 4 shows the expected sizes of the individual amplified V regions, the linker, and also the assembled ScFv (step 6) with some remaining V region fragments. In steps 5 through 9, the stippled regions represent sequences that are homologous between the 5' end of the linker and the 3' end of VH, and the 3' end of the linker and the 5' end of VK (respectively). These homologous regions permit assembly of the ScFv by performing cycling without oligonucleotides.

2. Overlay the tubes with paraffin oil, heat to 94°C for 5 min using a cycling heating block. Then add 0.5 µl *Taq* polymerase (5 units/µl (Cetus)) under the oil.

3. Cycle 25 times: 94°C 1 min, 60°C 1 min, 72°C 1 min.

4. Purify the ScFv linker fragment on a 2% (low-melting-point) agarose gel. To avoid contamination it is essential to depurinate the electrophoresis apparatus, combs, etc., with 0.25 M HCl for at least 30 min before use. Excise the band and place in the upper chamber of a SPIN-X column (Costar), freeze in dry ice for 10 to 15 min. Place in microcentrifuge tube, thaw and spin tube for 5 min at 13,000 rpm. Alternative methods of gel purification can also be used.

5. Precipitate the filtrate by adding 1/10 vol 3 M sodium acetate, pH 5.2, and 2.5 vol ethanol. Chill on dry ice for 5 min, spin at 13,000 rpm for 10 min. Wash the pellet in 1 ml 70% ethanol and dry under vacuum. Dilute purified linker into 5 µl water or TE per original 50 µl PCR, measure concentration on gel, dilute to approximately 20 ng/µl and store in 5 µl aliquots at 20°C.

$10 \times$ PCR buffer is 100 mM Tris-HCl, pH 8.3, 500 mM KCl, 15 mM $MgCl_2$ (for the Cetus enzyme). Otherwise use the buffer provided by the manufacturer. 5 mM dNTP is an equimolar mix of dATP, dTTP, dCTP and dGTP, with a total concentration of 5 mM per nucleotide at pH 7.0. BSA is acetylated BSA (New England Biolabs) at 10 mg/ml. See Table 10–1 for the sequence for MO-LINK-FOR and MO-LINK-BACK. The double-stranded ScFv linker (93 bp) can be amplified with the primers MO-link-FOR and MO-LINK-BACK from any ScFv gene made with the primers described in Clackson et al.[16] We use plasmid pSW1-ScFvD1.3 (described in the legend to Figure 1 of McCafferty et al., yielding the $(G_4S)_3$ linker described by Huston et al.[41] (see sequence of the linker region in the appendix). MO-LINK-FOR is complementary to the VK2BACK primer. MO-LINK-BACK is complementary to the VH1FOR-2 primer. Similarly, mouse VHCH1 and $V_\kappa C_\kappa$ fragments can be assembled as mouse Fab fragments using a linker providing the 3′ end of the CH1 domain, the pelB leader sequence, and the 5′ end of V_κ.

Method 1.2.2: PCR Assembly of Primary VH and V_κ PCR Bands in ScFv configuration

1. Make up two 50 µl PCR mixes containing:

	1	2
H20	35.5 µl	25.5 µl
$10 \times$ PCR buffer	5 µl	5 µl
$20 \times$ dNTPs	2.5 µl	2.5 µl

BSA (10 mg/ml)	0.5 µl	0.5 µl
ScFv LINKER (~ 20 ng)	1 µl	1 µl
VH 1°PCR band (min. 100 ng)	—	5 µl
VK 1°PCR band (min. 100 ng)	—	5 µl

For PCR-assembly of hybridoma genes, unpurified primary PCR bands can be used, but it may be necessary to remove excess primers or primer-dimer complexes from the products of the first PCR before assembly, which would otherwise compete with linker for priming. This is achieved here by gel purification (see step 7). The starting amount of primary PCR bands seems to be critical for the assembly reaction: 50 ng or more may be required. A roughly equimolar amount of linker is used. This may necessitate more than one PCR reaction to produce the required quantity of material.

2. Overlay with paraffin oil, heat to 94°C for 5 min using the PCR-block.
3. At the end of this incubation, add 0.5 µl *Taq* polymerase (5 unit/µl) (Cetus) under the oil. Then cycle 7 times: 94°C 1.5 min, 72°C 2.5 min. To minimise spurious priming events, the overlaps should be made long (24 nucleotides) to allow annealing to be performed at a high temperature (72°C).
4. After this cycling add:

VH1BACKSFI	2.5 µl	2.5 µl
VK4FOR NOT-MIX	2.5 µl	2.5 µl

5. Cycle 25 times: 94°C 1.5 min, 55°C 1 min, 72°C 2.0 min.
6. Analyze the PCR-assembly on a 2% agarose gel.
7. Gel-purify the assembled fragment on a 2% gel, excise the band corresponding to the assembled product (approximately 720 bp), and further purify the DNA using, for example, Geneclean (Bio101 Inc) or Magic PCR Preps (Promega). Resuspend the product in 10 to 25 µl of water. To avoid contamination it is essential to depurinate the electrophoresis apparatus, combs, etc., with 0.25 M HCl for at least 30 min before use.

Method 3.2: Cloning of ScFv Gene Cassette into Phagemid Vector pHEN1

It has been shown that functional antibody fragments (ScFv or FAb) can be produced in *E. coli* by secreting the antibody domains into the periplasm where folding and the formation of disulphide bridges can occur. In the pHEN1 phagemid vector,[36] the ScFv cassette is driven by the lacZ promoter (induction by IPTG). Export of the ScFv protein to the periplasm is directed by the *pel*B signal sequence, provided at the 5′ end of the ScFv sequence (Fig. 10–4). At the 3′ end of the gene, a sequence encoding a portion of the

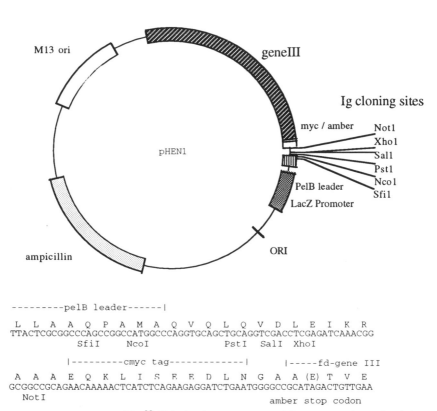

Figure 10–4. Map of pHEN.[36] This is the vector used for expression of soluble
ScFv fragments in bacteria. It contains origins of replication from ColEI and M13,
permitting growth in *E. coli* and packaging in filamentous phage, respectively;
ampicillinase as a selectable marker; the LacZ promoter to drive expression of the
ScFv; a PelB leader sequence; a polylinker for cloning immunoglobulins, a
sequence encoding a short stage derived from myc that is recognised by the mAb
9E10; an amber stop codon that allows the production of a soluble protein in both
suppressor and nonsuppressor strains, as well as a genelll-ScFv fusion protein
in suppressor strains; genelll from fd permitting phage display of the cloned ScFv
(see Chapter 2 for details). The sequence of the polylinker is also given.

c-myc protein is included, providing a C-terminal peptide on each soluble
antibody fragment, which can be recognised by the 9E10 antibody.[21] This
peptide tag facilitates detection in ELISA or Western blot and allows
purification of large amounts of antibody fragments directly from bacterial
culture medium by affinity chromatography with the 9E10 antibody. Finally,
at the 3′ end of the c-myc tag, an amber codon is inserted, allowing expression
of the cloned protein as a soluble fragment in nonsuppressing *E. coli* strains
such as HB2151. In those strains the amber is read as a stop codon and
soluble protein is secreted into the bacterial periplasm. When grown in a

suppressor strain, the amber is read as glutamine and read-through into the gene encoding the minor coat protein of filamentous phage fd (gene3) occurs. However, we have found that since suppression is at most 50%, sufficient soluble protein is also produced in suppressor strains to detect soluble protein by ELISA using the myc tag. The pHEN-1 phagemid vector was originally designed to allow the insertion of single-chain Fv sequences as fusions at the N-terminus of the gene III protein of filamentous phage fd.[36] The vector can therefore be used for the display of antibody fragments on the surface of phage fd, as well as for soluble expression. This is described in Chapter 2.

For cloning into pHEN1, the rare cutting enzyme SfiI and NotI sites are used, especially for cloning mouse ScFv and Fab cassettes, in which other enzymes like PstI, BstEII, and SacI are frequent cutters. Alternatively, human ScFv and FAb fragments can be cloned as NcoI-NotI cassettes (mouse V genes frequently have internal NcoI sites). Since the vector also contains codons for 5 amino acids derived from the N-terminus of the antibody heavy chain together with 6 amino acids derived from the C terminus of the light chain, sometimes restriction sites PstI and XhoI can also be used for cloning in combination with the primers VH1BACK and VK2FOR (see Table 10–1).

We describe here the cloning of an assembled ScFv fragment as an SfiI–NotI restriction fragment, its production by *E. coli* cells, and its detection in antigen-binding ELISA using the anti-myc tag 9E10 antibody as a reagent.

Method 3.3: Restriction and Cloning of Assembled PCR Fragment

1. For optimal digestion, the restriction of the PCR product is performed using the buffer supplied by the manufacturer. Large overdigestion is always necessary for efficient cutting. Mix the following together:

PCR-band	36 µl
10 × NEBuffer 2	5 µl
BSA (1 mg/ml)	5 µl
SfiI (10 units/µl)	4 µl

2. Overlay with paraffin. Incubate at 50°C for 3 to 6 hrs or overnight.
3. To this mix, add under the oil:

H$_2$O	31 µl
10 × NEB-NotI	5 µl
BSA (1 mg/ml)	5 µl
1 M NaCl	5 µl
NotI (10 units/µl)	4 µl

   (Alternatively, the mix can be precipitated in between the restriction digests.)

4. Leave 4 hrs or overnight at 37°C.
5. Recover the aqueous phase and gel-purify the digested product on a 1.5 to 2% (low-gelling-temperature) agarose gel. Recover the DNA by extraction using Geneclean or Magic PCR preps (Promega).
6. Finally, resuspend the DNA in 10 µl water, and determine its concentration by analysis on an agarose gel with markers of known size and quantity.
7. Set up the following ligation reaction:

	1	**2**	**3**
10 × NEB-Ligation Buffer	1 µl	1 µl	1 µl
H₂O	6 µl	5 µl	1 µl
Vector pHEN1 (30 ng/µ)	2 µl	2 µl	2 µl
Digested PCR band (20 to 59 ng/µl)	—	1 µl	5 µl

8. Spin for a few seconds in the microfuge. Then add:

	1	**2**	**3**
T4-DNA ligase (400 units/µl, NEB	1 µl	1 µl	1 µl

9. Leave overnight at 16°C; store at −20°C until further use.
10. Transform *E. coli* cells with the ligation mix, using standard calcium-phosphate competent cells (Hanahan) or electroporation.
11. Plate fractions on TYE plates containing 100 µg/ml ampicillin and 1% glucose and grow at 30°C.

NEBuffer 2 is 10 mM Tris-HCl, pH 7.9, 50 mM NaCl, 10 mM MgCl₂ and 1 mM dithiothreitol; NEB-NotI is 10 mM Tris-HCl, pH 7.9, 150 mM NaCl, 10 mM MgCl₂ and 0.1% Triton X-100. 10 × NEB-Ligation Buffer is 0.5 M Tris-HCl, pH 7.8, 0.1 M MgCl₂, 0.2 M DTT, 10 mM rATP and 500 µg/ml BSA. The phagemid pHEN1 is cut with SfiI and NotI, alkaline phosphatase treated and gel-purified using standard protocols.[76] For soluble expression of antibody fragments from pHEN1, the nonsuppressor strain HB2151 is used, although TG1 can be used if the vector is also to be used for phage display. TYE is 20 g Bacto-tryptone, 5 g Yeast extract and 0.5 g NaCl per liter.

The PCR amplification process can introduce mutations into the DNA, some of which could be deleterious for the binding characteristics of the antibody. Therefore, especially when PCR assembly has been used (another 32 PCR cycles), it is advisable to analyze many different clones for binding to antigen. We find that if no positive clone is found among 96 to 192 different transformants, a different approach should be taken (see Fig. 10–1). The next protocol describes the screening of 96 different transformants.

Method 3.4: Induction of ScFv Expression for ELISA

This method is based on that of De Bellis and Schwartz[19] and relies on the low levels of glucose present in the starting medium being metabolized by the time the inducer (IPTG) is added.

1. Inoculate 100 µl 2 × TY, 100 µg/ml ampicillin, 1% glucose in 96-well plates (cell wells, Nuclon) and grow with shaking (300 rpm) overnight at 37°C.

2. Use a 96-well transfer device to transfer small inocula from this plate to a second 96-well plate containing 150 µl fresh 2 × TY, 100 µl/ml ampicillin, 0.1% glucose per well. Grow at 37°C, shaking until $OD_{600\,nm}$ is approximately 0.9 (about 3 hrs). (Or transfer 4 µl to a well with 150 µl fresh medium). To the wells of the original plate, add 25 µl 60% glycerol per well and store at −70°C.

3. Add 50 µl 2 × TY, 100 µg/ml ampicillin, 4 mM IPTG (final concentration 1 mM IPTG). Continue shaking at *30°C* for a further 16 to 24 hrs.

4. Spin 4,000 rpm for 10 min and use 100 µl supernatant in ELISA the same day. During the second overnight growth, the outer membrane becomes leaky and antibody fragment can be detected in the culture supernatant as well as in the periplasm. For screening a large number of clones, we routinely use culture supernatant. The supernatant cannot be stored because the tag is lost with prolonged incubation.

Method 3.5: ELISA for Detection of Soluble ScFv Fragments

1. Coat plate (Falcon 3912) with 100 µl per well of protein antigen, usually at 1 to 10 µg/ml, but concentrations as high as 3 mg/ml are used in the case of certain proteins (e.g., lysozyme). Coating is usually in 50 mM sodium hydrogen carbonate, pH 9.6 (pH with NaOH) or in PBS. Leave overnight at room temperature.

2. Rinse wells 3 times with PBS, and block with 100 µl per well of 2% Marvel/PBS for 2 hrs at 37°C. Marvel is dried fat-free milk powder. PBS is 5.84 g NaCl, 4.72 g Na_2HPO_4 and 2.64 g $NaH_2PO_4 \cdot 2H_2O$, pH 7.2, in 1 liter.

3. Rinse wells 3 times with PBS, then add 50 µl 4% Marvel/PBS to all wells.

4. Add 50 µl culture supernatant containing soluble ScFv to the appropriate wells. Mix, leave 1.5 hrs at room temperature.

5. Discard test solution and wash out wells 3 times as in step 3. Pipette 100 µl of 4 µg/ml purified 9E10 antibody, in 2% Marvel/PBS, into each well. Incubate at room temperature for 1.5 hrs.

6. Discard 9E10 antibody and wash out wells 3 times as in step 3. Pipette 100 µl of 1 : 500 dilution of anti-mouse antibody (peroxidase-conjugated anti-mouse immunoglobulins, Dakopats/ICN, or peroxidase conjugated anti-mouse IgG, Fc-specific, Sigma A-2554). Incubate at room temperature for 1.5 hrs.

7. Discard second antibody and wash wells 3 times with PBS, 0.05% Tween-20, and 3 times with PBS.
8. Add one 10 mg ABTS (2,2′-azino bis(3-ethylbenzthiazoline-6-sulphonic acid), diammonium salt) tablet to 20 ml 50 mM citrate buffer, pH 4.5. (50 mM citrate buffer, pH 4.5 is made by mixing equal volumes 50 mM trisodium citrate and 50 mM citric acid).
9. Add 20 µl 30% hydrogen peroxide to the above solution immediately before dispensing.
10. Add 100 µl of the above solution to each well. Leave at room temperature 20 to 30 min.
11. Quench by adding 50 µl 3.2 mg/ml sodium fluoride. Read at 405 nm.

Points to Consider, Method 3

This procedure may not work for a number of reasons. In addition to the problems encountered in the amplification of V regions (discussed above), there are also problems associated with the formation of the ScFv itself. A significant proportion of hybridomas either will not fold as ScFvs, or the ScFvs when folded no longer recognise the antigen. It is clear that, in these cases, one has to proceed either by FAb cloning or by expression in eukaryotes. Another problem is related to the toxicity of antibodies when expressed in bacteria. The toxicity of any particular antibody is unpredictable. Toxicity is a possible cause of poor growth of clones on plates or in liquid culture. If it is suspected that this may be the case, it is worth changing the temperature of incubation (with a range 18 to 37°C). Usually a reduction in temperature helps, although some hybridomas are less toxic at 37°C.

Introduction to Method 4: Sequencing and Expression of Cloned V Regions in Mammalian Cells

If expression in mammalian cells is used to test the specificity, the V regions need to cloned into M13 based vectors, M13VKPCR and M13VHPCR (see Fig. 10–5), using standard techniques. These allow us to sequence them as well as providing the immunoglobulin enhancer, a eukaryotic leader sequence, the leader intron, and the first half of the intron downstream of the V region. The vectors already contain V regions which need to be replaced with the newly amplified PCR fragment. As indicated in Table 10–1, the primers used in the PCR reaction introduce restriction sites for six base cutters into extremities of the V-regions, thus allowing force cloning of the fragments. It is important to sequence more than one clone, to ensure that no PCR errors have been introduced; and once the V regions have been sequenced, they should be checked as described in points to consider below.

Following the appropriate sequence analysis in the M13 vectors, the

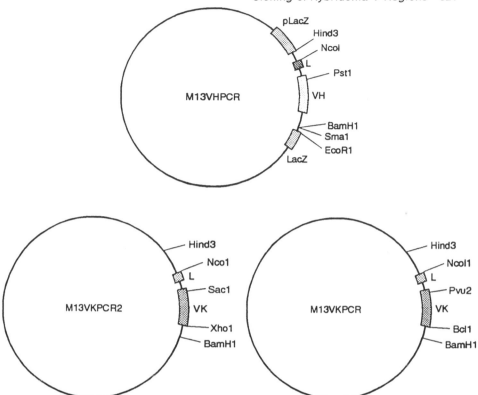

Figure 10–5. Maps of M13VHPCR, M13VKPCR1 and 2 inserts.[63] These are shuttle vectors for cloning V regions after amplification. They all contain a HindIII BamHI insert (sequences in the appendix), which is identical for the first 260hp. This part provides the immunoglobulin enhancer, an immunoglobulin leader exon and intron, and the last four amino acids of the leader region (which forms part of the V region exon). The rest of the insert provides in each case, the 5′ and 3′ ends of the appropriate V region, cloning sites for the V regions, and a 5′ splice site upstream of the BamHI. Although derived by roundabout routes, M13VHPCR is best considered to be the insert cloned into the HindIII BamHI sites of M13mp19; while M13VKPCR1 and 2 can be considered to be the respective inserts cloned as blunt functional HindIII/BamHI fragments into M13mp18 cut with PvuII. The difference between M13VKPCR1 and 2 lie in the V region cloning sites. Variations containing signals for different uses can be found in Figure 10–7. The inserts illustrated are found between the two PvuII sites (at 5960 and 6375) found in M13mp19, the rest of the sequence being M13mp19.

V-region and the relevant flanking sequences can then be cloned into the mammalian expression vector. We use pSV2 based vectors (see Fig. 10–6 and Method 4.2 for cloning details) which provide the second half of the V-C intron, the human IgG1/K constant regions, the immunoglobulin promoter, a SV40 polyadenylation site, the SV40 origin of replication, and selectable markers.

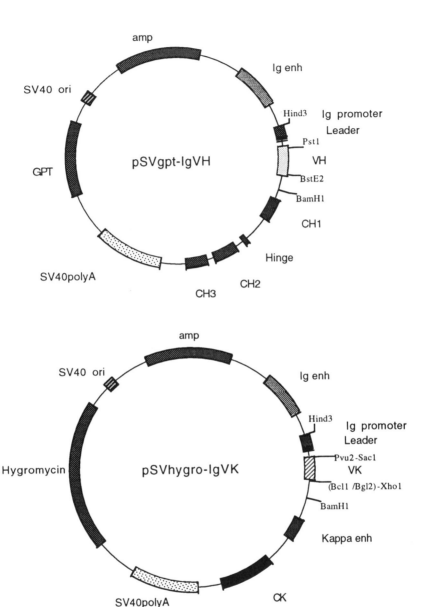

Figure 10–6. pSVhygro-IgVK and pSVgpt-IgVH.[63] These vectors for Ig expression are derived from the pSV2 series and contain, respectively, genes for hygromycin and gpt, in addition to sequences encoding the human kappa chain and IgG1 chain, respectively, under the control of the Ig promoter. The SV40 origin of replication permits high-level amplification of the plasmid in COS cells.

Method 4.1: Cloning into M13VPCR Shuttle Vectors

The heavy-chain V region primers incorporate a PstI site at the 5' end and a BstEII site at the 3' end of the V region (see Fig. 10–5 and Appendix). The PCR fragment and the shuttle vector M13VHPCR1 are digested with these enzymes; fragment and vector are subsequently gel purified and ligated together. Care should be taken to digest the vector completely, as it already contains a V region fragment that produces a functional polypeptide chain. Recombinant clones are identified by sequence analysis of the M13 vector. The sequence of the V region found in the parental M13VHPCR1 is given in the Appendix.

The light-chain V region may be amplified by two different primer sets, either the Pvu2/Bgl2 primers compatible with the vector M13VKPCR1, or the second generation Sac1/Xho1 primers compatible with M13VKPCR2. No difference in their capacity to amplify V region genes has been observed. The first-generation primers (Pvu2/Bgl2) and vector (M13VKPCR1) are associated with a few drawbacks: (1) and 3' restriction sites in the vector and the V region are Bcl1 and Bgl2, respectively, which yield compatible overhangs but result in loss of both restriction sites upon ligation; (2) the M13VKPCR1 vector must be grown in a Dam⁻–*E. coli* strain due to the Bcl1 site, which will otherwise not be cut; and (3) the restriction sites of the PCR fragment limit one to the mammalian M13 system, as they are incompatible with the bacterial system. These drawbacks have been overcome in the second generation of primers and vector in which all sites are retained and compatible with the bacterial system, and dam methylation is not a problem; however, this vector has not been used as extensively as the first.

If one is sure of the final destination of the antibody (intracellular, nuclear), then some M13 based vectors have been constructed (Fig. 10–7) that have the appropriate leaders already incorporated. The use of these ensures that the correct leader is already present when cloned into the pSV2-based vector.

Occasionally one isolates a V region that has an internal site identical to one of those used for cloning into the M13VPCR vectors (SacI, XhoI, BglII, PvuII, PstI, or BstEII). In the case of the light chain, the easiest solution is to change M13VPCR vector and primers. In the case of the heavy chain it is slightly more difficult. This is best resolved by doing an intermediate blunt cloning step of the PCR fragment into a general cloning vector and then extracting the fragment by a partial digestion (this cannot be done on the PCR fragment because one cannot be sure that the cloning site has been cut), or eliminating the site by *in vitro* mutagenesis (without affecting the amino acid sequence).

Approximately ten M13 clones should be sequenced in order to control for PCR-based errors. After the V regions have been translated, they should be checked as described in points to consider below.

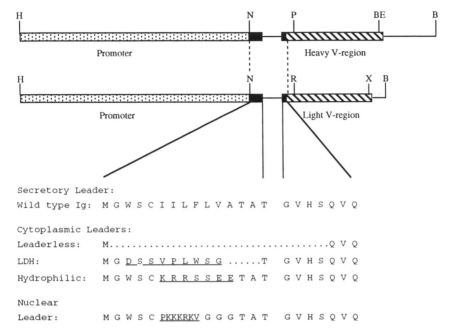

Figure 10–7. M13-based vectors that contain different targeting signals. These are identical to those in Fig. 10–5 with the exception of the N terminal end of the V regions. The normal secretory leader is depicted as well as three different cytoplasmic leaders (see text for details) and one nuclear localization leader. The single-letter amino acid code is used, dots indicate no amino acid, underlined amino acids are different to the normal secretory leader, and restriction enzyme sites are indicated as follows: H, HindIII; N, NcoI; P, OstI; BE, BstEII; B, BamHI. R and X are the sites found in the M13VKPCR series (PvuII, SacI and BcII, XhoI, respectively).

Points to Consider, 4.1: After Sequencing

1. Is it immunoglobulin? Translate the sequence in all three frames. Table 10–2 derived from Kabat et al.[44] Shows which amino acids are expected at different positions in rodent heavy- and light-chain V regions. It is relatively straightforward to take all three translated frames of a putative V region and compare them by eye. Al alternative method is to screen the translated sequence against the SWISS PROT database.

2. Is it productive? A productive V region will have no frameshifts (indicated by the loss of immunoglobulin specific residues in one frame and the appearance in another), no step codons, and no "junk" sequence. When frameshifts or stop codons are found, the original gels should be

Table 10–2. Amino Acids Found in Different Positions of Rodent V Regions

Kappa Light-Chain V regions				Heavy-Chain V regions			
1	DEQ	61	R	1	EDQ	**62**	**SAKT**
2	IVN	62	F	2	V	**63**	**LFVI**
3	VLQ	63	TS	3	QK	**63**	**KMIE**
4	MIL	64	G	4	L	**65**	**SGD**
5	TS	65	S	5	QKVEL	66	RK
6	Q	66	G	6	EQ	67	IAFT
7	STDA	67	S	7	SP	68	STI
8	PAET	68	GR	8	G	69	ILV
9	STLPAK	69	T	9	PAG	70	TS
10	SFIL	70	DS	10	SGE	71	RKVA
11	LNVMT	71	FY	11	L	72	DR
12	ATPSY	72	TS	12	VM	73	TNKD
13	VAT	73	L	13	KAQNR	74	STA
14	ST	74	TKRENS	14	PAT	75	KSQ
15	AVLSPI	75	I	15	SG	76	NS
16	GS	76	SHD	16	QAGSKR	77	QTIS
17	EQKDTS	77	SRPN	17	TS	78	YFVAL
18	KQSPRT	78	VML	18	LVM	79	FY
19	VAI	79	QE	19	SKRL	80	LFM
20	TS	80	AESPTQ	20	LIM	81	QKEN
21	MIL	81	ED	21	TS	82	LM
22	ST	82	D	22	C*	82A	NSR
23	C*	83	LVATI	23	STKAV	82B	SAKN
24	**KTRS**	84	AG	24	VAT	82C	VL
25	**SAV**	85	VTDMS	25	TS	83	TQR
26	**S**	86	Y	26	G	84	TSA
27	**EQKS**	87	YF	27	DYF	85	ED
27A	**S**	88	C*	28	STAND	86	D
27B	**LIV**	**89**	**Q**	29	ILF	87	TS
27C		**90**	**NQH**	30	TSK	88	AG
27D		91		**31**	**SDRN**	89	TMVIYL
27E	**S**	92		**32**	**GDYTF**	90	Y
27F		93		**33**	**YGWA**	91	YF
28		94		**34**	**WAVMI**	92	C*
29	**GSI**	95	P	35	NWHES	93	AT
30		95A		**35A**	**NS**	94	RT
31	**TSN**	95B		**35B**		95	
32	**Y**	95C		36	W	96	

(Continued)

325

Table 10–2. (*cont.*)

Kappa Light-Chain V regions				Heavy-Chain V regions			
33	**LM.V**	**95D**		37	IV	97	
34		**95E**		38	RK	**98**	
35	W	**95F**		39	KQN	**99**	
36	YFL	**96**		40	FPSRAT	**100**	
37	QL	97	T	41	PH	**100A**	
38	QK	98	F	42	GE	**100B**	
39	KR	99	G	43	NKQR	**100C**	
40	PSL	100	AGTS	44	KGSAR	**100D**	
41	GED	101	G	45	L	**100E**	
42		102	T*	46	E	**100F**	
43	SPTA	103	K	47	WY	**100G**	
44	PV	104	L	48	MLIV	**100H**	
45	KQR	105	E	49	GA	**100I**	
46	LRP	106	IL	**50**		**100J**	
47	LW	106A		**51**	IS	**100K**	**FM**
48	I	107	K	**52**	**WSNDR**	**101**	**DA**
49	Y	108	R	**52A**	**PNLS**	**102**	**YV**
50		109		**52B**		103	W
51	**AVT**	110		**52C**		104	G
52	**S**			**53**		105	QAN
53	**TNK**			**54**	**SGND**	106	G
54	**RL**			**55**	**GYS**	107	T
55	**EFAD**			**56**	**STGNAY**	108	**TSLV**
56	**S**			**57**	**TI**	109	VL
57	G			**58**		110	T
58	V			**59**	**Y**	111	V
59	P			**60**	**NDSTAP**	112	S
60	DAS			**61**	**PSQEAD**	113	S

Note: The table shows amino acids that are conserved at different positions in rodent heavy and light chains.[44] Asterisked amino acids are absolutely conserved (with only one or two exceptions). The presence of different amino acids at any of the positions indicated (except for those with asterisks) should not give cause for concern providing the sequence returns to the consensus. A radical departure indicates a likely frameshift. Amino acids indicated with letters, e.g., 100 A, B, etc., indicate additional amino acids, some or all of which may be present in different V regions.

examined carefully, as they may be due to compressions or other sequencing artifacts. Areas of frameshift should be sequenced on both strands and/or subjected to sequencing using either high-temperature polymerases, dITP, or 7-deaza-dGTP to eliminate artifacts. Such V regions derive from nonproductive rearrangements which do not produce functional protein, but continue to produce mRNA which can nevertheless act as a template for V region PCR.

3. Is the V region derived from the parental myeloma? The sequences of most parental myelomas can be found in Kabat et al.[44]

4. Is the V region a contaminant? It is important to check the sequence of the V region obtained against all the other V regions that have been cloned in the laboratory, since contamination in the PCR reaction is all too easy.

If then cloned V region passes all the above checks, it can be cloned into an expression vector, described in the next section.

Method 4.2: Cloning into the Expression Vectors

Once the V regions have been sequenced in the M13VPCR vectors, we use pSV2-based vectors (see Fig. 10–6), which provide human C regions downstream of the cloning site for the V region, as well as a eukaryotic promoter, the SV40 origin of replication, and selectable markers for expression. We use COS cells in transient transfections to demonstrate the presence of functional immunoglobulin. This exploits the SV40 origin of replication which, in the presence of the large T antigen, allows the plasmid to be amplified to high levels. The advantage of this system is that sufficient functional antibody to perform ELISAs can be produced after only three days, whereas the accumulation of equivalent amounts of antibody from stable clones can take weeks. The day before testing antibody specificity, we routinely check for the production of immunoglobulin by immunofluorescence. This allows us to exclude that a negative result is due to problems in the transfection efficiency.

This vector system can also be used to produce stable cell lines by transfecting into cells such as NSO and incubation in the appropriate selectable markers (G418 and hygromycin). This is very useful when the hybridoma is a rat-mouse hybrid, since it allows the production of ascites. Double-stranded DNA is made from the M13VPCR constructs. The V regions (with the associated elements) are extracted from the M13VPCR vectors by digestion with HindIII and BamHI and cloned into the pSV2-based vectors, which have also been digested with HindIII and BamHI. Bacteria harboring the pSV2 vectors grow very slowly, and often plates need to be left for two days to see colonies. Occasionally one isolates a V region that has an internal site identical to one of the second-step enzymes used for cloning into the expression vector (HindIII or BamHI). The solution to this is similar to that described above: either a partial digestion or *in vitro* mutagenesis to correct the site. The presence of diagnostic sites within the V regions is very useful to check that a V region has been cloned into the expression vector.

DNA preparation should be done by CsCl gradient using standard methods.[76] We have found that no method is superior to CsCl gradients (notwithstanding shortcut claims to the contrary) and, despite the inconvenience of this method, use it routinely prior to transfection.

Method 4.3: Transient Transfection in COS Cells with DEAE-Dextran [66]

Efficient transient transfection is absolutely dependent upon the quality of the cells and the quality of the DNA. Cells that have been allowed to overgrow may be irreversibly damaged (from a transfection point of view) however healthy they look afterwards. We suggest that when transfection efficiency decreases, one should return to (healthy) frozen stocks. Figures in parentheses in the following protocol indicate the volumes and compositions that should be used when performing parallel transfections in order to use the supernatant for assays. Step 10 uses much less DMEM in order to concentrate the antibody and includes HEPES to prevent acidification of the medium.

1. Plate the cells at semiconfluence in a 90 mm (60 mm) petri disk the day before the experiment.
2. Add 10 µg plasmid DNA to 2.7 ml DMEM.
3. Add 0.3 ml 5 mg/ml DEAE-dextran (Pharmacia, mol. wt. 500,000). DEAE-dextran should be autoclaved and stored at $-20°C$ in small aliquots.
4. Wash the cells twice in DMEM.
5. Add DMEM/DNA/DEAE-dextran to the dish (final concentration = 0.5 mg/ml).
6. Incubate 30 min at 37°C.
7. Add 15 ml 100 mm chloroquine in DMEM/10% FCS. Chloroquine (Sigma) should be filter sterilised and kept at $-20°C$ in small aliquots.
8. Incubate 3 hrs at 37°C in CO_2 incubator.
9. Wash with DMEM.
10. Feed with DMEM/10% FCS (2.5 ml DMEM/10% FCS containing 20 mM Hepes pH 7.2).
11. Harvest 2 or 3 days later.

Method 4.4: Indirect Immunofluorescence Detection of the Two Chains

1. Plate COS cells on polylysine-coated glass coverslips that have placed at the bottom of the petri dish and transfect them following Method 4.3.
2. Wash 3 times by dipping into a beaker containing PBS.
3. Fix the cells with 3.7% paraformaldehyde in PBS for 10 min at room temperature.
4. Wash once with PBS.
5. Permeabilize the cells with Tris-Cl 0.1 M pH 7.6/0.2% Triton X100 for 4 min at room temperature.
6. Wash 3 times by dipping into a beaker containing PBS.
7. Incubate with the appropriate dilution of anti-human isotype primery antibody (1 to 2 hours at room temperature) and subsequently with the fluorescein-labeled secondary antibody (1 to 2 hours at room temperature), with washes in PBS between incubations. Incubations should be done in a moist chamber (an airtight box with a piece of damp tissue paper at the bottom) to prevent drying out the cells. Biotin avidin based systems can also be used.

Method 4.5: Specific Assays for Checking the Binding Activity of the Recombinant Antibody in the Supernatant

Different assays can be done to check the binding activity of the cloned antibody, depending on the availability and nature of the antigen. In some cases (e.g., neurotrophins), biological assays can be done. In these cases, controls for the effect of the (COS cell conditioned) serum should be done, since we have found cases of a biological activity that overcomes the effect of the antibody (even when added exogenously). ELISA assays are relatively straightforward if the antigen is available in sufficient quantities, since there is enough specific antibody in supernatant collected from the transiently transfected COS cells. If the antibody recognizes the denatured antigen, Western blot analysis can also be used. Generally, whichever assay was used to characterize the hybridoma can also be used to characterize the COS cell supernatant.

Points to Consider, Methods 4.2 through 4.5

Transient transfections are usually done in parallel, half for immunofluorescence, and half for supernatant harvesting. This allows one to be sure that the supernatant used for assays is derived from cultures that have positive cells present. The presence of singly positive cells as well as double positive cells serves as a good internal control.

If the percentage of positive cells is lower than 1%, it is not worth proceeding with specific assays. Poor transfection efficiencies are usually due to poor DNA quality or unhealthy COS cells (overgrowth, mycoplasma contamination, or late passages prior to transfection). When 1 to 10% of cells are positive, one can expect up to 50 ng/ml immunoglobulin in the supernatant 3 days after the transfection, which is sufficient for most tests of specificity.

The pSV2-based vectors we use work in COS cells even though they only have the Ig promoter (beta globin versions have been constructed). This is probably due to cryptic promoter sequences found upstream.

ECTOPIC EXPRESSION OF CLONED V REGIONS FOR INTRACELLULAR AND INTERCELLULAR IMMUNIZATION

Once the specificty of a cloned hybridoma has been confirmed, a number of choices (of promoter, of antibody form, of delivery system, of targeting sequences) (see Fig. 10–2) need to be made in order to proceed with the planned experiment. Many of these choices are related to one another. The first choice pertains to the cellular location of the antigen to be recognized and therefore to the general strategy of the experiment. If the target antigen is active in the extracellular environment, then the antibody will have to be secreted in the same tissue region as the antigen, even if not by the same cells. For intracellular antigens, the antibody will have to be targeted to the same intracellular compartment as that in which the antigen is found.

The choice of antibody form also depends upon the type of experiment one is planning (secreted or intracellular). We have successfully expressed whole antibody molecules and ScFv fragments as both secreted and intra-cellularly targeted proteins. Despite the lack of systematic evidence, we suggest the use of whole antibodies in intercellular immunization, and of smaller antibody forms intracellularly. However, there are reasons related to the choice of the delivery system (e.g., the use of retroviral or viral vectors) that may lead one to use different forms instead of those suggested. It is clear from the introduction that the crucial aspects of the whole experimental strategy are represented by the two levels of control of antibody expression, the transcriptional one and the intracellular targeting one.

In the following paragraphs the two cases of intercellular and intracellular immunization will be described separately, as they have different requirements.

Intercellular Immunization

In this scheme, the antibodies will be expressed as secreted proteins in the extracellular environment of an otherwise intact tissue. We shall consider as an example the rodent central nervous system (CNS), as this is the one in which we have first-hand experience, but many of the considerations that will be made can also apply, or be extended, to other systems (or species) as well.

Choice of Delivery System

It is important that the antibody be delivered to its site of action, i.e., to the site where its antigen is present. This is a problem with two halves. The first is the delivery of the gene encoding the antibody (and associated control sequences) to the tissue(s) where the gene is to be found, and the second is the nature of the promoter. In this section we discuss the problem of gene delivery, and in the next, that of the promoter.

As schematically illustrated in Fig. 10–8, different approaches can be envisaged:

1. The production of transgenic mice by microinjection of the DNA construct in the pronucleus of fertilized eggs.[64] This has the advantage that all cells will contain the DNA encoding the antibody, and that expression of the

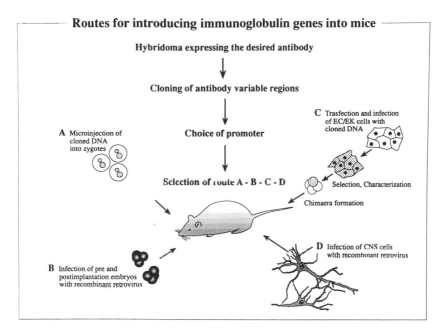

Figure 10–8. Choice of delivery system. Schematic illustration of the neuro-antibody approach: possible routes for expressing antibodies in cells of the nervous system.

antibody can be spatiotemporally controlled by the judicious choice of an appropriate promoter (see below). A disadvantage is the considerable work required to produce a mouse strain homozygous for a particular gene.

2. The production of transgenic mice by the use of embryonic stem cells (ES cells)[10,72] engineered *in vitro* to secrete the cloned antibody under the control of a suitable promoter. This technique is similar to the production of normal transgenic animals with the advantage that a tissue culture phenotype can be tested for prior to committing oneself to the creation of transgenic mice. Disadvantages are related to the difficulty of handling ES cells.

3. The use of retroviral[61] or viral (for the CNS, herpes virus, adenovirus) vectors to infect CNS cells. Retroviral vectors have not proven effective in gene delivery to the adult brain as they can only integrate into the genome of dividing cells, and most cells in the brain are not dividing. This property has, however, made them promising vectors for the delivery of "killer" genes to tumor cells in the brain.[18] If retroviral vectors are to be used in the normal brain, direct infection of the CNS has to be made early in development.[69] Alternatively, embryonic cells can be infected *in vitro* and then grafted into the CNS (see below). Herpes- and adenovirus-based vectors have been developed recently for gene delivery to the nervous system.[11,22,62] The principle of all of these is similar, in that the recombinant virus is replication deficient and able to infect postmitotic cells. The Herpes virus vectors can mediate gene expression in a substantial number of neurons, although stable expression (which occurs when the recombinant virus enters a latent state) can only be achieved in very few neurons. Adenorirus-based vectors, on the other hand, are striking in terms of the large number of cells that express the delivered gene product in the first week after infection, as well as for their apparent low pathogenicity. However, gene expression with these vectors falls to approximately 10% of peak levels by one to two months.

4. The transplantation into the CNS of cells engineered in vitro to secrete the recombinant antibody. These include the ES cells mentioned above, neuronal stem cells, fibroblasts, and primary neuronal or glial cells. DNA encoding for the recombinant antibody could be introduced in these cells by the use of the above-mentioned retroviral or viral vectors or by transfection, as appropriate. A similar technique, which does not require the cloning of the antibody genes, is the direct inoculation of the hybridoma into the ventricle.[2]

5. DNA mediated gene transfer. Preliminary work has shown that direct injection of DNA/liposome complexes into the brain results in some gene expression.[84] An alternative to liposomes is the formation of complexes between plasmid DNA and adenoviruses. In this case the DNA enters the cell following adenovirus receptor mediated endocytosis of the complex. It is still too soon to decide whether these methods are likely to become useful as general techniques.

Choice of Promoter

The use of antibodies in transgenic animals is, in principle, no different than the creation of transgenic animals containing other genes. In principle, it is possible to direct the expression of any gene to any specific cell type of an animal by using established transgenic methodology. This approach relies upon the fact that it is possible to combine the regulatory region(s) of a gene that is expressed in a cell-specific manner with any mRNA encoding a structural gene. The structural gene should then be transcribed according to the specificity dictated by the regulatory elements.

The possibility of achieving transcriptional control of the spatiotemporal pattern of expression of the transgenic antibody represents a key point of the whole approach. A regionally localized pattern of antibody expression can be achieved alternatively by transplanting engineered cells or by direct infection with retroviral or viral vectors (as described above). In this case a strong constitutive promoter should be utilized.

In most cases when the antibody is to be active extracellularly, it will be sufficient to have the antibody secreted into the bloodstream by a general purpose promoter (such as CMV[9] or EF-BOS).[59] There are some sites, however, such as the brain, that will be relatively inaccessible to antibody in the bloodstream (due to the presence of the blood brain barrier), and for these cases more specialized promoters are recommended.

CNS specific promoters. A number of promoters from genes specifically expressed by cells of the nervous system have been successfully used to direct the CNS-specific expression of heterologous transgenes and could be used to direct the expression of transgenic antibodies. These include the genes encoding for the myelin basic protein,[70] the neuron-specific enolase,[24] Thy-1,[48,83] and the neurofilament light chain.[54] The 5' flanking region of the NGF-regulated VGF8a gene[68] has been used by us to direct the expression of a monoclonal antibody against the tachykinin neuropeptide SP in discrete regions of the CNS of transgenic mice (our unpublished results and see below).

A general suggestion is to use promoters that have already been successfully used in transgenic mice and that express transgenes in the area of interest, unless testing a particular promoter is also part of the experimental design. More specific recommendations on the design of the plasmid construct for transgenic mice are given below.

Inducible promoters. The use of ectopic antibodies would greatly profit from systems that allow a stringent control of the expression of the transgenic antibody. Ideally, such systems would not only mediate an "on/off" situation of gene activity but also would permit limited expression at a defined level (dose response curves). Moreover, one would wish for an inducible promoter that is produced specifically only at the site of interest. Unfortunately, no such promoter exists yet. We have successfully utilized the promoter of the mouse metallothionein I gene[67] and the promoter of the Drosophila hsp70 heat shock protein[15] to drive the expression of secreted or intracellular antibodies in mammalian cell lines. The inducible promoters available tend to suffer from leakiness of the uninduced states (metallothionein) or from pleiotropic effects caused by the inducing agents themselves (heat shock). Another inducible promoter, which suffer from similar problems, is MMTV (induced by dexamethasone).

In search of regulatory systems that do not rely on endogenous control elements, several groups have been working to adapt prokaryotic regulatory systems, such as the (lac repressor/operator/inducer system of *E. coli*), to mammalian cells.[37] The most promising of such systems is without doubt that developed by Gossen ad Bujard,[29] which in HeLa cells allows regulation of expression of an individual reporter gene over five orders of magnitude. This system is based on control elements of the tetracycline-resistance operon encoded by Tn10, in which transcription of resistance-mediating genes is negatively regulated by the tetracycline repressor (tetR). In the presence of tetracycline, tetR does not bind to its operators located within the promoter region of the operon, so it allows transcription. By combining tetR with the C-terminal domain of HSV-VP16 (known to be essential for transactivation) a hybrid transactivator was generated that stimulates minimal promoters fused to tetracycline operator sequences, which activates the promoter in the absence of tetracycline (when the transactivator is able to bind to the operator).

The system is based on two plasmids: pUHD10-3, the cloning plasmid, contains a polylinker downstream of the tet operator sequences; while pUHD15-1 (or its derivatives) contains the transactivator under the control of the CMV promoter. We are presently attempting to utilize this system to obtain the inducbiel expression of ectopically expressed antibodies.

Immunoglobulin promoter and enhancers. This may appear as the most natural promoter to use if circulating antibody is required. However, the expression of transgenic antibodies,[79] transcriptionally driven by the immunoglobulin promoter/enhancer, is subject to a down regulation when the antibody recognizes a self-antigen, a finding that has a bearing on the development of immune tolerance.[27,28] Although the detailed mechanisms of this feedback have not yet been fully elucidated, the membrane form of the transgenic immunoglobulin is certainly implicated. We found that if the transgene encodes for the secretory form of the antibody only (in this case an antibody directed against the neuropeptide SP), good levels of the circulating antibody can be found in the adult as well as the newborn (Piccioli et al., unpublished results).

Other promoters. A theoretically ideal solution to deliver the antibody to exactly those cells where the antigen is produced, and hopefully inactivate it at source, would be to use the promoter of the gene whose protein product one is attempting to hit with the ectopically expressed antibody. If the antigen is a ligand acting on a receptor, the promoter of the gene encoding for the receptor could be used as well. A potential problem with this solution is that promoters for proteins expressed at low levels may not be sufficiently active to produce enough antibody to inhibit the activity of the antigen.

Choice of Antibody Form

We have successfully expressed whole antibody molecules and ScFv fragments as secreted proteins. In the case of intercellular immunization, i.e., the neutralization of extracellular (or secreted) antigens, the choice of antibody form depends mainly on the delivery system planned. In particular, the use of the retroviral or viral vectors calls ideally for the use of smaller antibody domains such as the single-chain Fv fragments. Work is presently being done in this direction.

On the other hand, in the intercellular scheme the effector functions carried out by the constant regions of the secreted antibody may be exploited to trigger a cascade of events downstream, for instance, complement activation or other effector functions by other isotypes. The half-life of circulating whole antibody is longer than that of ScFv fragments, which also makes it more suitable for intercellular immunization. Moreover, not all ScFv frequents appear to be efficiently secreted, some being retained in the ER (our unpublished observation and ref. 90).

Choice of Targeting Signals

Both soluble and membrane proteins pass through the cellular secretory apparatus before being secreted or retained at the cell surface. Therefore, the simplest way for antigen-antibody interaction to occur in these cases is for the antibody to be secreted. The secreted antibody will then be able to bind the antigen.

Secreted antibodies. The secretion of antibodies does not require the addition of any particular targeting signal, as antibodies are normally secreted. In addition to interacting with the antigen outside the cell, secreted antibodies may also be able to interact with their corresponding secreted antigen in the secretory apparatus, if they are coexpressed by the same cell. This intracellular interaction may lead to a block of antigen(-antibody) secretion, resulting in a reduction in the extracellular levels of the antigen. In this case, the antibody will be acting as a sort of anchor, which may be made more stringent by actively restraining the antibody itself (independently of its interaction with the antigen) within the cell.

ER-retained soluble antibodies. This may be achieved by adding at the C-terminus of the antibody the ER-retaining sequence SEKDAL.[60] We have engineered the SEKDEL sequence at the C-terminus of both the heavyand light immunoglobulin antibody chain, showing in both cases retention of the corresponding polypeptide in the lumen of the ER (Brocca and Ruberti et al., unpublished results), as already demonstrated for other secreted proteins.[60] Whether this will result in a corresponding retention of the otherwise secreted antigen will depend on whether the antigen is processed in a form recognized by the antibody in the ER, and this will have to be verified for each case. Sequences such as that present in the transmembrane region of the T-cell receptor alpha chain (NLSVMGLRILLLKVAGFNLLMTL), which confers retention and rapid degradation in the endoplasmic reticulum,[7,46] could, in principle, be used to utilize antibodies as "vectors for degradation" of the corresponding antigen. ScFv fragments retained in the ER have been used to block the appearance of the HIVgp120 protein to the plasma membrane.[30]

Membrane antibodies. The membrane form of IgM is expressed on the cell surface in B lymphocytes, whereas in plasma cells and nonlymphoid cells it is retained within the endoplasmic reticulum.[86] Since this retention can be overcome by mutation of the transmembrane segment of the heavy chain,[86] antibodies can be anchored to the cell surface by the use of these mutated transmembrane domains (or other transmembrane domains derived from, for example, MHC molecules[86] (see Table 10-3). In a more refined scheme, which is purely speculative, membrane antibodies may be used as "vectors for internalization" to deplete extracellular soluble antigen, by grafting the internalization signal of a rapidly recycling membrane protein (for instance, the sequence PGYRHV in the cytoplasmic tail of the lysosomal acid phosphatase) at the C-terminus.[52]

Intracellularly retained membrane antibodies. Since the transmembrane segment from the IgM heavy chain results in retention of the antibody in nonlympoid cells, this could be used (in these cells) to engineer an intracellularly retained antibody with the antigen recognition domain facing the lumen of the ER, thus acting as an intracellular anchor for secreted antigens. This may also be achieved by the use of ER-retaining sequences found in the cytoplasmic region of some viral membrane proteins. In this scheme antibodies would act as intracellular anchors for their corresponding antigen.

Table 10–3. Targeting Signals[a]

Compartment	Signal	Reference
secretory	normal leader at N terminal	[15]
cytoplasmic	methionine, hydrophilic leader or sequences of long lived cytoplasmic proteins at N terminal	[3] and Biocca et al., submitted)
nuclear	cytoplasmic[b] with a nuclear localisation signal (e.g., PKKKRKV for human and simian cells) at N or C terminal	[3], [85]
ER (endoplasmic reticulum)	normal leader at N terminal and SEKDEL sequence at C terminal	(S.D., F.R. et al., unpublished)
mitochondrial[c] mitochondrial[c]	cytoplasmic and N terminal presequence of mitochondrial proteins (e.g., aa-25 to -1 of human cytochrome c oxidase) MSVLTPLLLRGLTGSARRLPVPRAK at N terminal	(Biocca et al., unpublished)
extracellular membrane	normal leader and mutated μ transmembrane domain: NLWVVAAVFIVLFLLSLFFAVVVVLFKVK at C terminal	[86]
ER lumen	normal leader and normal μ transmembrane (in nonlymphoid cells): NLWTTASTFIVLFLLSLFYSTTVTLFKVK at C terminal	[86]

[a] The table shows which signals should be used (and in which position) to direct antibodies to different cellular compartments. All of these signals have been used. In some cases signals have to be combined (e.g., nuclear and cytoplasmic), in others, they are mutually exclusive (e.g., cytoplasmic and secretory).
[b] The cytoplasmic which has been used in this case is the deletion of the normal secretory leader.
[c] The amino-terminal presequences of imported mitochondrial proteins can direct foreign proteins to mitochondria and into specific intramitochondrial compartments, since these sequences contain information for both intracellular targeting to mitochondria and intramitochondrial sorting (matrix, inner membrane, intermembrane space, outer membrane). The sequence reported in the table is just one possible example that is being used to idrect antibody domains to the mitochondrial matrix.

Vectors for Transgenic Mice

Not all gene constructs work well in transgenic mice. The two most common problems are inappropriate expression patterns and failure to achieve adequate expression levels. The first problem is ideally solved by using promoters that have already proven to be successful in driving the expression of reporter genes. Inadequate expression is often associated with attempts to express cDNA constructs. The most baffling aspects of this phenomenon is that cDNA constructs that fail to work in transgenic mice are often expressed efficiently when transfected into tissue culture cells. This is likely to be due to the absence of introns in cDNA constructs.[65] In an attempt to decipher which introns to use and where to locate them in the construct, a number of different constructs have been compared,[65] with the general conclusion that genomic constructs should be used whenever possible for transgenic experiments. For this reason we include introns in our constructs. The next best strategy may be to insert a heterologous intron between the promoter and coding region, which may improve expression. Alternatively, the cloned gene can be inserted into the first exon of a gene that is known to be expressed well, between the promoter and the initiation codon, such that the inserted gene represents the first open reading frame.

DNA for microinjection should be linearized, after removing as much vector sequence as possible, and highly purified. The two transcriptional units for the heavy and the light chain need not be on the same DNA construct. We have found that coinjection results in cointegration (as multiple head-to-tail units) in approximately 50 to 75% of the cases. Mice positive for one of the two chains are very useful as controls for subsequent experiments.

Intracellular Immunization

In this scheme, the antibody molecules are redirected away from the secretory apparatus by removing the leader sequence for secretion and by adding targeting signals that are autonomous and dominant. The discussion of this section will be centered on experiments performed in systems *in vitro* (cells, oocytes) rather than in transgenic organism, although there is no reason why such experiments could not also be performed in transgenic organisms.

Choice of Promoter

The two crucial issues in this case are to achieve the highest levels of expression possible (disregarding problems of tissue specificity) and/or to achieve an inducible expression of the antibody. The first point helps to counteract the fact that the half-life of cytosolic immunoglobulins or immunoglobulin domains is shorter than that of their secreted counterparts (although this does not seem to be the case for single-chain Fv fragments in *Xenopus* oocytes or in plants). The second point may help to prove formally that an observed phenotype is indeed due to the expression of the intracellular antibody and not to a selection of cells. Inducible promoters are discussed in the section above. Strong constitutive promoters that can be used to drive intracellular antibody expression include general-purpose promoters such as CMV[9] and EF-BOS.[59]

Choice of Antibody Form

We have demonstrated that whole antibody molecules can be expressed successfully in the cell cytoplasm, where they assemble in a functional form, as shown by antigen binding and preservation of the idiotype (see Fig. 10–10).[3] In one case, with intracellular antibodies against the p21-ras protein, *in situ* interaction of the intracellularly expressed antibodies with the endogenous antigen was also demonstrated (refs. 87, 88, Fig. 11). From the cytoplasm, whole antibodies can be targeted to other compartments, such as the nucleus. However, antibodies expressed in the cytoplasm can undergo degradation. The exact of this degradation depends very much on the leader sequence used (see below). Interestingly, we have suggestive evidence that the variable domains may get proteolytically cleaved from their corresponding constant domains, to yield function Fv fragments carrying the idiotype and the antigen specificity.[3] For this reason, but without any systematic comparative evidence, we recommend the use of single-chain Fv fragments for intracellular expression. We have shown good expression of ScFv fragments targeted to secretory, cytoplasmic, nuclear and mitochondrial compartments in mammalian cells (in preparation). A caveat in the use of ScFv fragments is that their affinity for antigen may be lower than that of the whole antibody, so that what one gains in expression levels may be lost in affinity and or specificity. A comparison between the affinity and the specificity of the single-chain Fv fragment to that of the corresponding whole antibody molecule should be routinely performed after cloning the V regions.

Choice of Targeting Signal

For intracellular antibodies the leader to be used has to be chosen on the basis of the compartment to which one wants to direct the antibody. Leaders and signals that have been (or can be) used to target antibodies or antibody domains to different compartments are described in Table 10–3.

Cytoplasmic leaders. To retain the antibodies in the cell cytoplasm, the hydrophobic core of their N-terminal leader sequence for secretion has to be mutated or deleted. We have shown that while substitution of a small number of hydrophic amino acids from the secretory leader (one or two at a time) with charged residues is not effective in retaining the corresponding chains in the cell cytoplasm, mutating 6 or 7 of these amino acids into an equal number of charged residues (hydrophilic leader) leads to a complete cytoplasmic localization of the antibody chains. However, the corresponding proteins have a very short half-life, probably because the sequence we introduced at the N-terminus inadvertantly had a composition quite similar to that of the PEST sequences found at the N-terminus of short-lived cytoplasmic proteins.[73] We find that deleting the whole leader sequence (leaderless), leaving only the N-terminal methionine, is probably better in terms of expression levels. The hydrophobic nature of amino acids often found at the beginning of most variable regions does not appear to be a problem for cytoplasmic localization. An alternative is to place the N-terminus of a long lived cytoplasmic protein (e.g., lactate dehydrogenase) at the N-terminus of the antibody. This is also successful, but we have not made a systematic comparison of the different possible cytoplasmic constructs.

Nuclear localization. We have successfully used[3] the nuclear localization signal (NLS) from the large T antigen of SV40 (PKKKRKV).[45] By substituting the hyrophobic core of the V-region leader sequence with this NLS signal, one obtains a polypeptide that is not translocated across the ER membrane but is synthesized in the cytoplasm, from which it is translocated to the nucleus. Three glycine residues can be introduced downstream of the NLS as a spacer to ensure exposure of the nuclear leader. Although cintext effects on the efficacy of the NLS have been described, on the whole, the position of this signal in the molecule is not important, providing that it is exposed. The function of the SV40 large T antigen NLS, when it is out of its natural amino acid context, is cell-line and species dependent (being mostly efficient in monkey and human cells). Efficient nuclear targeting with such a signal in rodent cells can only be achieved either by reiteration of the signal itself[23] or by having a NLS on each of the two antibody chains. The failure of nuclear targeting with this signal (when it is in a single copy) results in the expression of antibody within the cytoplasm, and in fact in these conditions the NLS is a very good "cytoplasmic signal". Nuclear targeting in frog oocytes requires the use of species-specific NLSs, such as that of nucleoplasmin.[20] It is also clear that the nuclear localization of proteins can be controlled, for instance by cytoplasmic anchoring proteins,[39] and this could, in principle, be extended to the nuclear localization of antibodies.

Other signals. The N-terminal presequence of many mitochondrial proteins has been shown to contain the targeting information for mitochondria,[40] as can be demonstrated by fusing it to reporter proteins. In Table 10–3 we report, by way of example, the targeting presequence of the subunit VIII of human cytochrome c oxidase, which was recently shown, upon appropriate fusion, to target the photoprotein aequorin to mitochondria.[71] By fusing this sequence in frame with the N-terminal portion of antibody variable regions, antibody domains have been targeted to mitochondria (Biocca et al., submitted).

Targeting of antibodies to the cytoplasmic side of the plasma membrane should be possible by adding sequences for myristilation and palmytilation, such as those found in proteins involved in intracellular signaling (p21-ras, pp60-src, etc.)[31,49,55] to one or both chains of a cytoplasmic antibody.

As cell biology teaches us more about how proteins normally find their targets, this information can be exploited to engineer the targeting of antibody chains around the cell in more subtle ways. It should be noted, however, that not all targeting signals are made up of easily transplanted linear sequences; e.g., the signal to target proteins into the regulated secretory pathway has not yet been found and is thought to be a conformational signal.

While our aim is to have a series of vectors that will allow the addition of the appropriate leader sequence to a cloned antibody by a single cloning step (see below), these signal and leader sequences are presently added by *in vitro* mutagensis or PCR-based mutation approaches.

Points to Consider

A few crucial issues need to be considered when planning or discussing an experiment with intracellularly expressed antibodies.

1. Are the two chains (if two chains are expressed) correctly assembled? This is best monitored *in situ* by the use of anti-idiotypic antibodies (see later). Such sera (or monoclonals) should recognise an idiotype formed by the association of the heavy and light variable regions of the recombinant antibody, but not either chain on its own. Such anti-idiotypes may or may not be inhibited by the binding of antigen. Anti-idiotyopic antibodies can also be used for the detection of ScFv fragments, even if in this case association is less of a problem.

2. Disulphide bonds. The cell cytoplasm is reducing. It has not yet been clarified whether immunoglobulin disulphide bonds (inter- and intrachain) are formed. Even cytoplasmic proteins in which biochemical and structural evidence reveals the presence of juxtaposed cysteine residues do not show disulphide bridges in the cytoplasm of living cells. However, when separated immunoglobulin chains are mixed and brought to neutral pH, they reassociate to form intact Ig molecules even if the original disulphide linkages cannot reform. Moreover, in one naturally occurring case where the antibody is functional,[75] the second-half cysteine in the variable region of the heavy chain has been shown to be absent, having being replaced by a tyrosine residue. This suggests that the presence of a disulphide bridge in the heavy-chain variable region does not appear to be necessary for the function of this antibody and may not be obligatory for antibody function in general, as has been assumed so far. Therefore, the formation of inter- and intrachain disulphide bonds in the cell cytoplasm, while an interesting issue that deserves future investigation, does not appear to be fundamental for the ectopic assembly of functional antibodies or antibody domains in the cytoplasm.

3. Degradation and half life. As discussed below, degradation in the cytoplasm certainly occurs, but may not affect all parts of an antibody equally. The variable domains appear to rather more resistant than the linker between the constant and the variable region. Furthermore, the presence of the relevant antigen in the cell appears to stabilize the intracellular antibody, in keeping with old *in vitro* experiments on the reconstitution of antibody chains. Interestingly, in the oocyte system it appears that the half-life of the chtoplasmic ScFv is of the same order of magnitude as that of its secretory counterpart. A similar finding is reported for plant cells (see Chapter 11).

Choice of Expression System *(Xenopus Oocytes)*

As stated above, the first (and one of the most promising) fields of application of the intracellular immunization scheme is represented by mammalian cells in culture (but see also intracellular antibodies in plant cells, in the chapter by Tavladoraki et al.). The stable expression of the engineered antibodies will then be achieved by standard methods such as DNA transfection or retroviral vectors. As the production and analysis of stable transfectants can be slow and time consuming, we found that it is very important to be able to check the intracellular expression constructs with some sort of transient assay in order to show that the recombinant antibody chains (or antibody domains) are efficiently expressed (with a satisfactory half-life) and correctly targeted, and that the two chains (if present) are assembled. Moreover, for the intracellular immunization strategy to be effective, it is of utmost importance that the intracellular localization of the recombinant antibody matches that of its corresponding antigen in order that it may be inhibited. While this can (and has) been verified in the COS system or in other cell culture models, we have also developed an alternative system based on *Xenopus laevis* oocytes.[87,88]

Introduction to Method 5: Use of *Xenopus* Oocytes to Assay the Intracellular Expression of Antibodies

This method presents many advantages, particularly if the experimental setup for the oocyte system is already running in the laboratory. The advantages of this system include: (1) oocytes can be easily microinjected with plasmid DNA or mRNA; (2) the expression of intracellular antibodies in occytes is more efficient than in other transient systems tested so far (see below); and (3) the large dimension of oocytes, and their complex spatial organization, allow the localization of the antibodies or antibody domains to be accurately determined.

Stage 6 *Xenopus* oocytes are microinjected with plasmid DNA or *in vitro* transcribed mRNA encoding the immunoglobulin (domains), and their intracellular distribution is studied by confocal microscopy after indirect immunofluorescence. When the corresponding antigen is not present endogenously, it can be coexpressed. This allows one to verify whether the antigen and the antibody colocalize and, in particular, whether the intracellular distribution of the antibody is modified by the concomitant expression of the antigen (or vice versa).

In some cases the oocyte system may also offer a well-characterized experimental model to study the antibody of interest, as in the case of the p21-ras protein (see the working examples below). This approach is particularly recommended in those cases in which the oocyte system is routinely running. General references for microinjection can be consulted.[35, 56, 58]

Method 5.1: Preparation of Oocyte Sections for Confocal Analysis

After the oocytes have been microinjected with DNA or mRNA, they can be prepared for confocal analysis as follows:

1. Oocytes are frozen for a few seconds in liquid isopentane at $-161°C$ and are then dehydrated in 100% ethanol at $-80°C$ for three days.
2. After warming the ethanol to room temperature, transfer the oocytes through tertiary butanol to liquid paraffin at 60°C.
3. After embedding, sections are cut with a glass knife at 4 to 5 μm and floated onto 10 mM $MgCl_2$.
4. Sections are placed on glass slides and the paraffin is removed with xylene.
5. Slides are transferred to 100% ethanol and dehydrated through a series of ethanol washes into PBS + 10 mM $MgCl_2$ (PBS/Mg).
6. Slides are treated with 3 M urea for 3 min and washed with PBS/Mg.
7. Slides are "blocked" with 10% horse serum in TBS for 15 min, after which the primary antibody is applied. Slides are washed with PBS/Mg between antibodies.

Method 5.2: Labeling and Purification of Single-Chain Fv Fragments from Microinjected Oocytes

This protocol depends upon the presence of the myc tag at the COOH terminus of the ScFv.

1. Incubate the oocytes in Barth medium with 0.1 to 1 mCi/ml ^{35}S methionine for 4 hours. Barth medium is 88 mM NaCl, 1 mM KCl, 2.4 mM $NaHCO_3$, 15 mM Hepes, pH 7.4, 0.3 mM $Ca(NO_3)_2$, 0.41 mM, $CaCl_2$, 0.8 mM $MgSO_4$ with antibiotic solution and BSA at 0.4 mg/ml.

2. Wash in Barth and immediately freeze in dry ice or proceed with the preparation of the extracts.

For 5 oocytes:

3. Add 10 μl EB per oocyte and homogenize by pipetting up and down with a yellow tip. Extraction buffer (EB) is Tris/Cl 20 mM pH 7.4, $MgCl_2$ 10 mM, EGTA 10 mM, NP40 0.5% PMSF, leupeptin, chymostatin, and aprotinin at the usual concentrations.

4. Spin 5 min at 12,000 rpm.

5. Incubate the supernatant with 25 μl mAb 9E10-Sepharose slurry (diluted 1:1) for 2 hours at 4°C with gentle mixing. (The preparation of 9E10-Sepharose is carried out following the manufacturer's—Pharmacia—instructions).

6. To 50 μl supernatant add 200 μl EB without 0.5% NP40.

7. Spin briefly to pellet the Sepharose, remove the supernatant, and wash 4 times in Tris/Cl 20 mM pH 7.6 + 150 mm NaCl + 0.1% NP40 with brief spins in between. Longer centrifugations may damage the Sepharose matrix.

8. Add 10 μl 4× SDS Sample buffer to the Sepharose pellet, boil and perform 12% SDS polyacrylamide gel electrophoresis.

9. Dry the gel down and perform autoradiography.

Points to Consider, Method 5

1. Labeled antibody from a single oocyte can be seen.

2. The immunoprecipitation of the ScFv can also be analysed for the coprecipitation of the corresponding antigen. This can be done either in the same autoradiography (however, in this case, particularly if the antigen is endogenously expressed by the oocytes, the labeling time with ^{35}S-methionine should be longer—up to overnight), or by Western blot analysis of the immunoprecipitate.

3. If the antigen is localized in the cytoplasm, the parental antibody, microinjected as a protein, can serve as a positive control in the experiments. If, on the other hand, the antigen is localized in the secretory apparatus, the microinjected antibody protein will not have access to it. In this case, microinjection of the hybridoma mRNA[17] can be used as a positive control.

4. The expression of intracellular antibodies in *Xenopus* oocytes represents, in our opinion, a very convenient cellular assay to check the constructs for intracellular expression, even if the oocyte does not represent the experimental system for which the intracellular immunization experiments were planned. For this reason, if a biological/functional assay is not available in this system (perhaps because the

antigen is not endogenously expressed) it may be worthwhile putting an effort into developing such an assay by expressing recombinant forms of the antigen as well.

5. One obvious extension/application of the expression of intracellular antibodies in *Xenopus* oocytes is represented by the possibility of doing the same in fertilized eggs, and following the expression and the effects derived during development. This also applies to the expression of secreted antibodies.

WORKING EXAMPLES

The working examples given below are two experimental systems under study in our laboratory. In their presentation we hope to suggest, by way of example, the means by which such systems can be studied.

In transgenic systems, the first step is to obtain mice that are positive for integration of the DNA. If such mice have been obtained, and there is no obvious phenotype, we suggest that they are bred to derive homozygous positives (tested by dot or Southern blot). The increased gene dosage may reveal a phenotype if one was not obvious before. The next step is the demonstration first, that the antibodies are being made, and second, that they are in close vicinity to the antigen that is to be inhibited. In our experience this is best done by immunoprecipitation, immunofluorescence, and immunohistochemistry (using second-layer antibodies that recognise human constant regions), but will depend upon the system used. Simultaneous staining of both antigen and antibody will give an indication of whether they are close enough to interact.

There are a number of reasons why antibody expression may not be demonstrated. The most likely is an absence of mRNA, which may be caused by chromosomal integration within a transcriptionally silent area or problems in the transcription unit itself which result in low levels of mRNA expression. Unfortunately, expression in tissue culture cells is not a guarantee for good expression in transgenic animals. Although low mRNA levels are a common cause for low protein levels, other (poorly understood) mechanisms also apply, since we have had examples of normal (or even elevated) mRNA levels with low protein levels. mRNA levels are best studied by RNAse protection assays or by Northern blots. An anti-idiotypic serum (produced and purifed as described in Vaux et al.[82] or an anti-idiotypic mAb is very useful in the study of antibody expression, since it allows one to be confident that the antibodies that one sees by anti-human sera are also functional. Such anti-idiotypic reagents should be selected so that they do not compete with antigen binding (otherwise anrtigen binding *in vivo* may be mistaken for a lack of heavy/light chain assembly).

Ectopic Expression of Sdcreted Whole Ig in Transgenic Mice

Our initial experimental survey of the secretion of immunoglobulins by nonlymphoid cells of different origin showed that the efficiency of antibody secretion by cells related to the nervous system is particularly high, and comparable to that of lymphoid cells.[15] In the following we briefly describe results obtained on the expression of a recombinant monoclonal antibody in the central nervous system of transgenic mice (Piccioli et al., submitted).

The variable regions of the monoclonal antibody NC1/34HL directed against the tachykinin neuropeptide, Substance P, were cloned by the methods described above into vectors for the secretion of whole chimaeric immunoglobulins under the control of a number of different promoters.[67] In the case described below, the transcription was driven by the promoter of rat *vgf* (1, 5 kb of 5' flanking sequences of the *vgf* gene, including the first untranslated exon, the intron, and the first part of the second exon prior to the translation start), a gene whose expression is induced in PC12 cells in response to NGF.[53] The *vgf*-encoded protein is expressed only in cells of neuronal and endocrine origin,[68] and it was of interest to assess whether this is due to tissue-specific transcriptional control. Transgenic mice were generated with these constructs, and homozygous lines for the two transgenes (heavy and light chain) were utilized for expression studies. The presence of the human constant regions facilitated the detection of the transgenic antibody against the background of mouse Igs. Transgenic antibodies binding to SP can be detected in the serum, peaking between 2 to 3 weeks after birth. The highest levels, however, were found in the central nervous system, with a regional distribution that paralleled very closely that of the endogenous *vgf* gene, including spinal cord, hippocampus, hypothalamus, and cortex. This was studied on brain sections from the transgenic mice by immunohistochemistry, with biotinylated–anti-human constant region antibodies followed by Streptavidin-peroxidase. Figure 10–9 shows the distribution of cells positive for the transgenic antibodies in two such regions (hippocampus 9A and 9B and cortex, 9C and 9D) of a 16-day-old transgenic mouse. It is apparent from the higher magnification pictures that the staining is almost exclusively neuronal. Whether the neuronally expressed transgenic antibodies are able to inhibit synaptic transmission at SPergic nerve terminals is presently under study.

Intracellular Expression of Whole Immunoglobulins and ScFv Fragments

In order to demonstrate intracellular expression, assembly, and targeting of whole Ig, we present in Fig. 10–10 the results from an experiment in which CS simian fibroblast cells were cotransfected with plasmids encoding a heavy chain with cytoplasmic leader and a light chain with a NLS leader sequence.

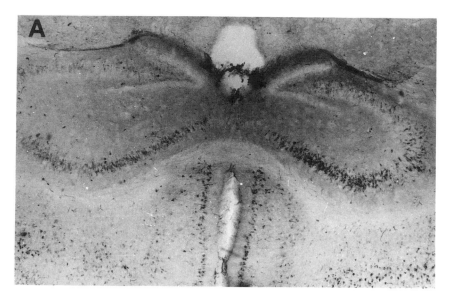

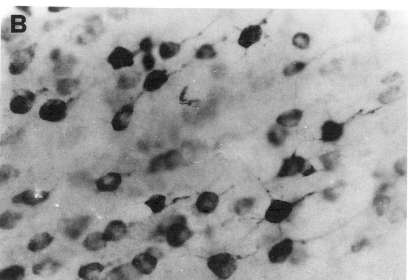

Figure 10–9. Expression of recombinant monoclonal antibodies in the CNS of transgenic mice. Brains were dissected from 16-day-old transgenic mice, homozygous for the heavy- and the light-chain genes encoding a human/rat chimaeric antibody against the neuropeptide Substance P. After fixation in 4% paraformaldehyde, sections were incubated with biotinylated antibodies against the human gamma constant region, followed by Streptavidin-peroxidase. After washing,

(continued)

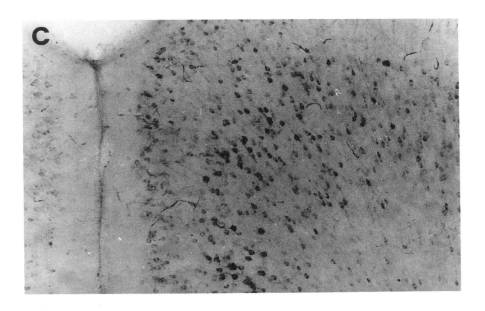

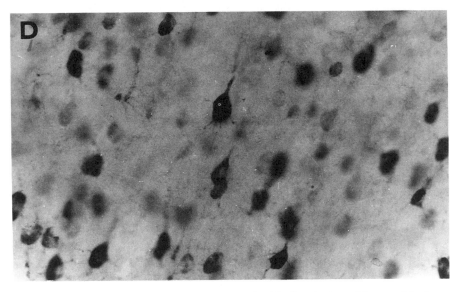

Figure 10–9. *(cont.)* sections were developed with diaminobenzidine and hydrogen peroxide. The figure shows sections through the hipocampus and the cortex, with higher magnification fields from both regions. A very similar spatial distribution of the trangenic antibody chain is obtained for the light chain (not shown).

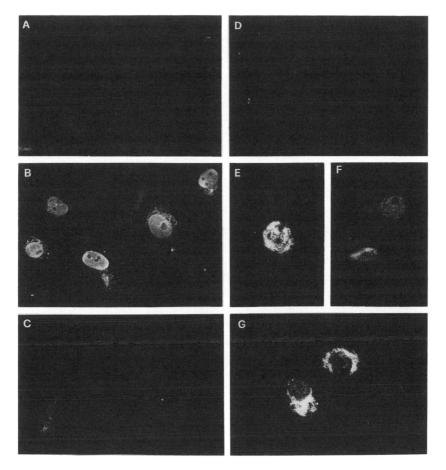

Figure 10–10. The nuclear localization of assembled and functional antibody. Immunofluorescence of cells expressing nuclear antibodies.[3] Untransfected CS simian fibroblast cells (A, D) and CSλNμ* cells (B, C, E, F, G) were stained with the anti-idiotypic monoclonal antibody AC38 (A, B, C) and with anti-λ antibodies (D, E, F, G). CSλNλ* cells are CS transfectants stably expressing a light chain with a nuclear localization signal and a heavy chain purely cytoplasmatic. The two intracellularly expressed immunoglobulin chains, when assembled to form an antibody, recognize the hapten 4-hydroxy-3-nitrophenacetyl (NP). The anti-idiotypic monoclonal antibody AC38 only recognizes the combination of the VH and VL and not either chain in the absence of association with the other. This particular idiotypic binding is competed by antigen (NP). In (B) nuclei light up very clearly, with the exception of nucleoli. Some staining of the cytoplasm is also visible. (C) Same cells and same antibodies as in (B), except that before addition of the anti-idiotypic antibodies, cell s were preincubated with 5 μg/ml of NP coupled to bovine serum albumin; the nuclear staining by the anti-idiotypic monoclonal antibody AC38 is abolished. (E, F, G) CS λNμ* cells stained with

(continued)

The individual chains, when transfected in isolation, show a cytoplasmic and a nuclear localization respectively. The antibody chains, when assembled, recognize a hapten (NP). Staining of cotransfected cells with anti-idiotypic antibodies lights up cell nuclei very clearly, revealing also the cytoplasm more weakly (Fig. 10–10B). All the positive cells in the transfected population (5 to 10%) show such nuclear staining. As these anti-idiotypic antibodies only recognize the combination of VH and VL, this result suggests that the heavy and light chains have assembled in the cytoplasm prior to translocation of the assembled Ig into the nucleus, since only the light chain carried the nuclear localization signal. The antibody molecules targeted to the nucleus retain their antigen-binding properties, as also shown by competition experiments in which fixed and permeabilized cells were preincubated with antigen prior to challenge with anti-idiotypic antibodies (Fig. 10–10C). When immunofluorescence is carried out on sister cultures with anti-isotype antibodies (for example anti λ) alongside cells in which the strongest signal is in the nucleus (Fig. 10–10E and F), one can see many cells in the population with an intense perinuclear distribution, a weaker dot-like staining within the nucleus itself, and a general more diffuse background staining in the cytoplasm (Fig. 10–10G). We interpret this as due to the varible domains being proteolytically separated from their corresponding constant domains, to yield Fv fragments carrying the idiotype and the antigen specificity of the parental antibody, as well as the nuclear localization signal (provided in this case by the light chain). Western blot analysis of intracellularly expressed light and heavy chains confirms this (unpublished results).

As an example of a strategy of intracellular immunization, we shall describe the expression of recombinant forms of the Y13-259 anti-p21ras monoclonal antibody in the cytoplasm of *Xenopus laevis* oocytes and show that this leads to a marked inhibition of the H1 kinase activity and of the meiotic maturation induced by insulin.[87,88] The *Xenopus* oocyte system provides one well-characterized assay for a signal transduction process involving the p21ras protein. Thus, *Xenopus* oocytes can be induced to mature meiotically *in vitro* by incubation with insulin, a process in which the protein p21ras has been shown to be involved.[4] This is consistent with the observation that microinjection of anti-p21ras antibodies in stage-6 oocytes markedly inhibits insulin-mediated meiotic maturation. Following activation of the p21ras protein, oocyte meiotic maturation requires the further

anti-λ antibodies. Some cells (E, F) show a nuclear staining that is similar to that obtained with the anti-idiotypic monoclonal antibody AC38, while the remaining positive cells in the population (G) show a cytoplasmic staining that is particularly intense in perinuclear regions and absent in the cell periphery. Fluorescence dots are also clearly visible in cell nuclei. See text for interpretation.

activation of the so-called maturation promoting factor (MPF) complex, which leads to germinal vesicle breakdown (GVBD). This active complex contains the serine threonine p34cdc2 kinase (also termed histone H1 kinase), whose activity can be readily assayed in oocyte lysates.

The variable regions of the antibody Y13-259 (which recognizes the p21ras proteins from all species[26] were cloned in the vectors described above, for its intracellular and extracellular expression as whole antibodies or as single-chain Fv fragments (ScFv). The Y13-259 ScFv fully retains the antigen-binding activity of the parental antibody.[89] These vectors were utilized to perturb the function of the cellular p21ras protein in *Xenopus* oocytes after microinjection of plasmid DNA or messenger RNA. Controls used for all the experiments described included the secreted version of the Y13-259 ScFvs and the intracellular version of a nonrelevant ScFv. Interestingly, the intracellular ScFvs are expressed at levels comparable to those of their secretory counterparts. Immunoprecipitation with the mAb 9E10[21] of ^{35}S labeled extracts from oocytes microinjected with mRNA encoding the cytosolic anti-p21ras ScFv revealed, together with a prominent band of 32 kDa, corresponding to the ScFv protein, a band of 21 kDa, suggesting that the intracellular anti-p21ras ScFvs interact *in vivo* with the endogenous p21ras protein (data not shown).

The intracellular distribution of the antibody chains, after plasmid microinjection in the nucleus or mRNA microinjection, was studied by indirect immunofluorescence viewed by confocal microscopy. Fig. 10–11(a) shows the distribution of the intracellular Y13-259 ScFvs and Fig. 10–11(b) shows the distributionnof whole Y13-259 molecules after microinjection of the encoding nucleic acid. The prominent feature of the staining is the distinct localization of a consistent proportion of ScFv and Ig molecules in the submembrane compartment of the animal pole, a pattern very similar to that of endogenous p21ras (Fig. 10–11(c)), and absent from secretory Y13-259 or irrelevent intracellular ScFv or antibodies (not shown).

These experiments, together with the coprecipitation experiments, suggest that the intracellular antibody chains of anti-p21ras specificity interact with the endogenous p21ras protein and, by virtue of this interaction, localize in the same submembrane compartment of the animal pole. That this specific interaction has a biological effect is shown by the fact that in the oocytes expressing intracellular anti-p21ras antibody chains or ScFv fragments, histone H1 kinase activity (induced by insulin and requiring p21ras activity) is markedly reduced (80%) with respect to those expressing a nonrelevant intracellular antibody, and furthermore that oocyte meiotic maturation (which is dependent upon histone H1 kinase activity) is significantly inhibited (80%) in those oocytes expressing anti-p21ras SvFvs.

The inhibition of insulin-induced histone H1 kinase activation and meiotic maturation by intracellularly expressed anti-p21ras recombinant antibodies

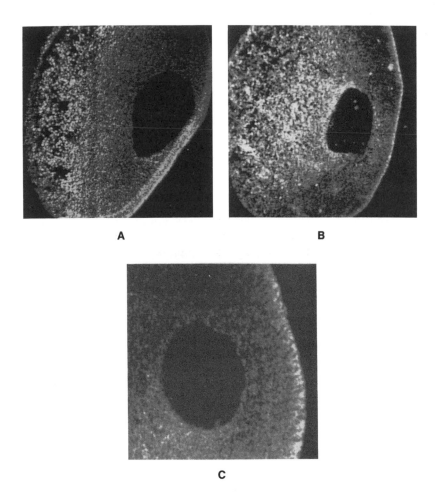

Figure 10–11. Confocal microscopy of intracellular antibodies in oocytes. Sections from oocytes microinjected with (A) mRNA encoding cytoplasmic ScFv Y13-259, and (B) a plasmid encoding cytoplasmic whole Y13-259. (C) staining for endogenous p21ras. Sections were viewed with a confocal microscope (Sarastro 2000). Images are in false colors (white yellow; maximum to blue; background).

and antibody domains gives further support to the concept of using antibody-mediated intracellular immunization in vertebrate cells as a general technique. The successful inhibition of the p21ras activity obtained with cytosolic ScFvs is particularly noteworthy, suggesting the possibility of expressing intracellular repertoires of variable regions in mammalian cells, to isolate new antibody specificities on the basis of selectable phenotypes.

THE INTEGRATED VECTOR SYSTEM

The use of cloned antibodies for inter- and intracellular immunization is likely to be a technique that will be used more frequently as the techniques for cloning and expression of immunoglobulin V regions improve. Unfortunately, the cloning of V regions from hybridomas is not as straightforward a task as it should be. While we expect that many antibodies in the future will be derived from large phage antibody libraries, a great many potentially useful hybridomas are already available for cloning. We are working on a aystem that will considerably simplify the cloning and expression of V regions, whether they are derived from hybridomas or from phage ntibody libaries. This is being done by the creation of an integrated vector system, which will have useful restriction sites bordering all components one may desire to change. This will allow the use of interchangeable cassettes to exchange promoters, targeting signals, constant regions, selectable markers, and antibody forms. Furthermore, in a single vector we expect to be able to sequence the V region and then express it is either transient or stable systems. This should considerably simplify the use of ectopic cloned antibodies as an experimental tool.

CONCLUSIONS

In this paper we have given an outline of all the steps involved in using cloned hybridoma V regions for ectopic *in vivo* expression in animal cells or organisms. This emerging technology, promising as it may be, is still in its infancy. While results with intracellular ScFv antibodies against p21ras, and those obtained in transgenic plants (see Chapter 11), conclusively demonstrate that a biological effect can be exerted by the ectopic expression of intracellular antibody domains, only with the demonstration of the effect in transgenic animals will it be possible to pursue this as a general technique for the inhibition of protein function *in vivo* in a controllable fashion.

Acknowledgments

We would like to thank Pietro Calissano for discussions and laboratory space, Ettore D'Ambrosio for oligonucleotides, and Gabriella Rossi for technical help. The work described in this chapter has been financed at various times by the following sources: EC BRIDGE (project number PL910008) awarded to SISSA, SIRS and CAT, NATO (CRG 890949), CNR (progetti finalizzati Ingegneria Genetica, Oncologia, Biotecnologia e Biostrumentazione), INFM (Biophysics Section), Associazione Italiana per le Richerche sul Cancro, Ministers della Sanità (AIDS) and MURST. Andrew Bradbury was supported by sequential EMBO and HFSP fellowships for part of the period of this work; Thomas Werge was supported by sequential fellowships from the EC and the Grosserer L. F. Fogth Foundation; and Patrizia Piccioli was supported by sequential fellowships, one from CNR and another supported by Montedison/Ferruzzi.

APPENDIX

The Sequences of inserts MADVKPCRA and 2; M13VHPCR; and ScFv linker region. For the M13VPCR vectors, the sequence between the HindIII/BamHI sites is shown. The M13VPCR backbone is common to all. Underlined regions represent oligonucleotides used to amplify the V regions. The amino acid sequence is indicated above the nucleic acid sequence. Restriction sites are in italics and indicated. The ScFv linker region encompasses the 3' end of the VH region and the 3' end of the VK region with oligonucleotides indicated by underling.

M13VPCR backbone

```
AAGCTTATGA ATATGCAAAT CCTCTGAATC TACATGGTAA ATATAGGTTT GTCTATACCA 60
HindIII

CAAACAGAAA AACATGAGAT CACAGTTCTC TCTACAGTTA CTGAGCACAC AGGACCTCAC 120

    M  G  W   S  C  I   I  L  F  L   V  A  T    A  T (leader)
CATGGGATGG AGCTGTATCA TCCTCTTCTT GGTAGCAACA GCTACAGGTA AGGGGCTCAC 180

AGTAGCAGGC TTGAGGTCTG GACATATATA TGGGTGACAA TGACATCCAC TTTGCCTTTC 240

          G   V  H  S  (3' end leader)
TCTCCACAGG TGTCCACTCC
```

M13VKPCR

```
                    D   I  Q   L  T  Q  S   P  S  S   L  S  A
                GACATCCAGC TGACCCAGAG CCCAAGCAGC CTGAGCGCCA 300
                        PvuII
S  V  G  D   R  V  T   I  T  C   R  A  S  G   N  I  H   N  Y  L
GCGTGGGTGA CAGAGTGACC ATCACCTGAT GAGCCAGCGG TAACATCCAC AACTACTGG 360

A  W  Y  Q   Q  K  P   G  K  A   P  K  L  L   I  Y  Y   T  T  T
CTTGGTACCA GCAGAAGCCA GGTAAGGCTC CAAAGCTGCT GATCTACTAC ACCACCACCC 420

L  A  D  G   V  P  S   R  F  S   G  S  G  S   G  T  D   F  T  F
TGGCTGACGG TGTGCCAAGC AGATTCAGCG GTAGCGGTAG CGGTACCGAC TTCACCTTCA 480

T  I  S  S   L  Q  P   E  D  I   A  T  Y  Y   C  Q  H   F  W  S
CCATCAGCAG CCTCCAGCCA GAGGACATCG CCACCTACTA CTGCCAGCAC TTCTGGAGCA 540

T  P  R  T   F  G  Q   G  T  K   V  V  I  K   | splice site
CCCCAAGGAC GTTCGGCCAA GGGACCAAGG TGGTGATCAA ACGTGAGTAG AATTTAAACT 600
                       GGGACCAAGG TGGTGATCAA
                                  BclI

TTGCTTCCTC AGTTGGTACC 620
              BamHI
```

M13VKPCR2

```
               D   E   L   T   Q   S   P   S   S   L   S   A
          GACATTGAGC TCACCCAGTC TCCAAGCAGC CTGAGCGCCA 300
                    SacI

  S   V   G   D   R   V   T   I   T   C   R   A   S   G   N   I   H   N   Y   L
GCGTGGGTGA CAGAGTGACC ATCACCTGAT GAGCCAGCGG TAACATCCAC AACTACCTGG 360

  A   W   Y   Q   Q   K   P   G   K   A   P   K   L   L   I   Y   Y   T   T   T
CTTGGTACCA GCAGAAGCCA GGTAAGGCTC CAAAGCTGCT GATCTACTAC ACCACCACCC 420

  L   A   D   G   V   P   S   R   F   S   G   S   G   S   G   T   D   F   T   F
TGGCTGACGG TGTGCCAAGC AGATTCAGCG GTAGCGGTAG CGGTACCGAC TTCACCTTCA 480

  T   I   S   S   L   Q   P   E   D   I   A   T   Y   Y   C   Q   H   F   W   S
CCATCAGCAG CCTCCAGCCA GAGGACATCG CCACCTACTA CTGCCAGCAC TTCTGGAGCA 540

  P   P   R   T   F   G   Q   G   T   K   L   E   I   A    |splice site
CCCCAAGGAC GTTCGGCCAA GGGACCAAGC TCGAGATAAA ACGTGAGTGG ATCC 594
                              XhoI                        BamHI
```

M13VHPCR

```
               Q   V   Q   L   V   E   S   G   P   G   L   V   R
          CAGGTCCAAC TGCAGGAGAG CGGTCCAGGT CTTGTGAGAC 300
                    PstI

  P   S   Q   T   L   S   L   T   C   T   V   S   G   S   T   F   S   S   Y   W
CTAGCCAGAC CCTGAGCCTG ACCTGCACCG TGTCTGGCAG CACCTTCAGC AGCTACTGGA 360

  M   H   W   V   R   Q   P   P   G   R   G   L   E   W   I   G   R   I   D   P
TGCACTGGGT GAGACAGCCA CCTGGACGAG GTCTTGAGTG GATTGGAAGG ATTGATCCTA 420

  N   S   G   G   T   K   Y   N   E   K   F   K   S   R   V   T   M   L   V   D
ATAGTGGTGG TACTAAGTAC AATGAGAAGT TCAAGAGCAG AGTGACAATG CTGGTAGACA 480

  T   S   K   N   Q   F   S   L   R   L   S   S   V   T   A   A   D   T   A   V
CCAGCAAGAA CCAGTTCAGC CTGAGACTCA GCAGCGTGAC AGCCGCCGAC ACCGCGGTCT 540

  Y   Y   C   A   R   Y   D   Y   Y   G   S   S   Y   F   D   Y   W   G   Q   G
ATTATTGTGC AAGATACGAT TACTACGGTA GTAGCTACTT TGACTACTGG GGCCAAGGGA 600

  T   T   V   V   S   S    |splice site
CCACGGTCAC CGTCTCCTCA GGTGAGTCCT TACAACCTCT CTCTTCTATT CAGCTTAAAT 660
     BstEII

AGATTTTACT GCATTTGTTG GGGGGGAAAT GTGTGTATCT GAATTTCAGG TCATGAAGGA 720

CTAGGGACAC CTTGGGAGTC AGAAAGGGTC ATTGGGAGCC CGGGCTGATG CAGACAGACA 780

TCCTCAGCTC CCAGACTTCA TGGCCAGAGA TTTATAGGAT CC 822
                                      BamHI
```

Linker Region of pSW1-scFvD1.3

```
             Q  G  T  T  V  T  V  S  S  G  G  G    G  S  G    G  G  G
5'VH.......CAAGG GACCACGGTC ACCGTCTCCT CAggtggagg cggttcaggc ggaggtggct
           MO-LINK-BACK —>
                  BstEII                    (gly4ser)3 linker

S  G  G  G    G  S  D  I  E  L    T  Q  S  P    A  S
ctggcggtgg cggatcgGAC ATTGAGCTCA CCCAGTCTCC AGCCTC......VK3'
                      CTG TAACTCGAGT GGGTCAGAGG TC
                      SacI        <— MO-LINK-FOR
```

REFERENCES

1. Baccala, R., Quang, T.V., Gilbert, M., Ternynck, T., and Avrameas, S. 1989. Two murine natural polyreactive autoantibodies are encoded by nonmutated germ-line genes. *Proc. Natl. Acad. Sci. USA* 86:4624–28.

2. Berardi, N., Cellerino, A., Domenici, L., Fagiolini, M., Pizzorusso, T., Cattaneo, A., and Maffei, L. 1994. Monoclonal antibodies to NGF affect the postnatal development of the rat visual system. *Proc. Natl. Acad. Sci. USA.* In press. 91:684–8.

3. Biocca, S., Neuberger, M.S., and Cattaneo, A. 1990. Expression and targeting of intracellular antibodies in mammalian cells. *EMBO J.* 9:101–8.

4. Birchmeier, C., Broek, D., and Wigler, M. 1985. RAS proteins can induce meiosis in *Xenopus* oocytes. *Cell* 43:615–21.

5. Bird, R.E., Hardman, K.D., Jacobson, J.W., Johnson, S., Kaufman, B.M., Lee, S.M., Lee, T., Pope, S.H., Riordan, G.S., and Whitlow, M. 1988. Single-chain antigen-binding proteins [published erratum appears in *Science* 1989 Apr. 28; 244(4903):409]. *Science* 242:423–26.

6. Bird, R.E., and Walker, B.W. 1991. Single chain antibody variable regions. *Trends Biotech.* 9:132–38.

7. Bonifacino, J.S., Suzuki, C.K., and Klausner, R.D. 1990. A peptide sequence confers retention and rapid degradation in the endoplasmic reticulum. *Science* 247:79–82.

8. Bookstein, R., Lai, C.-C., To, H., and Lee, W.-H. 1990. PCR-based detection of a polymorphic BamHI site in intron 1 of the human retinoblastoma (RB) gene. *Nucl. Acids Res.* 18:1666.

9. Boshart, M., Weber, F., Jahn, G., Dorsch-Häsler, K., Fleckenstein, B., and Schaffner, W. 1985. A very strong enhancer is located upstream of an immediate early gene of human cytomegalovirus. *Cell* 41:521–30.

10. Bradley, A. 1987. Production and analysis of chimeric mice. In *Teratocarcinomas and Embryonic Stem Cells: A Practical Approach.* Robertson, E. J., ed. (Oxford: IRL press).

11. Breakefield, X.O. 1993. Gene delivery into the brain using virus vectors. *Nat. New Genet.* 3:187–89.

12. Burke, B., and Warren, G. 1984. Microinjection of mRNA coding for an

anti-golgi antibody inhibits intracellular transport of a viral membrane protein. *Cell* 36:847–56.

13. Capecchi, M.R. 1989. The new mouse genetics: Altering the genome by gene targeting. *Trends Genet.* 5:70–76.

14. Carlson, J.R. 1988. A new means of inducibly inactivating a cellular protein. *Mol. Cell. Biol.* 8:2638–46.

15. Cattaneo, A., and Neuberger, M.S. 1987. Polymeric immunoglobulin M is secreted by transfectants of non-lymphoid cells in the absence of immuno-globulin J chain. *Embo J.* 6:2753–58.

16. Clackson, T., Hoogenboom, H.R., Griffiths, A.D., and Winter, G. 1991. Making antibody fragments using phage display libraries. *Nature* 352:624–28.

17. Colman, A., Besley, J., and Valle, G. 1982. Interactions of mouse immuno-globulin chains within *Xenopus* oocytes. *J. Mol. Biol.* 160:459–74.

18. Culver, K.W., Ram, Z., Wallbridge, S., Ishii, H., Oldfield, E.H., and Blaese, R.M. 1992. *In vivo* gene transfer with retroviral vector-producer cells for treatment of experimental brain tumors. *Science* 256:1500–52.

19. De Bellis, D., and Schwartz, I. 1990. Regulated expression of foreign genes fused by lac: control by glucose levels in growth medium. *Nucleic Acids Res.* 18:1311.

20. Dingwall, D., Robinns, J., Dilworth, S.M., Roberts, B., and Richardson, W.G. 1988. The nucleoplasmin nuclear location sequence is larger and more complex than that of SV40 large T antigen. *J. Cell Biol.* 107:841–49.

21. Evan, G.I., Lewis, G.K., Ramsay, G., and Bishop, J.M. 1985. Isolation of monoclonal antibodies specific for human c-myc proto-oncogene product. *Mol Cell Biol.* 5:3610–16.

22. Federoff, H.J., Geschwind, M.D., Geller, A.I., and Kessler, J.A. 1992. Expres-sion of NGF *in vivo* from a defective *Herpes simplex* virus 1 vector prevents effects of axotomy on sympathetic ganglia. *Proc. Natl. Acad. Sci. USA* 89:1636–40.

23. Fisher-Fantuzzi, L., and Vesco, C. 1988. Cell-dependent efficiency of reiterated nuclear signals in a mutant simian virus 40 oneoprotein targeted to the nucleus. *Mol. Cell. Biol.* 8:5495–503.

24. Forss-Petter, S., Danielson, P.E., Catsicas, S., Battenberg, E., Price, J., Nerenberg, M., and Sutcliffe, J.G. 1990. Transgenic mice expression β-galactosidase in mature neurons under neuron-specific enolase promoter control. *Neuron.* 5:187–97.

25. Frohman, M.A., Dush, M.K., and Martin, G. 1988. Rapid production of full length cDNAs from rare transcripts: Amplification using a single gene-specific oligonucleotide primer. *Proc. Natl. Acad. Sci. USA* 85:8998–9002.

26. Furth, M.E., Davis, L.J., Fleurdelys, B., and Scolnic, E.M. 1982. Monoclonal antibodies to the p21 products of the transforming gene of Harvey murine sarcoma virus and of the cellular *ras* gene family. *J. Virol.* 43:294–304.

27. Goodnow, C.C., Crosbie, J., Adelstein, S., Lavoie, T.B., Lavoie, G.S., Smith, G.S., Brink, R.A., Pritchard, B.H., Wotherspoon, J.S., Loblay, R.H., Raphael, K., Trent, R.J., and Basten, A. 1988. Altered immunoglobulin expression and functional silencing of self-reactive B lymphocytes in transgenic mice. *Nature* 334:676–82.

28. Goodnow, C.C., Crosbie, J., Jorgensen, H., Brick, R.A., and Basten, A.

1990. Induction of sel-tolerance in mature peripheral B lymphocytes. *Nature* 342:385–87.

29. Gossen, M., and Bujard, H. 1992. Tight control of gene expression in mammalian cells by tetracycline-responsive promoters. *Proc. Natl. Acad. Sci. USA* 89:5547–51.

30. Gu, H., Zou, Y.-R., and Rajewsky, K. 1993. Independent control of immunoglobulin switch recombination at individual switch regions evidenced through Cre-*loxP*-mediated gene targeting. *Cell* 73:1155–64.

31. Hancock, J.F., Magee, A.I., Childs, J.E., and Marshall, C.J. 1989. All *ras* proteins are polyisoprenylated but only some are palmitoylated. *Cell* 57:1167–77.

32. Haseloff, J., and Gerlach, W.L. 1988. Simple RNA enzymes with new and highly specific endoribonuclease activities. *Nature* 334:585–91.

33. Herrera, R.E., Shaw, P.E., and Nordheim, A. 1989. Occupation of the c-fos response element *in vivo* by a multi-protein complex is unaltered by growth factor induction. *Nature* 340:68–70.

34. Herskowitz, I. 1987. Functional inactivation of genes by dominant negative mutations. *Nature* 329:219–22.

35. Hitchcock, M.J.M., Ginns, E.I., and Marcus-Sekura, C.J., 1987. Microinjection into *Xenopus* oocytes: equipment. *Methods Enzymol.* 152:276–83.

36. Hoogenboom, H.R., Griffiths, A.D., Johnson, K.S., Chiswell, D.J., Hudson, P., and Winter, G. 1991. Multi-subunit proteins on the surface of filamentous phage: methodologies for displaying antibody (Fqb) heavy and light chains. *Nucl. Acids Res.* 19:4133–37.

37. Hu, M.C.T., and Davidson, N. 1987. The inducible iac operator-repressor system is functional in mammalian cells. *Cell* 48:555–66.

38. Hung, T., Mak, K., and Fong, K. 1990. A specificity enhancer for polymerase chain reaction. *Nucl. Acids. Res.* 18:4953.

39. Hunt, T. 1989. Cytoplasmic anchoring proteins and the control of nuclear localization. *Cell* 59:949–51.

40. Hurt, E.C., and van Loon, A.P.G.M. 1986. How proteins find mitochondria and intramitochondrial compartments. *Trends Biochem.* 11:204–207.

41. Huston, J.S., Levinson, D., Mudgett, H.M., Tai, M.S., Novotny, J., Margolies, M.N., Ridge, R.J., Bruccoleri, R.E., Haber, E., Crea, R., and Opperman, H. 1988. Protein engineering of antibody binding sites: recovery of specific activity in an anti-digoxin single-chain Fv analogue produced in *Escherichia coli*. *Proc. Natl. Acad. Sci. USA* 85:5879–83.

42. Izant, J.G., and Weintraub, H. 1985. Constitutive and conditional suppression of exogenous and endogenous genes by anti-sense RNA. *Science* 229:345–52.

43. Jones, S.T., and Bendig, M. 1991. Rapid PCR-cloning of full-length mouse immunoglobulin variable regions. *Bio/Technology* 9:88–89.

44. Kabat, E.A., Wu, T.T., Perry, H.M., Gottesman, K.S., and Foeller, C. 1991. In *Sequences of Proteins of Immunological Interest.* U.S. Department of Health and Human Services, U.S. Government Printing Office.

45. Kalderon, D., Roberts, B.L., Richardson, W., and Smith A. 1984. A short amino acid sequence able to specify nuclear location. *Cell* 39:499–509.

46. Klausner, R.D., and Sitia, R. 1990. Protein degradation in the endoplasmic reticulum. *Cell* 62:611–14.

47. Kohler, G., and Milstein, C. 1975. Continuous cultures of fused cells secreting antibody of predefined specificity. *Nature* 256:495–97.

48. Kollias, G., Spanopoulou, E., Grosveld, F., Ritter, M., Beech, J., and Morris, R. 1987. Differential regulation of a Thy 1 gene in transgenic mice. *Proc. Natl. Acad. Sci. USA* 84:1492–96.

49. Lacal, P.M., Pennington, C.T., and Lacal, J.C. 1988. Transforming activity of ras proteins translocated to the plasma membrane by a myristoylation sequence from the src gene product. *Oncogene* 2:533–37.

50. Larrick, J.W., Danielsson, L., Brenner, C.A., Abrahamson, M., Fry, K.E., and Borrebaeck, C.A. 1989a. Rapid cloning of rearranged immunoglobulin genes from human hybridoma cells using mixed primers and the polymerase chain reaction. *Biochem. Biophys. Res. Commun.* 160:1250–56.

51. Larrick, J.W., Danielsson, L., Brenner, C.A., Wallace, E.F., Abrahamson, M., Fry, K.E., and Borrebaeck, C.A.K. 1989b. Polymerase chain reaction using mixed primers: Cloning of human monoclonal antibody variable region genes from single hybridoma cells. *Bio/Technology* 7:934–38.

52. Lehmann, L.E., Eberle, W., Krull, S., Prill, V., Schmidt, B., Sander, C., von Figura, K., and Petres, C. 1992. The internalization signal in the cytoplasmic tail of lysosomal acid phosphatase consists of the hexapeptide PGYRHV. *EMBO J.* 11:4391–99.

53. Levi, A., Eldridge, J.D., and Patterson, B., 1985. Molecular cloning of a gene sequence regulated by nerve growth factor. *Science* 229:393–95.

54. Lewis, S.A., and Cowan, N.J. 1986. *Mol. Cell Biol.* 6:1529–34.

55. Magee, A.I., and Hanley, M.R. 1988. Protein modification: sticky fingers and CAAX boxes. *Nature* 335:114–15.

56. Markus-Sekura, C.J., and Hitchcock, M.J.M. 1987. Preparation of oocytes for microinjectionnof RNA and DNA. *Methods Enzymol.* 152:284–87.

57. McCafferty, J., Griffiths, A.D., Winter, G., and Chiswell, D.J. 1990. Phage antibodies: filamentous phage displaying antibody variable domains. *Nature* 348:552–54.

58. Melton, D.A. 1987. Translation of messenger RNA in injected frog oocytes. *Methods Enzymol.* 152:288–95.

59. Mizushima, S., and Nagata, S. 1990. pEF-BOS, a powerful mammalian expression vector. *Nucl. Acids Res.* 18:5322.

60. Munro, S., and Pelham, H.R.B. 1986. An Hsp-like protein in the ER: Identity with the 78kd glucose regulated protein and immunoglobulin heavy chain binding protein. *Cell* 46:291–300.

61. Nabel, G.J., and Plautz, G. 1990. Site specific gene expression *in vivo* by direct gene transfer into the arterial wall. *Science* 249:1285–88.

62. Neve, R.I. 1993. Adenovirus vectors enter the brain. *Trends Neurosci.* 16:251–53.

63. Orlandi, R., Gussow, D.H., Jones, P.T., and Winter, G. 1989. Cloning immunoglobulin variable domains for expression by the polymerase chain reaction. *Proc. Natl. Acad. USA* 86:3833–73.

64. Palmiter, R.D., and Brinster, R.L. 1986. Germ-line transformation of mice. *Ann. Rev. Genet.* 20:465–99.

65. Palmiter, R.D., Sandgren, E.P., Avarbock, M.R., Allen, D.D., and Brinster, R.L.

1991. Heterologous introns can enhance expression of transgenes in mice. *Proc. Natl. Acad. Sci. USA* 88:478–82.

66. Pelham, H.R.B. 1984. Hsp70 accelerates the recovery of nucleolar morphology after heat shock. *EMBO J.* 3:3095–3100.

67. Piccioli, P., Ruberti, F., Biocca, S., Di Luzio, A., Werge, T., Bradbury, A., and Cattaneo, A. 1991. Neuroantibodies: molecular cloning of a monoclonal antibody against substance P for expression in the central nervous system. *Proc. Natl. Acad. Sci. USA* 88:5611–15.

68. Possenti, R., Di Rocco, G., Nasi, S., and Levi, A. 1992. Regulatory elements in the promoter region of vgf, a nerve growth factor inducible gene. *Proc. Natl. Acad. Sci USA* 89:3815–19.

69. Price, J., Turner, D., and Cepko, C. 1987. Lineage analysis in the vertebrate nervous system by retrovirus mediated gene transfer. *Proc. Natl. Acad. Sci. USA* 84:156–60.

70. Readhead, C., Popko, B., Takahashi, N., Shine, H., Saavedra, R.A., Sidman, R.L., and Hood, L. 1987. Expression of a myelin basic protein gene in transgenic shiverer mice: correction of the dysmyelinating phenotype. *Cell* 48:703–12.

71. Rizzuto, R., Simpson, A.W.M., Brini, M., and Pozzan, T. 1992. Rapid changes of mitochondrial Ca++ revealed by specifically targeted recombinant aequorin. *Nature* 358:325–27.

72. Robertson, E., Bradley, A., Kuehn, M., and Evans, M. 1986. Germline transmission of genes introduced into cultured pluripotential cells by retroviral vector. *Nature* 323:445–48.

73. Rogers, S., Wells, R., and Rechsteiner, M. 1986. Amino acid sequences common to rapidly degraded proteins: the PEST hypothesis. *Science* 234:364–68.

74. Rosa, P., Weiss, U., Pepperkok, R., Ansorge, W., Niehrs, C., Stelzer, E.H.K., and Huttner, W.B. 1989. An antibody against secretogranin I (chromogranin B) is packaged into secretory granules. *J. Cell Biol.* 109:17–34.

75. Rudikoff, S., and Pumphrey, J.G. 1986. Functional antibody lacking a variable region disulfide bridge. *Proc. Natl. Acad. Sci. USA* 83:7875–78.

76. Sambrook, J., Fritsch, E.F., and Maniatis, T. 1989. In *Molecular Cloning: A Laboratory Manual.* Cold Spring Harbour Laboratory Press.

77. Sarker, G., Kapelner, S., and Sommer, S.S. 1990. Formamide can dramatically improve the specificty of PCR. *Nucl. Acids Res.* 18:7465.

78. Smith, A.J.H., and Kalogerakis, B. 1991. Detection of gene targeting by co-conversion of a single nucleotide change during replacement recombination at the immunoglobulin μ heavy chain locus. *Nucl. Acids Res.* 19:7161–70.

79. Storb, U. 1987. Transgenic mice with immunoglobulin genes. *Ann. Rev. Immunol.* 5:151.

80. Tang, Y., Hicks, J.B., and Hilvert, D. 1991. *In vivo* catalysis of a metabolically essential reaction by an antibody. *Proc. Natl. Acad. Sci. USA* 88:8784–86.

81. Valle, G., Bhamra, S.S., Martin, S., Griffiths, G., and Colman, A. 1988. Effect of anti-ER antibodies within the ER lumen of living cells. *Exp. Cell Res.* 176:221–23.

82. Vaux, D.J.T., Helenius, A., and Mellman, I. 1988. Spike-nucleocapsid interaction in Semliki Forest virus reconstructed using network antibodies. *Nature* 336:36–42.

83. Vidal, M., Morris, R., Grosveld, F., and Spanopoulou, E. 1990. Tissue specific control elements of the Thy 1 gene. *EMBO J.* 9:833–40.

84. Wahlestedt, C., Pich, E.M., Koob, G.F., and Helig, M. 1993. Modulation of anxiety and neuropeptide Y-Y1 receptors by antisense oligodeoxynucleotides. *Science* 259:528–31.

85. Werge, T., Biocca, S., and Cattaneo, A. 1990. Intracellular immunisation: cloning and intracellular expression of a monoclonal antibody to the p21ras protein. *FEBS Lett.* 274:193–98.

86. Williams, G.T., Venkitaraman, A.R., Gilmore, D.J., and Neuberger, M.S. 1990. The sequence of the mu transmembrane segment determines the tissue specificity of the transport of immunoglobulin M to the cell surface. *J. Exp. Med.* 171:947.

87. Biocca, S., Pierandrei-Amaldi, P., Campioni, N., and Cattaneo, A. 1994. Intracellular immunisation with cytosolic recombinant antibodies. *BioTech.*, in press.

88. Biocaa, S., Pierandrei-Amaldi, P., and Cattaneo, A. 1993. Intracellular expression of anti-p21ras single chain Fv fragments inhibits meiotic maturation of Xenopus oocytes. *B.B.R.C.* 197:422–27.

89. Bradbury, A., Persic, L., Werge, T., and Cattaneo, A. 1993. From gene to antibody: the use of living columns to select specific phage antibodies. *Biotech.* 11:1565–69.

90. Marasco, W.A., Haseltine, W.A., and Chen, S. 1993. Design, intracellular expression and activity of a human anti-human immunodeficiency virus type 1 gp120 single-chain antibody. *Proc. Natl. Acad. Sci. USA* 90:7889–93.

91. Ruberti, F., Bradbury, A., and Cattaneo, A. 1993. Cloning and expression of an anti-nerve growth factor antibody for studies using the neuroantibody approach. *Cell Mol. Neurobiol.* 13:559–68.

92. Ruberti, F., Cattaneo, A., and Bradbury, A. 1993. The use of the RACE method to clone hybridoma cDNA when V region primers fail. *J. Imm. Meth.*, in press.

CHAPTER 11

The Expression of "Single Chain" Antibodies in Transgenic Plants

Paraskevi Tavladoraki, Rosella Franconi,
Andrew Bradbury, Antonino Cattaneo,
and Eugenio Benvenuto

Since the first demonstration that plant cells are able to synthesize and secrete antibodies,[9,17] plants have been considered as alternative hosts for immunoglobulin expression in heterologous systems. Hence plants join the established list of nonlymphoid hosts such as bacteria,[29] yeasts,[18] insect cultured cells,[14] and mammalian cells,[3,7,36] each of which has advantages and disadvantages for the production of secreted antibodies or their domains.

Part of the difficulty with every recombinant host lies in the transfer and expression of two genes (heavy and light chains) and the subsequent assembly of the synthesized polypeptides. Other drawbacks are represented by the presence of distinct glycosylation patterns in each host and the necessity to adapt signals to drive processing, assembly, and secretion.

The pioneering work developed by Hiatt and colleagues[17] involved the sexual crossing of two separate plants expressing single γ or κ chains and the selection of F_1 progeny expressing assembled molecules. It has since been demonstrated that plants can express complete antibody molecules if the two genes are mounted on a single transformation vector[9] or after simultaneous transformations with two different heavy- and light-chain constructs.[8]

However, the data on the assembly and/or assembly patterns of plant-derived antibody are far from exhaustive. Moreover, although secretion has been obtained,[15,19] unusual targeting patterns have also been observed, such as the chloroplast[9] and nucleolus[8] where, according to prevailing dogma,[5] no antibody should be detected. These findings demonstrate that the fidelity with which the plant cell machinery can process an evolutionarily distant polypeptide is not always predictable. However, the yield of synthesized whole antibody may be as much as 1% of the total soluble proteins in stably transformed plants. This offers exciting perspectives for low-cost/large-scale production of bioactive diagnostic or therapeutic immunoglobulin.[16]

Expression of antibodies in plants is not only useful for mass-scale production, but could represent a powerful alternative to the "antisense" technology to "immunomodulate" specific plant functions *in vivo*, an approach first suggested for animal cells.[3,40,41] A recent boost for immuno-chemistry came from gene technology that permits the construction of new antibody reagents and the creation of novel variants of the immune-repertoire *in vitro*.[37] It is widely believed that such an approach will be extremely useful in plants, where simpler forms of immunoglobulin possess major advantages compared to the whole antibody, especially when the final aim is the construction of "genetically immunized" transgenic plants. Ideally, one would wish to avoid problems such as the transfer of two gene constructs and problems in assembly, folding, and stability. Since immunoglobulin effector functions (which anyway are likely to be inactive in plants) can be separated from binding domains, we focused our efforts on the expression of antibody fragments in plants. Plants have been shown to be capable of expressing functional engineered forms of antibody such as "single domain antibodies" (dAbs, VH fragment alone)[2] or "single chain antibodies" (ScFv, VH and VL joined by a linker).[27,33] Efficient expression of dAbs would be particularly attractive when targeting these molecules to exert molecular interference in different cellular compartments. There they would be able to act as very small recognition units and, due to their molecular size, may interfere effectively with key protein functions. The success of the approach is strictly dependent on the antigen-binding activity and specificity of the dAb which, at the moment, requires extensive engineering, particularly to modify the exposed hydrophobic surface.[34]

The need to build the smallest molecule endowed with the highest binding capacity and easy access to a potential target led us to consider ScFvs as the most versatile antibody derivatives for regulated expression in different tissue or cellular compartments of transgenic plants. This derives from their ability to fold correctly in the absence of specific assembly requirements, their high binding specificity, and the ease which they can be engineered. The ablation of a plant[27] and a viral[33] protein function, obtained with intracellular cytosolic ScFvs, demonstrates the feasibility of the strategy of intracellular immunization in plants. This has been recently demonstrated also for animal cells.[42,43] It is likely that this technique can be used for the molecular interference of many cellular functions and seems particularly appropriate as the leaderless heavy and light immunoglobulin chains fail to assembly in the plant cell cytoplasm.[15,17] Data on the yield and function (see below) forecast a possible use of plants as a source of single-chain based immunoreagents, especially if contaminants in industrial production based on bacteria or tissue-cultured based methods have implications for human health.[24]

The amplification of individual hybridoma V regions from either cDNA or genomic DNA using V region primers (and the problems which may be encountered) is described in Chapter 10. After amplification, the VH and VL fragments must be either assembled as a ScFv fragment by PCR (as described in Chapter 10), or inserted by cloning into a vector that will allow the expression of ScFv fragments. A number of vectors are available.[29] Different linker peptides to connect the variable domains have been designed that differ in the length (the minimum being 12 residues) and in the number of hydrophilic residues.[28,38] In some cases, naturally occurring linkers (VL-CL elbow, VH-CH$_1$ elbow) have been used that may possess reduced immunogenicity for *in vivo* applications.[1,4] In addition, ScFv antibodies have been devised in which the VH and VL domains have different orientations with respect to the linker,[4,21,29] With VH-linker-VL orientations,[1] the secretion of ScFv in bacteria has been demonstrated to be 20-fold lower than that obtained with VL-linker-VH orientation. However, VH-linker-VL orientations give a 10-fold higher affinity than VL-linker-VH orientations.[1] It may therefore be worth considering the factors influencing the biochemical properties of the ScFv antibodies before starting the cloning steps for a recombinent ScFv construct.

We have used the pJMScFv vectors (courtesy of MRC Laboratory of Molecular Biology, Cambridge) in our experiments for bacterial expression of ScFv. These vectors include the strong, tightly regulated $\lambda(P_L)$ promoter,[30] a ribosome-binding site, a hydrophilic linker peptide ([Gly$_5$Ser]$_3$)[19] that connects the VH carboxyl to the VL amino terminus, as well as the human c-myc Tag peptide,[25] which can be used to recognize the expression products with the 9E10 monoclonal antibody,[11] (Fig. 11–1). The secretory form [pelB] of the pJMScFv vector contains sequences encoding the bacterial pelB signal peptide[23] that direct the ScFv to be secreted into the periplasmic space. The intracellular form lacks this peptide and directs ScFv synthesis into the cytoplasm.

The method we used to clone ScFv fragments from hybridomas was first to create a cDNA library, to probe it with construct region probes by colony hybridization, and then to sequence the isolated chains. Functional immunoglobulins (on the basis of sequence) were recloned by V region PCR (see the

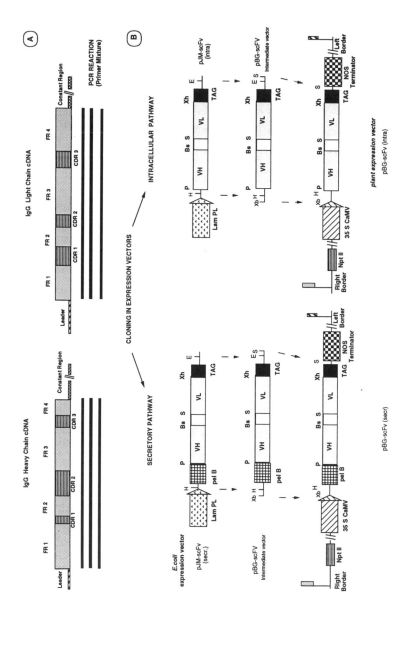

Figure 11–1. Vectors for expression for ScFv antibody in *E. coli* and in plants. (A) PCR amplification of variable domains. (B) Flow diagram of the cloning steps. Restriction sites used for cloning of the VH and VL fragments are shown. H:HindIII, P:PstI, Bs:BstII, S:SacI, E:EcoRI, Xb: Xba I. LAM PL: heat inducible promoter, pelB: bacterial signal peptide for secretion, TAG: human c-myc Tag peptide, 35S CaMV: Promoter derived from Cauliflower Mosaic Virus 35S RNA, NOS: terminator from Nopaline Synthase gene. Left and right border: 25 bp directed repeats, boundaries of *Agrobacterium* T-DNA insertion.

chapter by Bradbury et al., for details of methods and primers used) and then subcloned by restriction digestion (using PstI, BstEII, SacI, and XhoI) into the secretory and intracellular forms of the pJMscfv vectors (Fig. 11–1). It is clear that any of the methods available to create ScFv fragments are appropriate.

All the cloning steps utilizing the pJMScFv vector are carried out in the EMG578 strain of *E. coli*. This bacterial host strain expresses a thermolabile λ repressor. Fresh comptent EMG578 cells are prepared according to the basic CaCl$_2$ protocol.[32] No other protocol to produce higher transformation efficiencies seems to work with these cells. EMG578 bacteria harboring the pJMScFv vector should always be incubated at 28°C to ensure that repressed state of the promoter.

INTRODUCTION TO METHOD 1: EXPRESSION IN BACTERIA

Cloning of the ScFv construct in the right frame and the antigen-binding capacity of the engineered antibody is tested by expression of the ScFv construct in *E. coli*. This is very important, since it allows one to test the specificity of the cloned ScFv prior to embarking upon the time-consuming procedure of plant transformation. Several different *E. coli* expression vectors have been developed over the last years.[29] We have used the pJMScFv vectors described, above with the protocols described below being appropriate for the secretory form of the pJMScFv vector. Accumulation of the antibody fragment in the periplasmic space facilitates it purification, in contrast to the production of immunoglobulin fragments as intracellular inclusion bodies in *E. coli*, which requires long solubilization and refolding procedures that generally result in some loss of activity. After secretion into the periplasmic space, antibody fragments may also be found in the culture supernatant, particularly if the induction period is prolonged. This is thought to be due to leakage through the outer membrane, perhaps as a result of decreased cell viability during the expression of the antibody fragments (which are known to be toxic to bacteria).[29] Since purification from the culture supernatant can be laborious due to the large volumes, we prefer to purify it from the periplasmic space after a short heat induction when it can be obtained at high concentrations.

Method 1: Bacterial Induction

1. Inoculate a single colony of EMG578 carrying the pJMScFv [pel B] plasmid into LB medium + 100 µg/ml ampicillin.
2. Grow overnight at 28°C by continuous shaking at 250 rpm.

3. Dilute $OD_{600} = 0.05$ with fresh LB medium plus ampicillin and incubate at 28°C until $OD_{600} = 0.5$ is reached.
4. Heat the culture in a waterbath at 42°C by gentle shaking for 1 hr.
5. Harvest bacteria by centrifugation at $3,000 \times g$ for 15 min and resuspend the bacterial pellet in 100 µl of cold TES (0.2 M Tris-HCl pH 8.0, 0.5 mM EDTA, 0.5 M sucrose) per 10 ml of culture.
6. Add 150 µl of cold TES (diluted 1:4 with H_2O) to the bacterial suspension for each 10 ml of original culture and, after gentle mixing, transfer on ice for 30 min (osmotic shock).
7. Centrifuge at $20,000 \times g$ for 30 min to obtain the soluble periplasmic fraction in the supernatant. This is further cleaned up by a second centrifugation at $20,000 \times g$ for 30 min.
8. Check 20 µl of the periplasmic extract for expression of ScFv by Western blot analysis and/or use 100 µl of the same extract for antigen-binding activity by ELISA.

Points to Consider

We find that the yield of ScFv antibody should be 300 µg per liter of bacterial culture, and that longer periods of incubation do not increase the yields of ScFv produced. Once antigen-binding activity of the ScFv has been confirmed, the plant expression construct can be made.

INTRODUCTION TO METHOD 2: EXPRESSION IN PLANTS

Several vectors for expression of heterologous proteins in plants have been developed in an effort to get high levels of expression. Various strategies to obtain transgenic plants have also been developed, the *Agrobacterium*-based procedure being the most widely used.[35] The choice of a suitable vector and transformation procedure to obtain transgenic plants is very important.

For intracellular immunization the choice of the targeting signal to direct the engineered antibody to a particular plant compartment is fundamental, obviously being influenced by the location of the antigen to be inhibited. To date, the protein-localizing mechanism in plants is not fully understood. Targeting of immunoglobulin chains to the endoplasmic reticulum (ER) and for secretion has been obtained by the native immunoglobulin signal peptides[15] for whole immunoglobulin molecules and with the bacterial pelB leader for ScFvs, although at lower levels in the latter case.[2] Strangely, the N-terminal signal peptide from barely α-amylase directed the expression of whole immunoglobulin chains to the ER and the chloroplasts,[9] but not to more downstream parts of the secretory apparatus, even though it derives from a well-characterized extracellular protein of barley aleurone[31] and directs secretion of another heterologous protein in potato.[10] Even more surprisingly, FAb molecules were identified in the nucleolus of meristematic shoot tissue when the signal peptide of the 2S albumin of *Arabidopsis thaliana*[22] was used.[8] Other targeting signals to direct immunoglobulin (as a whole or fragments) to different cellular compartments have not been fully characterized yet. In Fig. 11–1, the cloning strategy and the vectors constructed in our laboratory for expression of ScFvs in plants are shown. The 35S CaMV (Cauliflower Mosaic Virus) promoter and NOS (Nopaline synthase) terminator regulatory elements as well as the selectable marker NPT II (Neomycine Phosphotransferase II) are derived from the pBI 121.1 binary vector.[20] The constructs for plant transformation used in our experiments lack a signal peptide and so direct the expression of the ScFv antibody to the cytoplasm. Relatively high levels of ScFv expression (0.1% of the total soluble proteins) are observed in stably transformed plants, and results from ELISA and Western blot analysis demonstrated that the total amount of the expressed ScFv antibody is functional. The successful expression in the plant cytoplasm of a functional ScFv antibody, also reported by Owen et al.,[27] highlights the versatility of the ScFv molecules for expression in plants, and the ease with which they are accumulated in the plant cytoplasm contrasts with whole immunoglobulins that are not assembled.

Method 2.1: Extraction of Total Soluble Plant Proteins

To test intracellular antibody expression in transgenic plants, total soluble proteins are first extracted.

1. A minimum of 0.3 g of leaves are homogenized by Polytron homogenizer in about 0.3 ml of 50 mM Tris-HCl pH 8, 0.5 M NaCl, 0.25% Nonidet P-40, 1.3% Polyvinylpolypyrrolidone (PVPP), 1 mM Phenylmethylsulfonylfluoride (PMSF), 3 µg/ml pepstatin, 1 µg/ml leupeptin, 5 mM ascorbic acid.
2. Centrifuge the homogenate at $20,000 \times g$ for 30 min at 4°C. The supernatant is centrifuged a second time in the same conditions to remove further debris.
3. The plant extract can be kept at -20°C or immediately processed for affinity chromatography purification.

Method 2.2: Affinity Chromatography Purification and Western Blot Analysis of the Expressed Proteins

1. Incubate the crude plant extract with 9E10 Sepharose (prepared from CNBr-activated Sepharose 4B (Pharmacia) following the

manufacturer's instructions) in an Eppendorf tube overnight at 4°C by constant rotation.

2. Wasn the Sepharose resin four times with 50 mM Tris-HCl pH 8.0, 0.5 M NaCl, 0.1% NP-40 and once with 200 mM Glycine pH 5.8. In each washing step centrifuge the resin at 1000 × *g* for 30 seconds, just enough to pellet it without damaging it.

3. Elute the antibody fragment with 200 mM Glycine pH 3.0 and immediately neutralize to pH 7.5 by adding a titrated volume of 1 M Tris-HCl pH 8.0.

4. An aliquot of the eluted protein is subjected to SDS-PAGE electrophoresis and electrophoretically transferred to nitrocellulose membrane (e.g., Hybond-C Super, Amersham).

5. Preincubate blotted filters overnight at 4°C in 3% BSA, 0.3% gelatine, 10 µg/ml total goat IgG, in PBS.

6. Before the addition of antibodies, wash the blots three times (ten minutes each) with vigorous shaking as follows: once in PBS, 0.1% Tween 20 (PBST), once in PBST containing 0.5 M NaCl; and once in PBST.

7. Incubate blots for 2 hr with an appropriate dilution of purified 9E10 monoclonal antibody in 0.5% BSA, 0.1% Tween-20, 1 × PBS at room temperature with constant shaking.

8. Repeat step 6.

9. Incubate with a biotinylated sheep anti-mouse whole Ig antibody (Amersham) in PBST containing 0.5% BSA for 1.5 hr at room temperature.

10. Wash as before.

11. Incubate with streptavidin-conjugated horseradish peroxidase (Amersham) in PBST containing 0.5% BSA for 45 min at room temperature.

12. After washing, peroxidase activity is determined by the ECL detection system (Amersham).

Method 2.3: ELISA Analysis of ScFv Antibody Expressed in Plants

1. Coat ELISA microtitre plates (Dynatech Immunol) with the known antigen concentration in $Na_2CO_3/NaHCO_3$ 0.2 M pH 9.6 or in 1 × PBS overnight at 4°C.

2. Wash 3 times with PBST and once with PBS.

3. Block with 2% low-fat dry milk, 0.1% gelatin in 1 × PBS for 2 hr at 37°C.

4. Wash as before.

5. Crude plant extract or elutate from the affinity column is added in serial dilutions to microtitre wells. The plate is incubated overnight at 4°C.

6. Wash as before.

7. Incubate with 9E10 monoclonal antibody in PBS containing 2% milk for 2 hrs at 37°C.
8. Repeat washing steps.
9. Add peroxidase-conjugated goat anti-mouse antibody in PBS containing 2% milk. Incubate for 1 hr at 37°C.
10. Bound ScFv fragment is detected using the KPL detection system (ABTS and H_2O_2) according to the manufacturer's instructions.

Method 2.4: Antibody Secretion in Plants: Testing the Protoplast Culture Supernatant for ScFv Secretion

When a signal peptide for secretion is used, antibody export should be checked. This can be performed by testing the supernatant of a protoplast culture and/or the intercellular washing fluid (IWF) from leaves for the presence of antibody.

1. Prepare protoplast from plant leaves according to Zhu and Negrutiu.[39]
2. Transfer into a 6-well culture plates 3 ml culture of protoplasts (10^6 protoplasts/ml), and incubate in low light conditions for 96 hrs.
3. Check cell viability with 0.1 µg/ml FDA (Fluorescein diacetate 1 mg/ml stock solution in acetone stored in the freezer) and calculate the percentage of live protoplasts.
4. Gently pellet protoplasts at $150 \times g$ and keep the supernatant separate from the pellet.
5. Wash the protoplast pellet once with culture medium and then freeze to help lysis.
6. Check the culture supernatant for the presence of cytosolic contaminants by assaying for glucose-6-phosphate dehydrogenase activity.
7. Desalt through PD-10 Sephadex G-25 column (Pharmacia).
8. Concentrate the eluate to a final volume of about 400 µl by ultrafiltration using the centrifuged ultrafree-20 filtration units (Millipore). Add $10 \times$ PBS to a final concentration of $1 \times$.
9. Purify the eluate through 9E10-Sepharose affinity chromatography and visualize secreted ScFv antibody by Western blot (see Method 2.2). Protoplasts are also analyzed for the presence of ScFv antibody.
10. Thaw frozen protoplasts and resuspend in about 400 µl of buffer ($1 \times$ PBS, 2% polyvinylpyrrolidone, PVP, 0.05% Tween-20).
11. Incubate protoplast suspension on ice for 10 min and disrupt protoplasts using a teflon homogenizer and by vigorous vortexing.
12. Centrifuge protoplast suspension at $20,000 \times g$ for 15 min and purify the ScFv antibody by affinity chromatography.

Comparison of the amount of the ScFv antibody purified from both protoplast pellet and culture supernatant gives an indication of the degree of ScFv secretion. A high yield of living protoplasts and the complete absence of cytosolic contaminants in the culture supernatant is crucial for the success of this procedure. To avoid cytosolic contaminants due to cell death during prolonged incubation, phase-chase labeling experiments may be performed according to Hein et al.[15]

Method 2.5: Testing Intercellular Washing Fluid (IWF) for the Presence of Secreted ScFv Antibody

IWF from plant leaves is prepared according to Düring et al.[10] and Hammond-Kosack.[13]

1. Leaves from greenhouse-grown plants are cut to strips of 0.5 cm.
2. Strips are washed in water for 15 min to remove any cytosolic contaminants from the cut edges and infiltrated with 1 × PBS at 4°C under vacuum, three times for 10 min.
3. Leaf pieces are gently blotted dry and transferred to an Eppendorf tube having a hole in the bottom. Place this Eppendorf tube on top of another tube and centrifuge the assembly at 200 × g for 10 min at 4°C remove the adhering fluid.
4. Collect IWF by centrifugation at 1,000 × g.
5. Centrifuge the collected IWF at 20,000 × g for 10 min at 4°C.
6. Check supernatant cytosolic contaminants by assaying glucose-6-phosphate dehydrogenase activity.
7. Test the presence of secreted ScFv antibody by affinity chromatography followed by Western blot analysis.

CONCLUSIONS

The possibility of employing the extensive immune repertoire against almost every molecular structure offers plant biologists a powerful and versatile tool to endow plants with new properties.

Molecular interference mediated by antibody-protein interactions seems to be an expanding field where the exploitation of the intrinsic properties of antibodies can contribute to basic and applied research issues. Many antibodies raised against protein antigens block their functions. This is probably due to the fact that accessibility[26] and flexibility[12] influence antigenicity and that active or functional sites of many proteins reside in the most accessible and flexible domains. Hence the chances to select antibodies mediating inhibition of a protein function are relatively high. We have found the ScFv format to be particularly appropriate for expression in plants due to the high level of functional antibody produced. These may be derived from hybridomas or from phage display libraries.[6] With the latter method, the problems of V region cloning is considerably reduced.

The experimental procedure described here has been devised to block a viral function with the final aim to confer protection to plants.[33] A ScFv antibody has been constructed that retained to a great extent the affinity of the parental monoclonal antibody raised against the coat protein of the plant virus AMCV (Artichoke Mottled Crinkle Virus). Intracellular expression of this antibody fragment resulted in a drastic reduction of infection incidence and a marked delay in the development of pathological symptoms in transgenic plants. Although the mechanisms operating are not yet clarified, the factors influencing the success of our approach are probably due to: (1) the use of ScFv instead of whole antibody molecules; (2) the relatively high levels of expression obtained (0.1% of the total protein); (3) the high affinity of the parental antibody and its derived ScFv fragment for the virus; and (4) the high level of functional ScFv expressed. These findings have implications not only for molecular plant pathology but could be of general utility when other plant functions or activities should be permanently modulated. Indeed, progress made in the plantibody (plant-derived antibodies) field leaves no doubt that important basic and applied achievements will be realized in the near future.

REFERENCES

1. Anand, N.N., Mandal, S., Mackenzie, C.R., Sadowska, J., Sigurskjold, B., Young, N.M., Bundle, D.R., and Narang, S.N. 1991. Bacterial expression and secretion of various single chain Fv genes encoding proteins specific for a *Samonella* serotype B O antigen. *J. Biol. Chem.* 266:21874.

2. Benvenuto, E., Ordas, R.J., Tavazza, R., Ancora, G., Biocca, S., Cattaneo, A., and Galeffi, P. 1991. Phytoantibodies: a general vector for the expression of immunoglobulin domains in transgenic plants. *Plant Mol. Biol.* 17:865–74.

3. Cattaneo, A., and Neuberger, M.S. 1987. Polymeric immunoglobulin M is secreted by transfectants of non-lymphoid cells in the absence of immunoglobulin J chain. *EMBO J.* 6:2753–58.

4. Chaudhary, V.K., Batra, J.K., Gallo, M.G., Willingham, M.C., FitzGerald, D.J., and Pastan, I. 1990. A rapid method of cloning functional variable-region antibody genes in *Escherichia coli* as single-chain immunotoxins [published erratum appears in Proc. Natl. Acad. Sci. USA 1990 Apr; 87(8): 3253]. *Proc. Natl. Acad. Sci. USA* 87:1066–70.

5. Chrispeels, M.J. 1991. Sorting of proteins in the secretory system. *Ann. Rev. Plant Physiol. Plant Mol. Biol.* 42:21.

6. Clackson, T., Hoogenboom, H.R., Griffiths, A.D., and Winter, G. 1991. Making antibody fragments using phage display libraries. *Nature* 352:624–28.

7. Colcher, D., Milenic, D., Roselli, M., Raubitscheck, A., Yarranton, G., King, D.,

Adair, J., Whittle, N., Bodmer, M., and Schlom, J. 1989. Characterization and biodistribution of recombinant/chimeric constructs of monoclonal antibody B72.3. *Cancer Res.* 49:1738.

8. De Neve, M., De Loose, M., Jacobs, A., van Houdt, H., Kaluza, B., Weidle, U., van Montagu, M., and Depicker, A. 1993. Assembly of an antibody and its derived antibody fragment in *Nicotiana* and *Arabidopsis. Transgenic Res.* 2:227.

9. Düring, K., Hippe, S., Kreuzaler, F., and Schell, J. 1990. Synthesis and self-assembly of a functional monoclonal antibody in transgenic *Nicotiana tabacum. Plant Mol. Biol.* 15:281.

10. Düring, K., Porsch, P., Fladung, M., and Lörz, H. 1993. Transgenic potato plants resistant to the phytopathogenic bacterium *Erwinia carotovora. Plant J.* 3:587.

11. Eban, G.I., Lewis, G.K., Ramsay, G., and Bishop, J.M. 1985. Isolation of monoclonal antibodies specific for human c-myc proto-oncogene product. *Mol. Cell Biol.* 5:3610–16.

12. Getzoff, E.D., Geysen, H.M., Rodda, S.J., Alexander, H., Tainer, J.A., and Lerner, R.A. 1987. Mechanisms of antibody binding to a protein. *Science* 235:1191.

13. Hammond-Kosack, K.E. 1992. Preparation and analysis of intercellular fluid. In *Molecular Plant Pathology, A Practical Approach.* Gurr, McPherson, and Bowes eds. IRL Press (Washington, D.C.: Oxford).

14. Haseman, C.A., and Capra, J.D. 199. High-level expression of a functional immunoglobulin heterodimer in a baculovirus expression system. *Proc. Natl. Acad. Sci. USA* 87:3492.

15. Hein, M.B., Tang, Y., McLeod, D.A., Janda, K.D., and Hiatt, A. 1991. Evaluation of immunoglobulins from plant cells. *Biotechnol. Prog.* 7:455.

16. Hiatt, A. 1990. Antibodies produced in plants. *Nature* 344:469.

17. Hiatt, A., Cafferkey, R., and Bowdish, K. 1989. Production of antibodies in transgenic plants. *Nature* 342:76–87.

18. Horwitz, A.H., Chang, C.P., Better, M., Hellstrom, K.E., and Robinson, R.R. 1988. Secretion of functional antibody and Fab fragment from yeast cells. *Proc. Natl. Acad. Sci. USA* 85:8678–82.

19. Huston, J.S., Levinson, D., Mudgett, H.M., Tai, M.S., Novotny, J., Margolies, M.N., Ridge, R.J., Bruccoleri, R.E., Haber, E., Crea, R., and Opperman, H. 1988. Protein engineering of antibody binding sites: recovery of specific activity in an anti-digoxin single-chain Fv analogue produced in *Escherichia coli. Proc. Natl. Acad. Sci. USA* 85:5879–83.

20. Jefferson, R.A., Kavanagh, T.A., and Bevan, M.W. 1987. GUS fusions: beta-glucuronidase as a sensitive and versatile gene fusion marker in higher plants. *EMBO J.* 6:3907.

21. Knappik, A., Krebber, C., and Plückthun, A. 1993. The effect of folding catalysts on the *in vivo* folding process of different antibody fragments in *Escherichia coli. Bio/Technology* 11:77.

22. Krebbers, E., Herdies, L., De Clercq, A., Surinck, J., Leemans, J., Van Damme, J., Segura, M., Gheysen, G., Van Montagu, M., and Vandekerckhove, J. 1988. Determination of the processing sites of an *Arabidopsis* 2S albumin and characterization of the complete gene family. *Plant Physiol.* 87:859.

23. Lei, S.-P., Lin, H.-C., Wang, S.-S., Callaway, J., and Wilcox, G. 1987. Characterization of the *Erwinia cartovora* pelB gene and its product pectase lyase. *J. Bacteriol.* 169:4379.
24. Mariani, M., and Tarditi, L. 1992. Validating the preparation of clinical monoclonal antibodies. *Bio/Technology* 10:394.
25. Munro, S., and Pelham, H. 1986. An Hsp70-like protein in the ER: identity with the 78 kd glucose-regulated protein and immunoglobulin heavy chain binding protein. *Cell* 46:291/300.
26. Novotny, J., Handshumaker, M., Haber, E., Bruccoleri, R.E., Carlson, W.B., Fanning, D.W., Smith, J.A., and Rose, G.D. 1986. Antigenic determinants in protein coincide with surface regions accessible to large probe (antibody domains). *Proc. Natl. Acad. Sci. USA* 83:226.
27. Owen, M., Gandecha, A., Cockburn, B., and Whitelam, G. 1992. Synthesis of a functional anti-phytochrome single chain Fv protein in transgenic tobacco. *Bio/Technology* 10:790.
28. Pantoliano, M.W., Bird, R.E., Johnson, S., Asel, E.D., Dodd, S.W., Wood, J.F., and Hardman, K.D. 1991. Conformational stability, folding and ligand-binding affinity of single-chain Fv immunoglobulin fragments expressed in *Escherichia coli*. *Biochem.* 30:1017.
29. Plückthun, A. 1991. Antibody engineering: advances from the use of *Escherichia coli* expression systems. *Bio/Technology* 9:545–51.
30. Remaut, E., Stanssens, P., and Fiers, W. 1981. Plasmid vectors for high-efficiency expression controlled by the pL promoter of coliphage lambda. *Gene* 15:81.
31. Rogers, J.C., and Milliman, C. 1983. Isolation and sequence analysis of a barley alpha amylase cDNA clone. *J. Biol. Chem.* 258:8169.
32. Sambrook, J., Fritsch, E.F., and Maniatis, T. 1989. In *Molecular Cloning: A Laboratory Manual*. Cold Spring Harbor Laboratory Press.
33. Tavladoraki, P., Benvenuto, E., Trinca, S., De Martinis, D., Cattaneo, A., and Galeffi, P. 1993. Transgenic plants expressing a functional single chain Fv antibody are specifically protected from virus attack. *Nature* 366:469–72.
34. Ward, E.S., Gussow, D., Griffiths, A.D., Jones, P.T., and Winter, G. 1989. Binding activities of a repertoire of single immunoglobulin variable domains secreted from *Escherichia coli* [see comments]. *Nature* 341:544–46.
35. Weising, K., Schell, J., and Kahl, G. 1988. Foreign genes in plants: transfer, structure, expression and applications. *Ann. Rev. Genet.* 22:421.
36. Whittle, N., Adair, J., Lloyd, C., Jenkins, L., Devine, J., Schlom, J., Raubitschek, A., Colcher, D., and Bodmer, M. 1987. Expression in COS cells of a mouse-human chimeric B72.3 antibody. *Prot. Engin.* 1:499.
37. Winter, G., and Milstein, C. 1991. Man-made antibodies. *Nature* 349:293–99.
38. Wu, X.-C., Ng, S-C., Near, R.I., and Wong, S.-L. 1993. Efficient production of a functional single chain anti-digoxin antibody via an engineered *Bacillus subtilis* expression secretion system. *Bio/Technology* 11:71.
39. Zhu, X.-Y., and Negrutiu, I. 1981. Isolation and culture of protoplasts. In *A Laboratory Guide for Cellular and Molecular Plant Biolohy*. Negrutiu and Gharti-Chetri, eds. (Basel: Birkhäuser Verlag).
40. Biocca, S., Neuberger, M.S., and Cattaneo, A. 1990. Expression and targeting of intracellular antibodies in mammalian cells. *EMBO J.* 9:101–08.

41. Piccioli, P., Ruberti, F., Biocca, S., Di Luzio, A., Werge, T., Bradbury, A., and Cattaneo, A. 1991. Neuroantibodies: molecular cloning of a monoclonal antibody against substance P for expression in the central nervous system. *Proc. Natl. Acad. Sci. USA* 88:5611–15.
42. Biocca, S., Pierandrei-Amaldi, P., and Cattaneo, A. 1993. Intracellular expression of anti-p21ras single chain Fv fragments inhibits meiotic maturation of Xenopus oocytes. *B.B.R.C.* 197:422–27.
43. Biocca, S., Pierandrei-Amaldi, P., Campioni, N., and Cattaneo, A. 1994. Intracellulae immunisation with cytosolic recombinant antibodies. *BioTech.*, in press.

Index

377